Toxicology

MINISTRY OF AGRICULTURE
FISHERIES AND FOOD

ISSUED BY PRINTING AND PUBLICATIONS SECTION

Room 624,
Great Westminster House

AF843437

615.9/BRO

Information Centre

-15126

BROWN, SS

Chemical toxicology

Chemical Toxicology and Clinical Chemistry of Metals

Organizing Committee

IUPAC Clinical Chemistry Division, Commission on Toxicology

D. B. Tonks (Canada) (Chairman)
J. Savory (USA)
F.W. Sunderman Jr. (USA)
S.S. Brown (UK)
M. Stoeppler (FGR)
P. Schneider (Canada)
J.Y. Savoie (Canada)
Y. Tanaka (Canada)
W. Purdy (Canada)

INTERNATIONAL UNION OF PURE AND APPLIED CHEMISTRY

IUPAC Secretariat: Bank Court Chambers, 2-3 Pound Way,
Cowley Centre, Oxford OX4 3YF, UK

INTERNATIONAL UNION OF PURE AND APPLIED CHEMISTRY
Clinical Chemistry Division

Chemical Toxicology and Clinical Chemistry of Metals

Proceedings of 2nd International Conference held in Montreal, Canada, 19-22 July 1983

Edited by

Stanley S. Brown

Dudley Road Hospital, Birmingham, England

John Savory

University of Virginia Medical Center, Charlottesville, Virginia, USA

1983

ACADEMIC PRESS

A Subsidiary of Harcourt Brace Jovanovich, Publishers

London New York

Paris San Diego San Francisco

São Paulo Sydney Tokyo Toronto

546.3 : 615.9 BAT

ACADEMIC PRESS INC. (LONDON) LTD.
24/28 Oval Road,
London NW1

United States Edition published by
ACADEMIC PRESS INC.
111 Fifth Avenue
New York, New York 10003

Copyright © 1983 by
INTERNATIONAL UNION OF PURE AND APPLIED CHEMISTRY

IUPAC Secretariat: Bank Court Chambers, 2-3 Pound Way,
Cowley Centre, Oxford OX4 3YF, UK

All Rights Reserved
No part of this book may be reproduced in any form by photostat, microfilm, or any other means, without written permission from the copyright holder

British Library Cataloguing in Publication Data
Chemical toxicology and clinical chemistry of metals.
1. Metals – Toxicology – Congresses
I. Brown, Stanley S.   II. Savory, John
615.9'25'3   RA1231.H/

ISBN 0-12-137520-X

LCCCN 83-72587

Printed in Great Britain by
Whitstable Litho Ltd., Whitstable, Kent

ACKNOWLEDGEMENTS

The Chairman and Members of the Organizing Committee acknowledge support from the following organizations and exhibitors.

International Union of Pure and Applied Chemistry (IUPAC)
Canadian National Committee for IUPAC
International Federation of Clinical Chemistry (IFCC)
Association of Clinical Scientists

Alcan Aluminum Limitée, Montréal, Québec
Behringwerke Aktiengesellschaft, Marburg, FRG
Canadian Institute of Mining and Metallurgy, Montreal
C.I.L. Inc., North York, Ontario
Commission of the European Communities, Luxembourg
Dupont Canada, Inc., Mississauga, Ontario
Gouvernement du Québec, Cabinet du Ministre délégué à la Science et à la Technologie, Québec
Health and Welfare Canada, National Health Research and Development Program, Ottawa
Imasco Limitée, Montréal, Québec
Institut de recherche en santé et en sécurité du travail du Québec, Montréal, Québec
McGill University, Faculty of Graduate Studies and Research, Montreal, Quebec
National Bank of Greece, Montreal, Quebec
National Research Council of Canada, Ottawa, Ontario
National Sciences and Engineering Research Council of Canada, Ottawa, Ontario
The Noranda Foundation, Toronto, Ontario
The Samuel and Saidye Bronfman Family Foundation, Montreal, Quebec
Sherritt Gordon Mines Ltd., Toronto, Ontario
Société Canadienne du Cancer, Division du Québec, Montréal, Québec
Technicon Canada Inc., Montreal, Quebec

Beckman Instruments Ltd. Toronto, Ontario
Canadian Cabinets Company Ltd. Nepean, Ontario
CANLAB, Division de Distribution McGraw Ltée, Town of Mount Royal, Quebec

EG & G Princeton Applied Research, Downsview, Ontario
Fisher Scientific, Ottawa, Ontario
Hitachi Scientific Instruments-Nissei Sangyo Canada Inc., Rexdale, Ontario
Hospal Medical, East Brunswick, New Jersey, USA
Perkin-Elmer (Canada) Limitée, Montréal, Québec
Scintrex, Concord, Ontario
Varian Canada, Georgetown, Ontario
VG Isotopes c/o Analytech Instrumentation and Service Inc., Montreal, Quebec

FOREWORD

These Proceedings contain the texts of the Plenary Lectures and a selection of the communications which were presented to the Second International Conference on the Clinical Chemistry and Chemical Toxicology of Metals held in Montreal, 19-22 July 1983.

As with the First Conference,* the Plenary Lectures reflect the concern of the sponsoring organizations to bring together physicians and scientists from different disciplines to present a broad, balanced and up-to-date perspective of metal toxicology. In this context, the scope of the 140 contributed papers ranged from clinical and epidemiological case histories to studies of cellular and molecular mechanisms. To achieve a Proceedings of manageable size, however, it was necessary to make a selection of about 50 papers on the basis of the merits of the submitted abstracts and their relevance to the themes of the Plenary Lectures. We hope that this volume truly refects the state-of-the-art of metal toxicology and that it conveys a faithful picture of a dynamic subject which is of world-wide interest and importance.

We are grateful to the Chairman and members of the Organizing Committee for their advice and encouragement; to the sponsors, financial contributors and exhibitors who made it possible to hold the conference; and, not least, to all of the participants for their support and lively interest.

Stanley S. Brown
Dudley Road Hospital
Birmingham, UK

John Savory
University of Virginia
Charlottesville, USA

July, 1983

* Clinical Chemistry and Chemical Toxicology of Metals (Ed. S. S. Brown). Elsevier/North-Holland, Amsterdam 1977.

CONTENTS

CLINICAL AND EXPERIMENTAL STUDIES

ANALYSIS AND QUALITY ASSURANCE

CONTAMINATION CONTROL IN ROUTINE ULTRATRACE ANALYSIS OF TOXIC METALS

Evert Nieboer and Algis A. Jusys

Department of Biochemistry, McMaster University
Health Sciences Centre,
1200 Main Street West
Hamilton, Ontario
Canada L8N 3Z5

1. INTRODUCTION

The general availability of sensitive instrumental methods for the routine analysis of elements at the ng g^{-1} and μg L^{-1} levels in environmental and biological samples has revealed that contamination by containers, analytical reagents and airborne dust constitutes a serious source of both systematic and random errors. Interlaboratory comparisons (e.g. Parr, 1977; Brown *et al.*, 1981; Sunderman Jr. *et al.*, 1982), and the use of standard reference materials (e.g. Parr, 1977; Kosta and Byrne, 1982) clearly show contamination as a limiting factor in trace and ultratrace analyses (Kosta, 1982).* This realization has resulted in marked downward adjustments in the normal levels of elements in human tissues (AIN, 1981) and body fluids (Versieck and Cornelis, 1980). Establishment of reliable normal levels of toxic metals is critical in exposure and risk assessment studies, and in the evaluation of genuine biological variation.

Two specific examples will help to illustrate the importance of controlling contamination. The data in Fig. 1 indicate that the normal serum (plasma) nickel concentration has been revised downward by a factor of 30-40. Currently determined serum values for unexposed individuals are 2.5 ± 1.0 μg L^{-1}, while those exposed occupationally typically have concentrations in the range 3 - 10 μg L^{-1} (Sunderman Jr., 1977; Torjussen and Andersen, 1979; Tola *et al.*, 1979). Although improved detection limits are in part responsible for today's accepted lower nickel concentrations, the elimination of contamination during collection, storage and analysis has been shown by the IUPAC interlaboratory comparisons for nickel in urine and serum specimens to have contributed in a major way (Adams *et al.*, 1978; Brown *et al.*, 1981).

*Throughout this article, the designation 'trace' refers to concentrations at the μg g^{-1} and mg L^{-1} levels, and 'ultratrace' at ng g^{-1} and μg L^{-1}.

Chemical Toxicology and Clinical Chemistry of Metals
ISBN 0.12.137520.X
Copyright © 1983 by IUPAC
All rights of reproduction in any form reserved

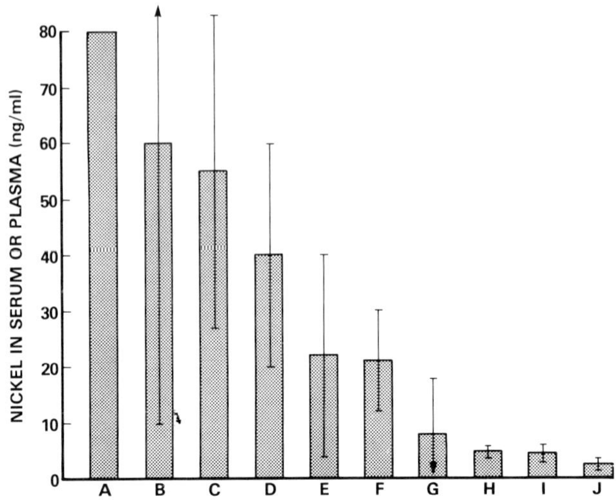

FIG. 1 "Normal" nickel concentrations in human serum or plasma for individuals with no occupational exposure to nickel (mean ± SD). The arrangement from left to right is roughly in chronological order. Analytical procedures involved emission spectroscopy (histograms A-D, G), flame atomic absorption spectrometry, AA (F, I, J), electrothermal AA (H), and spectrophotometry (E); wet-acid digestion (B, D, E), ashing (C, H), TCA deproteinization (F, G, I, J) and extraction (E-J). The concentrations depicted by histograms H and I are for populations environmentally exposed to nickel fallout from nickel refineries. Current normal levels assigned by the electrothermal AA IUPAC Reference Method, which involves acid digestion and extraction (Brown *et al.*, 1981), are as shown in J (2.5 ± 1.0 μg L^{-1}; see Sunderman Jr., 1977). Reproduced with permission from Cecutti and Nieboer (1981).

Recent compilations of chromium levels in human tissues and body fluids tell a similar story. Normal serum and urine chromium concentrations have decreased markedly with time; from ~200 μg L^{-1} to 0.1 - 0.2 μg L^{-1}, and from ~150 μg L^{-1} to ~1 μg L^{-1}, respectively (Versieck and Cornelis, 1980; Guthrie, 1982). Gross contamination during collection, handling, storage and analysis can again be incriminated (Versieck and Cornelis, 1980; Torgrimsen, 1982). Chromium, like nickel, is ubiquitous because of its presence in stainless steels. As with most other trace and ultratrace metals, the chromium concentrations reported for human tissues are likely to be in gross error because of extraneous additions during sampling and analysis (AIN, 1981; Guthrie, 1982). These uncertainties in normal levels, especially for body fluids, have rendered it difficult to assess the nutritional role of chromium (the Cr/glucose/insulin inter-relationships), and to evaluate the extent of exposure in the workplace by biological monitoring.

Much has been written about the impact of contamination and the need and methods for its elimination. An entire issue of the analytical journal *Talanta* was recently devoted to this topic; it featured 12 major articles (Vol. 29: 963-1055, 1982). An analyst seeking direction from the analytical literature is besieged by recommendations ranging from the necessity of working in a special trace-element clean laboratory, to the feasibility of working in a classical laboratory providing exhaustive cleaning of laboratory wares is practised. It is very difficult to assess the extent that precautionary measures are needed for a specific metal. The aim of the present article is to formulate a *modus operandi* or guidelines which will allow the design of a protocol that takes into consideration the degree of contamination, the precision and accuracy sought, and which allows analysis cost and time to be minimized. The principle adopted in this review is that the sophistication required in contamination control can be and should be tailored to specific analytic problems and needs.

2. CONSIDERATIONS OF ANALYTICAL QUALITY

Fundamental to our examination of contamination and its control are the basic concepts of detection limit, precision and accuracy. It is imperative that these terms be defined clearly before we proceed.

(a) *Limit of Detection*

The limit of detection may be defined as the smallest concentration (or amount) which can be reported with a specified degree of certainty for a given analytical procedure (IUPAC, 1976; Sunderman Jr. *et al.*, 1982). Expressed quantitatively (Currie, 1968; Kateman and Pijpers, 1981):

$$x = \bar{x}_o + 1.41k\sigma_o/n^{\frac{1}{2}} \qquad (1)$$

$$x = cS \qquad (2)$$

with x the limit of detection of the gross signal (blank + net); $\bar{x}_o$, mean of blank signals x_o; n, the number of determinations; k, a factor fixed by the probability level of the blank and net signals; σ_o, the standard deviation of the probability distribution of x_o, which in routine work may be assumed to be reasonably well known and invariant; c, concentration corresponding to the signal x; S, the sensitivity (defined as the change in measured signal resulting from a concentration change of one unit). In the derivation of (1), it is assumed that standard deviations corresponding to the blank and the sample signals are equal. When k has a value of 3.29, x (and thus c) denotes the lowest gross signal (concentration) that can be detected with a probability of 95%, providing that a Gaussian distribution applies for the measurements (Currie, 1968; Kateman and Pijpers, 1981).

(b) *Systematic and Random Errors*

Systematic errors introduce a consistent bias in experimental measurements, and may originate from incorrect sampling, contamination, losses by volatilization and adsorption, matrix effects, interferences, and errors in calibration, among others (e.g. Gretzinger *et al.*, 1982). On the other hand, random error corresponds to variance whose source(s) cannot be completely identified and controlled, and is subject to chance. It determines the precision of analytical fluctuations, and can arise, for example, from contamination, instrument fluctuations, and inconsistent performance by the analyst. Accuracy refers to the agreement between the measured concentration and the 'true value'. The principal limitations on accuracy are both random and systematic errors (IUPAC, 1976; Kateman and Pijpers, 1981).

3. THE PLANNING PHASE OF CONTAMINATION CONTROL

A preliminary assessment of the potential for contamination is perhaps the most beneficial first step when contemplating analytical work involving metals at trace and ultratrace concentrations. The most important issues to resolve are what the expected levels of the designated metal are, and its propensity for contamination. Helpful hints may be derived by consulting recent compilations for toxic metal concentrations in body fluids and tissues, water and food (e.g., Underwood, 1977; Friberg *et al.*, 1979; Versieck and Cornelis, 1980; Stokinger, 1981). In most instances, some indication is given about potential analytical problems especially contamination-related errors. A strong consensus exists that significant disparities in reported concentrations signal inadequate contamination control, with the lowest values cited most likely being the most reliable estimates.

It is obvious that the potential for extraneous addition increases when the level sought is low and when the metal in question is ubiquitous in the laboratory environment. A summary of the major sources of contaminants is provided in Table 1. Interestingly, many of the metals that are present in air at intermediate to high concentrations are known to be most problematic (e.g. Al,Co,Cr,Mo,Ni,Pb,Zn). As indicated in Table 1, most of these metals are also present in significant amounts in laboratory wares and tools. Except for Zn,which occurs in relatively high concentrations, these metals are present in serum (and it appears also in urine) at levels of several μg L^{-1} (Versieck and Cornelis, 1980; Everson and Patterson, 1980), and thus considerable care is necessary during collection, handling and analysis.

TABLE 1

Major Sources of Contamination

Potential Source	Typical Metals	Reference
Air	High concentration: Al, Ca, Cu, Fe, K, Mg, Na, Pb, Ti, Zn Intermediate: Ag, Cd, Co, Cr, Hg, Mo, Mn, Ni, V, Sn Low concentration: Au, Pt, Tl, U	Bowen, 1979
Industrial environment	Significant ambient levels of specific metals in working and adjacent laboratory environments	-
Laboratory construction materials and surfaces	Stainless steel and other metal-containing items (e.g, bench tops, hood linings, ovens, etc.)	-
Reagent grade chemicals	All except rare metals are present at ng g^{-1} or $\mu g\ L^{-1}$ levels	Zief & Mitchell, 1976; manufacturers' labels
Laboratory wares (plastics, Teflon, glass & silica)	Most metals can be leached from these items; borosilicate glass, polyethylene and Teflon wares are most suitable for routine analytical work after acid washing	Zief & Mitchell, 1976; Moody & Lindstrom, 1977; Laxen & Harrison, 1981; Heydorn & Damsgaard, 1982; Kosta, 1982
Laboratory tools	Stainless steel blades, knives, syringes (Fe, Cr, Mo, Ni): grinding and milling materials (e.g. Ca, Cu, Al, Fe, Ni, Co, Zn)	Zief & Mitchell, 1976; Kosta, 1982

The importance in ultratrace analysis of blank reproducibility and the selection of an adequate number of blank determinations can be deduced from Eqn (1). Thus the lowest gross concentration c that can be detected with a probability of 95% is as shown in Eqns (3) and (4).

$$c = (\bar{x}_o + 4.6\sigma_o/n^{\frac{1}{2}})/S \qquad (3)$$

$$c = (\bar{x}_o + 2.3\sigma_o)/S \quad \text{(for n=4)} \qquad (4)$$

It is our current practice in the determination of nickel in urine by atomic absorption spectrometry to include four blanks per experiment along with samples and standards (cf. Eqn 4). Previously when working in a laboratory environment subject to nickel fallout due to a local nickel smelter, blank values were typically 3.1 ± 1.7 μg L^{-1} (4 experiments, n = 13, S = 0.007 absorbance units · L μg^{-1}; Nieboer *et al.*, 1980). This work was done under ambient laboratory conditions, although a rigorous laboratory-ware cleaning regime was employed. According to Eqn (4), a gross nickel concentration of 7 μg L^{-1}, corresponding to a net or actual sample concentration of 4 μg L^{-1}, could thus be detected with a probability of 95%. These working conditions would therefore be inadequate to determine normal urine nickel levels which are considerably lower (2.0 ± 1.0 μg L^{-1}; Sunderman Jr., 1977). Nevertheless, in two of the four experiments alluded to, the magnitude and precision of the blank were suitable to attain this desired limit of detection. It is obvious that if normal urinary nickel levels were to be determined routinely in the laboratory described, stricter measures for contamination control would need to be adopted (see Section 4). For comparison, in our laboratory at McMaster which is not affected by industrial nickel fallout, the net detection limit is ∿1 μg L^{-1} (without the aid of clean air facilities).

The determination of zinc in serum illustrates that the absence of contamination is not guaranteed when expected concentrations are at the trace level (∿0.8 mg L^{-1} for Zn). Reimold and Besch (1978) found significant contaminating quantities of zinc in almost every procedural step in the analysis. Zinc is an ubiquitous contaminant of glass, plastics, rubber, water and laboratory reagents and apparatus. In contrast to zinc, serum copper determinations present no real methodological problems, even though it is present in serum at comparable concentrations (Versieck and Cornelis, 1980).

And finally, the need for contamination control is less severe when measuring elevated concentrations. This conclusion is inherent in the discussion of Eqns (1-4). For example, blood levels of gold (up to 1 mg L^{-1}) are considered by some a useful index in chrysotherapy (Finkelstein *et al.*, 1976; Pedersen and Graabaek, 1980), and plasma platinum (up to 10 mg L^{-1}) is monitored in cancer treatment with cis-platinum (Ribaud *et al.*, 1981). For these metals, both normal levels (in the μg L^{-1}

range) and contamination tendencies are low (see Table 1; Jones, 1976; Pedersen and Graabaek, 1980; Stokinger, 1981). The need for less than the most stringent contamination prevention measures is also encountered in the biological monitoring of individuals occupationally exposed to metals (e.g., Hagedorn-Götz *et al.*, 1977; Tossavainen *et al.*, 1980).

4. MEASURES TO ASSURE LOW BLANK VALUES

As stated in the Introduction, our approach to trace and ultratrace analysis is that the degree of sophistication required for measures to regulate contamination should be tailored to need. Unnecessary control measures are costly, especially as they tend to be labour intensive. Although a general approach to preventing adventitious additions is desirable, the fact remains that as contaminants each metal is idiosyncratic. Consequently, there are no *a priori* guarantees that the measures outlined in Table 2 will assure contamination-free analytical work. An appropriate criterion, of course, is whether the detection limit at the desired probability can be attained (see Eqns 3 and 4) for the measures adopted. If not, their application must be more exhaustive or new and more elaborate approaches are required.

a) *Cleanliness and Cleaning of Laboratory Wares*

Cleanliness of the working area can be improved by working on plastic sheets or trays, preferably placed on expoxy-coated bench tops (Zief and Mitchell, 1976). Koirtyohann and Hopps (1981) emphasize that laboratory personnel should wear disposable polyvinyl gloves that are presterilized and packaged without talc. Sweat, and presumably talc, is a known source of metals. Of the many methods recommended in the literature for cleaning plasticware, the simplest is suitable for most situations, namely soaking the containers in 10% HNO_3 for 48 h, then rinsing with distilled deionized water (DDI), followed by drain drying (Laxen and Harrison, 1981). Our own procedures for glassware such as digestion tubes employed in nickel analysis is similar and involves the following steps: wash with non-ionic detergent, water rinse, soak for 1 week in 6 M HCl (reagent grade), rinse with DDI, soak for 1 week in 1 M HCl (double distilled, DD; Nieboer *et al.*, 1980), rinse with DDI, dry (e.g. Al oven), and cover with Parafilm; vortex with 6 M HCl (DD) and DDI just prior to use. All acid-soaking is done in polyethylene containers with snap-on lids. Plastic containers, although handled individually, are subjected to the same cleaning regime in our laboratory. Since metal particulates appear to be entrapped in plastic wares during manufacturing, the acid-washing of newly purchased items should involve proven long-term acid-leaching procedures. In subsequent use, shorter protocols should suffice, such as the 1½ h leaching period employed in the IUPAC Nickel Reference Method (Brown *et al.*, 1981). HNO_3

TABLE 2

Measures to Assure Low Blank Values

Measure	Rationale
1. Assure cleanliness of work area	To avoid obvious contamination
2. Wearing of talc-free disposable gloves	Sweat and talc are known sources of metals
3. Acid-wash glass and plastic wares	To prevent leaching and adsorption of metals
4. Contamination-free specimen collection	To eliminate additions during handling
5. Sealed specimen storage	To avoid contact with ambient air
6. Minimum sample size	To minimize additions and contact with laboratory ware
7. Use high-purity reagents	To reduce addition of contaminants
8. Minimize reagent additions	To reduce addition of contaminants
9. Direct, small-scale analytical procedures	To avoid excessive handling and transfers; to minimize surface area contact with containers and ambient air
10. If required, work in clean-air installations.	To reduce blank values to acceptable levels

can be substituted for HCl if Cl^- or low pH interfere in subsequent detection steps (Laxen and Harrison, 1981). It appears that HCl diffuses into and adsorbs onto container walls to a greater extent than HNO_3.

b) *Specimen Collection and Analysis*

Extreme caution is warranted in specimen collection and handling to prevent gross contamination and transfer of metals from laboratory tools (Anand *et al.*, 1975; AIN, 1981; Kosta, 1982). Laminar-flow benches are recommended for the handling of specimens, especially body tissues. Proper storage is also essential. Subsequently, in the actual analysis, keeping the sample as small as possible has the benefit of minimizing the extraneous addition of unwanted metals. Even ultra-pure reagents have residual impurities. Because these specialty chemicals are very expensive, purification procedures for analytical reagents are available (e.g., Mitchell, 1982a,b; Moody and Beary, 1982). Of course, the design and adoption of direct small-scale analytical methods further reduces the potential of contamination by assuring that the surface area in contact with containers and with the ambient air is kept small. For this reason, considerable effort is currently focused on developing direct micro-methods of urine analysis for use in biological monitoring programmes involving laboratories subject to environmental fallout such as encountered in industry (e.g., Ader and Stoeppler, 1977; Adams *et al.*, 1978). This is especially useful in situations where elevated concentrations are likely, and where rapid analytical determinations are crucial from a medical viewpoint, such as in accidental exposure to nickel carbonyl (Hagedorn-Götz *et al.*, 1977; Adams, 1980).

c) *Clean-Air Facilities*

It is our contention that the rigorous application of measures 1-9 suggested in Table 2 will assure adequate blank levels and acceptable precision in many situations where adventitious addition of impurities is of major concern. Kosta (1982) expresses the view that only at the lowest range of concentrations and in special operations is the availability of clean air facilities a prerequisite. Normal serum levels of chromium and lead appear to be examples (Everson and Patterson, 1980; Guthrie, 1982). It is also obvious if ultratrace analyses are to be carried out in a laboratory known to be subject to ambient metal fallout, such as is often encountered in industry, clean-air facilities are warranted. Controlled environments range from the simple glove box, to laminar-flow clean exhaust hoods, and to clean rooms and clean laboratories (Zief and Mitchell, 1976; Mitchell, 1982a; Moody, 1982). The more elaborate of these dust-free installations are, of course, essential in the independent assignment of concentrations to standard reference materials such as by the US National Bureau of Standards.

5. VERIFICATION OF ANALYTICAL PROCEDURES

As pointed out in Section 2, negative analytical errors are possible. Thus it is conceivable that a contamination error is balanced by a systematic error of opposite sign. Verification of an analytical procedure is therefore advisable. Some of the major approaches and measures currently employed are summarized in Table 3.

In principle, the absence of random error such as contamination can be verified by comparing the observed variability with the *a priori* errors (i.e., all random sources of error that can be identified in the analytical procedure) (Heydorn and

TABLE 3

Verification of Analytical Procedures

Measure	Aspect Evaluated	Reference
Comparison of expected and observed variability	To identify the presence of random error (e.g. contamination)	Heydorn & Nørgård, 1973
Direct estimation of sample contamination	Identify and quantify extraneous additions during handling and preparation (e.g. by neutron activation)	Versieck *et al.*, 1982; Heydorn & Damsgaard, 1982
Analysis of Standard Reference Material	To identify systematic and random errors	Parr, 1977, 1980; Gretziner *et al.*, 1982; Kosta & Byrne, 1982; Versieck, 1983
Inter-laboratory comparisons		Sunderman Jr. *et al.*, 1982
a) Single experimental protocol (collaborative survey)	Method evaluation (accuracy and precision)	Brown *et al.*, 1981
b) Multi experimental protocols (cooperative survey)	Disclosure of unreliable methods	Tonks, 1982
Inter-method comparisons (independent or reference method)	Disclosure of error and assessment of relative variability	Brown *et al.*, 1981; Kosta & Byrne, 1982

Nørgård, 1973; Parr, 1977). If a residual contribution from random contamination is still suspected after such a statistical analysis, and if the measures summarized in Table 2 fail to eliminate it, direct estimation of the transfer of extraneous impurities from containers and laboratory tools may be warranted. Although the irradiation process can in some instances aggravate blank values by releasing metals from materials, neutron activation analysis is a suitable technique for this purpose (Zief and Mitchell, 1976; Versieck *et al.*,1982;Cornelis *et al.*, 1982; Heydorn and Damsgaard, 1982). Furthermore, the analysis of standard reference materials provides another powerful avenue for error testing. A disadvantage of this approach is the relative dearth of biological materials that are available. Furthermore, concentrations of many trace and ultratrace metals remain unassigned (Parr, 1980; Kosta and Byrne, 1982).

Of late, considerable effort has been devoted to the establishment of reference methods. Substantial disappointment has been expressed about *cooperative* interlaboratory surveys, in which samples are distributed but the choice of method is left to the individual analyst (e.g. Parr, 1977; Sunderman Jr. *et al.*, 1982). Results seem to be difficult to interpret. Nevertheless, cooperative studies would appear to permit the elimination of methods which give either imprecise and/or inaccurate results (Tonks, 1982). On the other hand, in a *collaborative* interlaboratory survey the experimental protocol is explicitly defined. Such studies constitute an essential exercise in the establishment of a reference method. Several round-robin trials may be required to eliminate or replace undersirable components. Both materials for which true values are known and unknown can be employed to assess accuracy and precision (e.g. Brown *et al.*, 1981). Once established, a reference method permits an authoritative evaluation of new methods. Most importantly, the performance of less rigorous, time-saving, and more direct analytical procedures can be assessed. If a suitable reference procedure is not available, analysts are urged to compare their results with those generated by an independent method (e.g., destructive *versus* non-destructive technique). Such an independent verification has the potential of disclosing errors, including contamination problems.

6. CONCLUDING REMARKS

Evidence is overwhelming that the general acceptance of measures to control contamination in trace and ultratrace analysis has lagged behind instrumental developments. Consequently, many of the normal concentrations reported for environmental and biological samples are suspect. For some metals, this uncertainty renders difficult the proper assessment of endogenous biological roles, the evaluation of genuine biological variation, and the confirmation of industrial exposure. It is apparent that the metals present in air at intermediate to high concentrations are the most problematic. In most cases, the situation is aggravated by the extremely low concentrations that need to be determined.

A general approach to the prevention of extraneous additions during collection, handling and analytical procedures is possible and is recommended. However, the fact remains that as contaminants, each metal is idiosyncratic. This requires that all potential contributions to the blank be carefully evaluated for each metal. Considerations of analytical quality suggest that a useful guide to the stringency of the contamination control required, and the effectiveness achieved, is the lowest concentration that can be detected at a specified probability such as 95%. Both low blank values (systematic error) and good precision (random error) are essential to attain this goal. In this manner, the sophistication of the control measures required can be linked to analytical need. It is concluded that only at the lowest range of concentrations and in special operations is the availability of clean-air installations and clean facilities an absolute prerequisite for reliable analytical work. Less stringent approaches when employed systematically and routinely appear to assure suitably low blank levels and acceptable precision in many instances. More extensive use of reference materials and reference methods is urged as a vehicle for disclosing both random and systematic inaccuracies introduced by contamination. International co-operation is also needed to establish additional reference methods by collaborative interlaboratory surveys.

7. REFERENCES

D.B. Adams (1980) in *Nickel Toxicology* (Eds., S.S. Brown and F.W. Sunderman Jr.), Academic Press, London, pp. 99-102.

D.B. Adams, S.S. Brown, F.W. Sunderman Jr. and H. Zachariasen (1978) *Clin. Chem.*, 24, 862-867.

D. Ader and M. Stoeppler (1977) *J. Anal. Toxicol.*, 1, 252-260.

AIN (American Institute of Nutrition) (1981) *Fed. Proc.*, 40, 2111-2158.

V.D. Anand, J.M. White and H.V. Nino (1975) *Clin. Chem.*, 21, 592-602.

H.J.M. Bowen (1979) *Environmental Chemistry of the Elements*, Academic Press, London, pp. 1-12.

S.S. Brown, S. Nomoto, M. Stoeppler and F.W. Sunderman Jr. (1981) *Pure Appl. Chem.*, 53, 773-781.

A.G. Cecutti and E. Nieboer (1981) in *Effects of Nickel in the Canadian Environment*, National Research Council of Canada, Ottawa, pp. 193-216.

R. Cornelis, J. Hoste and J. Versieck (1982) *Talanta*, 29, 1029-1034.

L.A. Currie (1968) *Anal. Chem.*, 40, 586-593.

J. Everson and C.C. Patterson (1980) *Clin. Chem.*, 26, 1603-1607.

A.E. Finkelstein, D.T. Walz, V. Batista, M. Mizraji, F. Roisman and A. Misher (1976) *Ann. Rheum. Dis.*, 35, 251-257.

L. Friberg, G.F. Nordberg and V.B. Vouk (Eds.) (1979) *Handbook on the Toxicology of Metals*, Elsevier/North-Holland Biomedical Press, Amsterdam, 709 pp.

K. Gretzinger, L. Kotz, P. Tschöpel and G. Tölg (1982) *Talanta*, 29, 1011-1018.

B.E. Guthrie (1982) in *Biological and Environmental Aspects of Chromium (Topics in Environmental Health, Vol. 5)* (Ed., S. Langård), Elsevier Biomedical Press, Amsterdam, pp. 117-148.

H. Hagedorn-Götz, G. Küppers and M. Stoeppler (1977) *Arch. Toxicol.*, 38, 275-285.
K. Heydorn and E. Damsgaard (1982) *Talanta*, 29, 1019-1024.
K. Heydorn and K. Nørgård (1973) *Talanta*, 20, 835-842.
IUPAC (International Union of Pure and Applied Chemistry) (1976) *Pure Appl. Chem.*, 45, 101-103.
A.H. Jones (1976) *Anal. Chem.*, 48, 1472-1474.
G. Kateman and F.W. Pijpers (1981) *Quality Control in Analytical Chemistry (Chemical Analysis, Vol. 60)*, John Wiley & Sons, New York, pp. 73-78.
S.R. Koirtyohann and H.C. Hopps (1981) *Fed. Proc.*, 40, 2143-2148.
L. Kosta (1982) *Talanta*, 29, 985-992.
L. Kosta and A.R. Byrne (1982) *J. Radioanal. Chem.* 69, 117-129.
D.P.H. Laxen and R.M. Harrison (1981) *Anal. Chem.*, 53, 345-350.
J.W. Mitchell (1982a) *J. Radioanal. Chem.*, 69, 47-105.
J.W. Mitchell (1982b) *Talanta*, 29, 993-1002.
J.R. Moody (1982) *Anal. Chem.*, 54, 1358A-1376A.
J.R. Moody and E.S. Beary (1982) *Talanta*, 29, 1003-1010.
J.R. Moody and R.M. Lindstrom (1977) *Anal. Chem.*, 49, 2264-2267.
E. Nieboer, P. Lavoie, D. Padovan, T.K. Venkatachalam and A.G. Cecutti (1980) in *Nickel Toxicology* (Eds., S.S. Brown and F.W. Sunderman Jr.), Academic Press, London, pp. 179-182.
R.M. Parr (1977) *J. Radioanal. Chem.*, 39, 421-433.
R.M. Parr (1980) in *Trace Element Analytical Chemistry in Medicine and Biology* (Eds., P. Bratter and P. Schramel), de Gryter, Berlin, pp. 631-655.
S.M. Pedersen and P.M. Graabaek (1980) *Ann. Rheum. Dis.*, 39, 576-579.
E.W. Reimold and D.J. Besch (1978) *Clin. Chem.*, 24, 675-680.
P. Ribaud, J. Gouveia, M. Bonnay and G. Mathe (1981) *Cancer Treatment Reports*, 65 (suppl. 3), 97-105.
H.E. Stokinger (1981) in *Patty's Industrial Hygiene and Toxicology* (Eds., G.D. Clayton and F.E. Clayton), John Wiley & Sons, New York. 3rd edn., Vol. 2A, pp. 1493-2060.
F.W. Sunderman Jr. (1977) *Ann. Clin. Lab. Sci.*, 7, 377-398.
F.W. Sunderman Jr., S.S. Brown, M. Stoeppler and D.B. Tonks (1982) in *IUPAC Collaborative Interlaboratory Studies in Chemical Analysis* (Eds., H. Egan and T.S. West), Pergamon Press, Oxford, pp. 25-35.
S. Tola, J. Kilpiö and M. Virtamo (1979) *J. Occup. Med.*, 21, 184-188.
D.B. Tonks (1982) *Clin. Biochem.*, 15, 67-68.
T. Torgrimsen (1982) in *Biological and Environmental Aspects of Chromium (Topics in Environmental Health, Vol. 5)* (Ed. S. Langård), Elsevier Biomedical Press, Amsterdam, pp. 65-99.
W. Torjussen and I. Andersen (1979) *Ann. Clin. Lab. Sci.*, 9, 289-298.

A. Tossavainen, M. Nurminen, P. Mutanen and S. Tola (1980) *Br. J. Ind. Med.*, 37, 285-291.

E.J. Underwood (1977) *Trace Elements in Human and Animal Nutrition*, Academic Press, New York, 4th edn., 545 pp.

J. Versieck (1983) *Trends Anal. Chem.*, 2, 110-113.

J. Versieck, F. Barbier, R. Cornelis and J. Hoste (1982) *Talanta*, 29, 973-984.

J. Versieck and R. Cornelis (1980) *Anal. Chim. Acta*, 116, 217-254.

M. Zief and J.W. Mitchell (1976) *Contamination Control in Trace Element Analysis (Chemical Analysis, Vol. 47)*, John Wiley & Sons, New York, 262 pp.

ACKNOWLEDGEMENT

Financial support by the Natural Sciences and Engineering Research Council of Canada is gratefully acknowledged.

ELECTROCHEMICAL DETECTION OF TOXIC METALS IN BIOLOGICAL MATERIALS

William C. Purdy

Department of Chemistry, McGill University, 801 Sherbrooke St. W., Montreal, Quebec, Canada H3A 2K6

1. INTRODUCTION

Since Clinical Chemistry and Toxicology rest firmly on an analytical base, it is vital that the clinical investigator be provided with reliable analytical methods. In addition to the final determinative step, these methods must include sampling protocols, sample-preparation procedures, quality-control steps, etc. and give due attention to the question of speciation (Brown *et al.*, 1981; Florence, 1982; Seltner *et al.*, 1981; Sullivan and Delfino, 1982). This is extremely important if the procedure involves destruction of organic matter prior to the determination of a metallic element. The digestion step has often been termed the "weak link" in the entire analytical procedure and the literature documents the difficulties encountered, especially when the element sought forms volatile compounds. Many elements are not recovered with 100% efficiency even though the final determinative step in the procedure may be accurate and precise and possess low detection limits. These low recoveries are principally due to losses in the digestion step.

The first comprehensive examination of digestion methods in recent times was made by Gorsuch (1959). He employed radiochemical techniques to study the recovery of fourteen trace elements from organic and biological matter. Many of his observations were quite startling. Depending upon digestion conditions, lead recoveries varied from 4 to 100%, mercury recoveries from 0 to 93%, selenium recoveries from 0.5 to 101% and arsenic recoveries from 47 to 97%. Gorsuch (1962) used the radiochemical technique to demonstrate that the losses, in the case of antimony, chromium, ferric iron, lead and zinc, were due to the volatilization of a chloride salt. A mixture of 50% hydrogen peroxide and sulfuric acid was recommended for the destruction of organic matter (Down and Gorsuch, 1967).

Chemical Toxicology and Clinical Chemistry of Metals
ISBN 0.12.137520.X
Copyright © 1983 by IUPAC
All rights of reproduction in any form reserved

Most of the elements examined could be recovered quantitatively, but arsenic, germanium, ruthenium and selenium suffered losses under some or all of the conditions employed. Janghorbani *et al.*, (1982) have shown that for complete oxidation of urinary selenium to the plus-4 state, a digestion mixture of nitric acid and perchloric acid is required. The trimethylselenonium ion found in urine resists oxidation even with hydrogen peroxide (Ting *et al.*, 1982).

Leyden and Wegscheider (1981) have discussed the preconcentration step in trace-metal determinations as a means of increasing the sensitivity of the method and minimizing sample-matrix effects. Others have evaluated reference materials for the complete determination of manganese (Clegg *et al.*, 1982) and vanadium (Heydorn and Damsgaard, 1982) in biological materials and studied the effect of preservation methods on trace-element concentrations of fish and mussel tissues (Klusek and Heit, 1982). In a review on environmental lead levels Bryce-Smith (1982) comments that older literature values for blood lead "... are suspect because of the use of hypodermic syringes having the needle soldered into the base ...". Sample contamination from plasticware vessels is reported in zinc analyses (Butrimovitz and Purdy, 1977) and can also be a problem in the analysis of cadmium, copper, iron, manganese and zinc in human hair if the proper washing procedures are not followed (Salmela *et al.*, 1981).

When all the proper pre-determinative-step precautions have been followed, the clinical investigator is left with an aqueous solution containing the metal ion of interest. The quantitation of this metal may be accomplished in a number of ways, some of which are the subject of plenary lectures in this conference. The present paper is devoted to electrochemical methods of analysis. Because of the great variety of these methods, the author has decided to concentrate his discussion on four (ion-selective electrodes, differential pulse polarography, stripping voltammetry and potentiometric stripping) and to make passing mention of some novel initiatives.

2. ION-SELECTIVE ELECTRODES

Prior to 1950 a review of ion-selective electrodes (ISEs) would have involved little more than a discussion of the glass electrode for pH measurements. Now such a review would have to describe a wide variety of electrochemical sensors, including field-effect transistors (Janata, 1982), used to determine the activity of ionized and non-ionized chemical species. The electrodes themselves are actually combinations of a primary electrode, such as a metal or metal/metal salt electrode, coupled with a solution of an electrolyte and a permselective membrane. The critical component of the ISE is the membrane for it determines the sensitivity and selectivity of the device. Membranes vary from specially formulated glasses to valinomycin in hexane on a cloth wick. Buck (1981) has prepared an excellent review on the electrochemistry of ISEs and the reader is referred to this source for a thorough, up-to-date, theoretical discussion.

An improved lithium ISE has been developed which yields selectivity coefficients for lithium ion over the alkali-metal ions of >100 and over the alkaline-earth metal cations of about 1000. The permselective membrane is based on the lipophilic N,N,N',N'-tetraisobutylcyclohexane-*cis*-1,2-dicarboxylic diamide (Zhukov *et al.*, 1981). By shaking a commercial calcium liquid ion exchanger in chloroform with aqueous 0.5 *M* beryllium sulfate, a beryllium ion exchanger can be made (Moody *et al.*, 1981). This can then be incorporated into a polyvinyl chloride (PVC) matrix to form a beryllium ISE. Bourgognon *et al.* (1981) have improved the characteristics of a cadmium ISE by changing the nature of the polycrystalline membrane. Instead of the classical mixture of Ag_2S and CdS, these authors tested a number of polymetallic chalcogenides and obtained excellent reproducibility, response time and Nernstian-response range. On the other hand a membrane based on HgS(black)/Hg(0) was found to be superior to mercury(II) chalcogenides as a mercury ISE for direct potentiometry or as an indicator in potentiometric titrations (East and Da Silva, 1983). Solid membranes of copper hexacyanoferrate(III) in Araldite gave linear, near Nernstian response for thallium(I) over the concentration range of 10^{-1} to 5×10^{-4} *M* (Jain *et al.*, 1982).

Pungor and his group have been active in ISE research for a number of years. Recently they developed an ISE with a PVC membrane containing the tetradecylphosphonium trichloromercurate(II) ion-pair complex (Kopytin *et al.*, 1983). This electrode gives nearly Nernstian response for trichloromercurate(II) ion down to about 10^{-6} *M* and is unaffected by pH over the range of 1.5 to 6. Another group (Pungor *et al.*, 1983) has studied changes in the surface morphology of ISEs. Using a scanning electron microscope they examined the surface of a Ag_2S-PbS membrane and found that lead passivation was due to chemical and electrochemical oxidation of PbS to $PbSO_4$ and PbO, respectively. Finally, ISEs have been used to assay pharmaceuticals for inorganic substances which are present either as contaminants or decomposition products (Pungor *et al.*, 1982).

3. DIFFERENTIAL PULSE POLAROGRAPHY

Polarography involves electrolysis at a dropping mercury electrode (DME) where the current at the working electrode is controlled by the rate of diffusion of the electroactive substance from the solution to the mercury surface. The current that flows as a function of an applied DC voltage ramp and the location of the polarographic wave on the current-voltage axes form the basis of a simultaneous quantitative and qualitative analysis. In polarography the measured current comes from two main sources, the electrolysis of the material at the electrode surface and the charging of the double layer around the mercury drop. The former, the faradaic current, is related by Faraday's Law to the concentration of the electroactive species in solution; the latter, the capacitative current, sets the detection limits for polarography.

Many different approaches have been tried for separating the faradaic and capacitative currents. One method that has proved useful is differential pulse polarography (DPP). DPP employs short duration, small amplitude, voltage pulses of constant height on top of a slowly rising DC potential ramp. These pulses occur near the end of the lifetime of the mercury drop when the surface area of the drop is changing least. The current is sampled just before pulse application and toward the end of the pulse when the capacitative current has decayed. A difference measurement is made and this is recorded as a function of applied potential. The resulting curve is a series of peaks with the peak current of each being directly proportional to the concentration of an electroactive species in the bulk solution. DPP gives a linear response between peak current and concentration from 10^{-4} to 10^{-8} *M*. de Silva and Brooks (1978) have reviewed voltammetric methods and discussed their application to pharmaceutical analyses.

Recently ppb levels of silicon in water samples have been determined by DPP (Wang and Wang, 1982). The heteropoly molybdosilicic acid complex produced five peaks in a citrate buffer containing 20% 2-butanone. The current for each of these peaks was proportional to the silicon concentration. In a perchlorate medium containing catechol, Ge(IV), Sn(IV), As(V), Sb(V), Se(IV) and Te(VI) all gave well-defined peaks which had currents proportional to the concentration of the metal ion (McCrory-Joy and Rosamilia, 1982). The detection limits for the various ions were 82, 28, 4000, 25, 115 and 17 ppb, respectively. Finally fast-scan DPP has been described and applied to the determination of amalgam-forming metals like cadmium (Gajda and Horák, 1982). The technique, which determines with good reproducibility concentrations above 10^{-8} *M*, permitted a ten-fold decrease in measurement time compared to conventional DPP.

4. STRIPPING VOLTAMMETRY

If the detection limits afforded by DPP are not low enough,

an initial preconcentration step may be interposed before the current measurement. One way to preconcentrate the metal ions in a sample is to deposit them electrolytically at a constant potential onto an inert electrode surface, typically a hanging mercury drop electrode (HMDE) or a thin-film mercury electrode (TFME). If the metal is one which is soluble in mercury, the product of this preconcentration step will be an amalgam. However, if the material is a substance which forms a precipitate with an oxidized form of mercury, the product will be that precipitate (i.e. Hg_2Cl_2 for the deposition of chloride ion).

Following the preconcentration step a voltage scan is applied to the working electrode. This voltage scan, which is most often applied in the DPP mode, causes an electrolytic dissolution, or stripping, which releases the material back into the solution. If the product of the preconcentration were an amalgam, the voltage scan is applied anodically and the technique is called anodic stripping voltammetry (ASV). An alternative is employed with mercury precipitates. Here the voltage scan which is used reduces the oxidized form of mercury back to the element and releases the anion. This technique is termed cathodic stripping voltammetry (CSV). If the voltage scan is applied in the differential-pulse mode, then a DP will precede the ASV or CSV.

Stripping voltammetry is currently the most sensitive electroanalytical technique available. It is capable of identifying and quantitating trace components from 10^{-5} to 10^{-9} *M* with excellent sensitivity and selectivity. Its use as an analytical tool has been reviewed by Nürnberg (1982) and Wang (1982b) who also evaluated DPASV at a TFME (Wang, 1982c). Nürnberg and Raspor (1981) have addressed the problem of speciation of heavy metals determined by DPASV. The use of a standard-additions technique has been suggested in those cases where ASV leads to the formation of intermetallic compounds which would cause significant errors in the determination of analyte concentration (Gerlach and Kowalski, 1982). These studies were carried out on the Cu-Zn-Hg system.

Several groups have studied ASV at different working electrodes. One electrode that has been investigated as a replacement for the TFME is the rotating glassy carbon ring-disc electrode. Various stripping parameters have been evaluated (Brihaye and Duyckaerts, 1982), a linear ASV technique for this electrode has been compared to DPASV (Brihaye and Duyckaerts, 1983), and the application of this linear ASV technique to the analysis of cadmium, lead, copper, antimony and bismuth in sea water has been described (Brihaye *et al.*, 1983). A semi-stationary mercury electrode has been suggested for the DPASV analysis of trace metals in natural waters (Andruzzi *et al.*, 1982) and a graphite-epoxy microelectrode for *in vivo* measurements of trace metals (Wang, 1982a).

One of the elements that has been frequently determined by ASV is lead. The total lead concentration in polluted estuarine waters was found to range between 0.07 and 5.4 μg kg^{-1} (Acebal and de Luca Rebello, 1983). Copper was shown to interfere with the analysis at pH≥2. The determination of lead and cadmium in bones gave comparable results by DPASV and flameless atomic absorption spectroscopy (Simon and Liese, 1981). The concentration of lead in the sample ranged from 10 to 70 ppb and for cadmium from 0.05 to 0.4 ppb.

Arsenic in the plus-3 oxidation state may be deposited on a rotating gold electrode (Bodewig *et al.*, 1982). Following a 4-min deposition the arsenic is stripped anodically from the gold surface using the DP mode. The limit of detection for As(III) in polluted sea-water samples is 0.2 μg L^{-1}. As(V) will not deposit on gold under the conditions used for As(III), permitting arsenic-speciation studies. If total As levels are desired, the sample is treated with SO_2 to reduce the plus-5 As. Other workers have employed a gold-plated tubular graphite electrode for the DPASV determination of arsenic in polluted waters (Leung *et al.*, 1982). A gold-film glassy carbon disc electrode was used to determine methylmercury in the presence of inorganic mercury (Ireland-Ripert *et al.*, 1982). A method combining double standard additions and DPASV was suggested to detect methylmercury with a limit of 2×10^{-8} *M*. Trace amounts of bismuth in urine and blood have been assayed by DPASV at a HMDE (Kauffmann *et al.*, 1981). As little as 5 μg L^{-1} bismuth in the sample may be quantitated in less than 2 hours total analysis time.

Selenium is one of the elements which may be determined by anodic or cathodic stripping voltammetry. Selenium(IV) is deposited at a rotating gold-disc electrode from a 0.1 *M* perchloric acid solution (Andrews and Johnson, 1975). At an applied potential of -0.3 V, selenium deposits in three distinct states, bulk deposited Se, adsorbed Se and an intermetallic Au-Se compound. Upon anodic stripping using a linear voltage scan to +1.2 V, three anodic stripping peaks are observed, the largest corresponding to the bulk deposited Se. The detection limit for this ASV assay is reported to be approximately 0.04 ppb of Se(IV) in 0.1 *M* $HClO_4$. The method has been applied to the determination of selenium in NBS bovine liver. The use of an *in situ* gold-plated glassy carbon disc electrode has been suggested (Posey and Andrews, 1981). A single anodic stripping peak was obtained for Se(IV) concentrations less than 10^{-6} *M*. The limit of detection was reported to be 2×10^{-9} *M* and ASV values compared well with certificate values for NBS SRM 1577 Bovine Liver.

CSV at a mercury electrode has been used for the determination of selenium. In one procedure the selenium was reduced to selenide by sodium borohydride (Dennis *et al.*, 1976). Acidification evolved hydrogen selenide which was passed into and trapped in a cell containing the HDME in an alkaline electrolyte. The mercuric selenide formed on the mercury drop was stripped off in the differential pulse mode using a voltage scan from -0.7 to -1.0 V. The limit of detection was 1 ppb with linear standard curves from 4-50 ppb. Others have employed electroreduction of selenium(IV) to mercuric selenide (Jarzabek and Kublik, 1982). Although the potential for this reduction varied with the nature of the supporting electrolyte, a potential range of 0.0 to -0.1 V proved effective. The stripping was carried out over the voltage range of -0.3 to -0.8 V. Detection limits were about 10^{-9} *M* and the entire procedure required about 30 min for a soluble selenium(IV) sample. Ebhardt and Umland (1982) studied the effects of pH and the nature of the mercury-drop surface on the determination of selenium and tellurium. Using a new AC controlled CSV technique they were able to quantitate 0.2 ppb of selenium in the presence of tellurium. Finally a simple and rapid stripping assay has been reported for the determination of selenium, copper, lead and cadmium (Adeloju *et al.*, 1983). The selenium was deposited at a potential of -0.3 V and stripped by DPCSV over a voltage range of -0.5 to -0.7 V. The three other elements were then deposited at -0.9 V. In this process the selenium was oxidized to the electro-inactive Se(VI). Copper, lead and cadmium were then quantitated by DPASV. Selenium was determined in the presence of interfering ions at 1000-fold excess in concentration. The detection limits were 0.1 μg L^{-1} for Se, 0.5 μg L^{-1} for Cu and 0.01 μg L^{-1} for Cd and Pb and the method was satisfactorily applied to NBS bovine liver and some fish samples.

Arsenic in drinking water contaminated with copper has been determined by DPCSV (Sadana, 1983). Arsenic(III) does not deposit at a HMDE but in the presence of copper it forms copper compounds which are soluble in mercury. The procedure is as follows. As(V) is reduced to the plus-3 state by the addition of conc. HCl and 48% HBr. The pH is then adjusted by the addition of hydrazinium chloride, copper ion is added and the arsenic is deposited at -0.6 V, presumably as copper arsenide, Cu_3As_2. Following deposition the material is stripped cathodically yielding hydrogen gas, arsine and copper amalgam. The height of the stripping peak is proportional to the As(III) concentration and permits detection of As(III) to a limit of 1 ng mL^{-1}. Since As(V) will not deposit unless first reduced to As(III), some idea of arsenic speciation can be gained from this analytical method.

5. POTENTIOMETRIC STRIPPING ANALYSIS

In 1976 Jagner and Graneli reported on potentiometric stripping analysis (PSA), a method which is analogous in many ways to ASV or CSV. The metal to be determined is preconcentrated by electrolysis on a working mercury-drop or mercury thin-film electrode. Once the electrolysis has been terminated, the time-potential behavior of the working electrode is studied as it undergoes chemical oxidation. The metals that have amalgamated at the mercury electrode are then oxidized at a rate dependent on the rate of transport of the oxidizing agent, usually mercury(II) ion, to the working-electrode surface. The resultant time-potential curve resembles a chronopotentiogram and has time steps (similar to transition times) which are proportional to the amount of each deposited metal. PSA should be well suited to *in situ*-plated thin-film mercury electrodes since mercury(II) ions are being used both for film formation and as an oxidizing agent.

Jagner (1978) developed a simple instrument package for PSA consisting of a battery-operated potentiostat with a timer/switcher unit for the preconcentration and a pH meter and x-y recorder for the recording of the time-voltage curves. This package was used for the analysis of bismuth, cadmium, thallium, lead, copper and zinc in tap water, mussels from the Swedish coast and Baltic sea water. Instead of recording the time-voltage curve, a derivative circuit can be placed before the recorder so that the plotted curves are actually the first derivative of the time-interval steps (Jagner and Årén, 1978). This derivative PSA was applied to the analysis of zinc, cadmium, lead and copper. All four elements had detection limits on the order of 0.05 ppb and an analytical range of 0.1 to 1000 ppb.

Others (Labar and Lamberts, 1981) have controlled the transport mechanism of the oxidant during the stripping step. They used a stationary glassy carbon working electrode onto which was plated a thin film of mercury. By controlling the transport of mercury(II) ions to the working-electrode surface these authors claimed enhancement by a factor of 50 for copper, cadmium, zinc, lead, bismuth, thallium and gallium. Christensen *et al.* (1982a) studied the effects of the formation of water-soluble species during the preconcentration step. These reduced species could give rise to interfering signals during the subsequent stripping step. The authors discussed the mechanism of the interference and indicated ways of avoiding problems. They illustrated the efficacy of their methodology for eliminating interferences by determining cadmium, lead and thallium (Christensen *et al.*, 1982b). The same three elements plus copper were determined in a flow PSA procedure (Anderson *et al.*, 1982). A six-step inlet valve permitted the stripping process to take place in a solution of optimum composition for the particular element to be quantitated. Flow conditions were evaluated and the system was successfully applied to the analysis of lead in urine.

Hu *et al.* (1983) have combined PSA with flow-injection analysis providing an automated method for the determination of heavy metals in groundwater. The usefulness of the method was illustrated by the quantitation of copper, cadmium and lead. Copper in sea water at environmental pH was measured at a thin-film mercury-coated glassy carbon electrode (Sheffrin and Williams, 1982). Copper gave a linear response over the range of 2-200 ppb. Several reports have been made on the PSA determination of lead in biological materials. Lead in urine was determined using the Radiometer Ion Scanning System 820, a commerical PSA apparatus (Watts *et al.*, 1981). The results compared well with those obtained by atomic absorption spectrophotometry. The same PSA apparatus was employed to determine lead in fruit juices and soft drinks (Mannino, 1982).

In the initial work and that described above mercury(II) ion was used as the oxidizing agent in the stripping step. Other potential oxidizing agents were removed from the solution. This necessitated deoxygenation of all solutions during the preconcentration and stripping steps. Jagner (1979) found that deaeration was unnecessary and used the oxygen in air as the oxidizing agent for the stripping step. For zinc, thallium, cadmium, bismuth, lead and copper concentrations above 25 $\mu g\ L^{-1}$, PSA in nondeaerated samples was more rapid than with mercury(II) ions as the oxidant. It is also possible to use other oxidizing agents in the stripping process. Jagner *et al.* (1982) employed the flow PSA technique (Anderson *et al.*, 1982) for the determination of mercury(II). Preconcentration was carried out at a gold electrode from a sample solution to which ammonia and iodide had been added. The valve was then switched and stripping was done in an acidified bromide solution containing chromium(VI) as the oxidant. The detecton limit was 0.4 $\mu g\ kg^{-1}$ and the analytical range for mercury was about three orders of magnitude. Copper(II) interfered when present at a 1000-fold excess and silver at a 5-fold excess over mercury. This technique was then applied to the analysis of mercury(II) in sediment and in acid-digested orchard leaves (NBS SRM 1571) and fish muscles (Jagner and Årén, 1982).

6. NEW ELECTROANALYTICAL INITIATIVES

Izutsu *et al.* (1983) have reported on a form of CSV at a trioctylphosphine oxide (TOPO)-coated glassy carbon working electrode. At a potential of 0 V uranyl ions in a 0.5 *M* NaCl solution adjusted to pH 4.0 were concentrated into the TOPO-coated electrode, presumably by an extraction process. Following preconcentration the uranyl ions were irreversibly reduced yielding a current peak which quantitated the analyte in the concentration range of 2×10^{-9} to 2×10^{-7} *M*. The method, which was reasonably selective for uranyl ions, was applied to artificial sea-water samples.

Methylmercury in fish has been determined by DPP following isolation and separation by liquid chromatography (Holak, 1982). The mobile phase for the separation was methanol-ammonium acetate solution at pH 5.5 containing mercapto-ethanol. Ion chromatography has been used to separate cyanide, sulfide, iodide and bromide ions (Rocklin and Johnson, 1983). Following separation on a HPIC-AS4 anion exchange column, the effluent was determined amperometrically at a silver working electrode. The reported detection limits were 2, 30, 10 and 10 ppb for cyanide, sulfide, iodide and bromide ions, respectively. Cyanide bound loosely in complexes with cadmium or zinc could be determined quantitatively as total "free" cyanide while that in nickel or copper complexes was only partially "free". Cyanide in strongly bound complexes like cobalt, gold or iron was not detected. Sevenich and Fritz (1983) used ion chromatography with a conductivity detector to separate and determine polyvalent metal ions. The addition of complexing (masking) agents to the sample increased selectivity. The technique was reported to be rapid and highly selective for the determination of calcium, magnesium and strontium and to work well for zinc, cobalt, manganese, cadmium and lead as well as some of the trivalent lanthanide ions.

Bruckenstein and his students have developed a technique which they call pneumatoamperometry (Beran and Bruckenstein, 1980b). A trace solute is reacted with a suitable reagent to produce a volatile electroactive product. A nonelectroactive purge gas is used to carry this volatile product to a gas-sensing electrode of the Clark or fuel-cell type. A typical sample train employs nitrogen as the purge gas and a gold gas-porous electrode (Au GPE) as the indicator (Gifford and Bruckenstein, 1980a). Mercury(II), reduced to the element with Sn(II), is oxidized back to the divalent state at 975 mV; arsenic is converted to arsine with sodium borohydride then electro-oxidized to As(III). Iodide or iodate ion can be converted to volatile iodine by reaction with the other and then be oxidized to iodate ion electrochemically. The detection limits for these four analytes are reported to be 5, 3, 6.5 and 0.5 ppb, respectively. Since the borohydride reduction works so well for the determination of arsenic, it has been extended to the determination of antimony and tin (Gifford and Bruckenstein, 1980b). The volatile hydrides produced are detected at the Au GPE. Se(II) up to 1 μg and nitrate ion up to 20 ppm do not interfere with the analysis. Mercury interferes at levels $\geq$ 0.5 ppm causing a decrease in arsenic response and an increase in antimony response. Cr(VI) above 1.0 ppm causes a 10% decrease in both arsenic and antimony response and copper(II) at 30 ppb causes a 10% decrease in antimony response. Copper does not interfere with arsenic response at levels $<$ 0.4 ppm. Detection limits for 5-mL samples were reported to be 0.2, 0.8 and 0.2 ppb for arsenic(III), tin(II) and antimony(III), respectively.

Nanogram amounts of cyanide were determined by reaction with iodine to form ICN followed by amperometric regeneration of I_2 at the Au GPE (Beran and Bruckenstein, 1980a). The detection limit was estimated to be 5 ng in a 4 mL sample; nitrate, chloride, silver, sulfide and sulfite did not interfere. Ni(II), Pb(II) and Zn(II) gave slight positive interferences, thiocyanate and nitrite strong positive interferences and Pd(II) a strong negative interference. It was not possible to determine the cyanide in complexes with iron or cobalt(III).

Finally, several groups have used electrochemical detectors in flow-injection analysis. Wang *et al.* (1983) have applied the flow-injection technique to ASV. Preconcentration was carried out in a flow cell at a mercury-coated wall-jet glassy-carbon disk electrode. Using sample volumes of 500 μL, ppb levels of trace metals were determined simultaneously at a rate of about ten samples per hour. Improvements in this system have been made by employing dual coulometric-voltammetric cells (Wang and Dewald, 1983). The "upstream" coulometric cell, set at a more positive potential than the analytical cell, removes more easily reduced interferences, eliminating problems caused by dissolved oxygen, intermetallic compounds and overlapping peaks. The "downstream" voltammetric analytical cell then determined the desired trace metal as above by flow-injection DPASV. Examples of cadmium, copper, indium, lead and zinc analyses were given.

7. CONCLUSIONS

The author has attempted to touch on some of the recent developments in the application of electrochemistry to the determination of toxic metals in biological materials. To be sure many other examples exist that have not been mentioned. Most of the methods discussed here have been used for the determination of metals in other sample matrices as well as in those cited. Thus they have a proven "track record". Since the practical range of usefulness for these electroanalytical techniques matches, in most cases, that of fluorometry and atomic spectroscopy, the researcher should give serious thought to electrochemical solutions to their biological problems involving trace metal analysis.

8. REFERENCES

S.A. Acebal and A. de Luca Rebello (1983) *Anal. Chim. Acta,* 148, 71-78.

S.B. Adeloju, A.M. Bond and H.C. Hughes (1983) *Anal. Chim. Acta,* 148, 59-69.

L. Anderson, D. Jagner and M. Josefson (1982) *Anal. Chem.,* 54, 1371-1376.

R.W. Andrews and D.C. Johnson (1975) *Anal. Chem.,* 47, 294-299.

R. Andruzzi, A. Trazza and G. Marrosu (1982) *Anal. Lett.,* 15, 1565-1584.

P. Beran and S. Bruckenstein (1980a) *Anal. Chem.,* 52, 1183-1186.

P. Beran and S. Bruckenstein (1980b) *Anal. Chem.,* 52, 2207-2209.

F.G. Bodewig, P. Valenta and H.W. Nürnberg (1982) *Fresenius Z. Anal. Chem.,* 311, 187-191.

H. Bourgognon, J.-J. Fombon, F. Lancelot, J. Paris, M. Roubin and J. Tacussel (1981) *Analusis,* 9, 385-388.

C. Brihaye and G. Duyckaerts (1982) *Anal. Chim. Acta,* 143, 111-120.

C. Brihaye and G. Duyckaerts (1983) *Anal. Chim. Acta,* 146, 37-43.

C. Brihaye, G. Gillain and G. Duyckaerts (1983) *Anal. Chim. Acta,* 148, 51-57.

S.S. Brown, S. Nomoto, M. Stoeppler and F.W. Sunderman, Jr. (1981) *Clin. Biochem.,* 14, 295-299.

D. Bryce-Smith (1982) *Trends in Anal. Chem.,* 1, 199-203.

R.P. Buck (1981) *Sensors and Actuators,* 1, 197-260.

G.P. Butrimovitz and W.C. Purdy (1977) *Anal. Chim. Acta,* 94, 63-73.

J.K. Christensen, L. Kryger and N. Pind (1982a) *Anal. Chim. Acta,* 136, 39-50.

J.K. Christensen, L. Kryger and N. Pind (1982b) *Anal. Chim. Acta,* 141, 131-146.

M.S. Clegg, C.L. Keen, B. Lonnerdal and L.S. Hurley (1982) *Biol. Trace Elem. Res.,* 4, 145-156.

B.L. Dennis, J.L. Moyers and G.S. Wilson (1976) *Anal. Chem.,* 48, 1611-1616.

J.A.F. de Silva and M.A. Brooks (1978) in *Drug Fate and Metabolism (Methods and Techniques)*, (Eds. E.R. Garrett and J.L. Hirtz), Marcel Dekker, Inc., New York, vol. 2, pp. 1-48.

J.L. Down and T.T. Gorsuch (1967) *Analyst,* 92, 398-402.

G.A. East and I.A. DaSilva (1983) *Anal. Chim. Acta,* 148, 41-50.

K.-B. Ebhardt and F. Umland (1982) *Fresenius Z. Anal. Chem.,* 310, 406-409.

T.M. Florence (1982) *Talanta,* 29, 345-364.

V. Gajda and K. Horák (1982) *Anal. Chim. Acta,* 134, 219-226.

R.W. Gerlach and B.R. Kowalski (1982) *Anal. Chim. Acta,* 134, 119-127.

P.R. Gifford and S. Bruckenstein (1980a) *Anal. Chem.,* 52, 1024-1028.

P.R. Gifford and S. Bruckenstein (1980b) *Anal. Chem.* 52, 1028-1031.
T.T. Gorsuch (1959) *Analyst*, 84, 135-173.
T.T. Gorsuch (1962) *Analyst*, 87, 112-115.
K. Heydorn and E. Damsgaard (1982) *J. Radioanal. Chem.*, 69, 131-146.
W. Holak (1982) *Analyst*, 107, 1457-1461.
A. Hu, R.E. Dessy and A. Graneli (1983) *Anal. Chem.*, 55, 320-328.
J. Ireland-Ripert, A. Bermond and C. Ducauze (1982) *Anal. Chim. Acta*, 143, 249-254.
K. Izutsu, T. Nakamura, R. Takizawa and H. Hanawa (1983) *Anal. Chim. Acta*, 149, 147-155.
D. Jagner (1978) *Anal. Chem.*, 50, 1924-1929.
D. Jagner (1979) *Anal. Chem.*, 51, 342-345.
D. Jagner and A. Graneli (1976) *Anal. Chim. Acta*, 83, 19-26.
D. Jagner and K. Årén (1978) *Anal. Chim. Acta*, 100, 375-388.
D. Jagner and K. Årén (1982) *Anal. Chim. Acta*, 141, 157-162.
D. Jagner, M. Josefson and K. Årén (1982) *Anal. Chim. Acta*, 141, 147-156.
A.K. Jain, R.P. Singh and G. Bala (1982) *Anal. Lett.*, 15, 1557-1563.
J. Janata (1982) *Anal. Proc.*, 19, 65-68.
M. Janghorbani, B.T.G. Ting, A. Nahapetian and V.R. Young (1982) *Anal. Chem.*, 54, 1188-1190.
G. Jarzabek and Z. Kublik (1982) *Anal. Chim. Acta*, 143, 121-130.
J.-M. Kauffmann, G.J. Patriarche and G.D. Christian (1981) *Anal. Lett.*, 14, 1209-1220.
C.S. Klusek and M. Heit (1982) *Bull. Environ. Contam. Toxicol.*, 28, 202-207.
A.V. Kopytin, P. Gábor-Klatsmányi, V.P. Izvekov, E. Pungor and G.A. Yagodin (1983) *Anal. Chim. Acta*, 148, 35-40.
C. Labar and L. Lamberts (1981) *Anal. Chim. Acta*, 132, 23-33.
P.C. Leung, K.S. Subramanian and J.C. Méranger (1982) *Talanta*, 29, 515-518.
D.E. Leyden and W. Wegscheider (1981) *Anal. Chem.*, 53, 1059A-1065A.
S. Mannino (1982) *Analyst*, 107, 1466-1470.
C. McCrory-Joy and J.M. Rosamilia (1982) *Anal. Chim. Acta*, 142, 231-238.
G.J. Moody, J.D.R. Thomas and M.A. Yarmo (1981) *Lab. Prac.*, 30, 1111-1112.
H.W. Nürnberg (1982) *Pure & Appl. Chem.*, 54, 853-878.
H.W. Nürnberg and B. Raspor (1981) *Environ. Tech. Lett.*, 2, 457-483.

R.S. Posey and R.W. Andrews (1981) *Anal. Chim. Acta*, 124, 107-112.
E. Pungor, Zs. Fehér, G. Nagy, E. Lindner and K. Tóth (1982) *Anal. Proc.*, 19, 79-82.
E. Pungor, K. Tóth, G. Nagy, L. Pólos, M.F. Ebel and I. Wernisch (1983) *Anal. Chim. Acta*, 147, 23-32.
R.D. Rocklin and E.L. Johnson (1983) *Anal. Chem.*, 55, 4-7.
R.S. Sadana (1983) *Anal. Chem.*, 55, 304-307.
S. Salmela, E. Vouri and J.D. Kilpiö (1981) *Anal. Chim. Acta*, 125, 131-137.
H.D. Seltner, H.R. Linder and B. Schreiber (1981) *Intern. J. Environ. Anal. Chem.*, 10, 7-12.
G.J. Sevenich and J.S. Fritz (1983) *Anal. Chem.*, 55, 12-16.
N. Sheffrin and E.E. Williams (1982) *Anal. Proc.*, 19, 483-487.
J. Simon and T. Liese (1981) *Fresenius Z. Anal. Chem.*, 309, 383-385.
J.R. Sullivan and J.J. Delfino (1982) *Talanta*, 29, 345-364.
B.T.G. Ting, A. Nahapetian, V.R. Young and M. Janghorbani (1982) *Analyst*, 107, 1495-1498.
Er-Kang Wang and Meng-Xia Wang (1982) *Anal. Chim. Acta*, 144, 147-153.
J. Wang (1982a) *Anal. Chem.*, 54, 221-223.
J. Wang (1982b) *Environ. Sci. Technol.*, 16, 104A-109A.
J. Wang (1982c) *Talanta*, 29, 125-128.
J. Wang and H.D. Dewald (1983) *Anal. Chem.*, 55, 933-936.
J. Wang, H.D. Dewald and B. Greene (1983) *Anal. Chim. Acta*, 146, 45-50.
M.T. Watts, B.J. Poulsen, L.L. Szabo, M.A. Kenny and W.Y. Lee (1981) *J. Anal. Toxicol.*, 5, 231-235.
A.F. Zhukov, D. Erne, D. Ammann, M. Güggi, E. Pretsch and W. Simon (1981) *Anal. Chim. Acta*, 131, 117-122.

PROCESSING BIOLOGICAL SAMPLES FOR METAL ANALYSES

Markus Stoeppler

Institute of Applied Physical Chemistry,
Nuclear Research Center (KFA) Juelich, GmbH.
P.O.Box 1913, D-5170 Juelich
Federal Republic of Germany

1. INTRODUCTION

It is still valid that sampling, sample preparation, sample storage and various treatment procedures prior to determination are the least spectacular but with respect to reliability most crucial steps of an analytical procedure. Therefore it is obvious that proper performance of this part of metal analysis requires a highly skilled staff. This is particularly the case for metal analyses in biological materials (i.e. urine, whole blood, blood derived fluids and tissues) where the contents are by far predominantly of the order of a few µg/l or µg/kg and can only reach the mg/kg-level in particular cases (see below) of frank exposure.

Therefore, this type of analysis requires properly planned and performed sampling, with subsequent storage and treatment steps which also include appropriate measures for the minimization of contamination and losses. Due to the fact that not only the total concentration of a metal or metalloid is of importance but also the type and amount of different chemical forms, sample processing should increasingly provide a material with composition as far as possible chemically unchanged. This adds another task to the usual conservation measures and is an important research field, for example, within biochemical studies and specimen banking programs (Stoeppler et al, 1982).

This paper thus sets out to summarize, after a definition of the most important analytical tasks, the state-of-the-art of all steps of procedures which can be summarized as processing prior to the determination step. Since all aspects of contamination are treated in detail elsewhere (Nieboer,1983), this item will only be mentioned here where appropriate and information on more or less new technical developments will be given priority.

Chemical Toxicology and Clinical Chemistry of Metals
ISBN 0.12.137520.X

Copyright © 1983 by IUPAC
All rights of reproduction in any form reserved

2. DEFINITION OF TASKS - SELECTION OF INDICATOR MATERIALS

Trace metal analyses are required mainly in three areas of significantly different concentration levels whose values pre-determine the efforts to be made for a reliable analytical approach at all relevant procedural stages.

- *Level 1:* Toxic concentrations in cases of acute poisoning, examples: Tl, As, Hg. Here on average the trace metal contents in the selected indicator specimens are at the mg/l or mg/kg level and contamination precautions do not play any significant role in sampling, sample preparation and sample treatment. The only important fact in this connection is the proper selection of indicator specimens, which very often are urine, whole blood or serum if recent poisoning is considered. In order to trace back particular exposure, body fluids as well as tissue samples (post mortem) and in some, e.g. poisoning, cases also hair can be used successfully for retrospective studies (Hagedorn-Götz and Stoeppler, 1975). Usually no significant risk of contamination is encountered in those investigations; however, for specimen collection from corpses, possible post-mortem changes, for example change in moisture content, of organs taken for analysis have to be considered.
- *Level 2:* Occupational and environmental exposure. In this case trace metal contents in biopsy materials such as urine, serum or whole blood range from slightly to significantly enhanced values possibly up to two or even three orders of magnitude higher than normal levels. Therefore sampling has to be performed with care and taking appropriate contamination precautions. Particularly in occupational exposure contamination from the working environment is likely to occur if, for instance, urine samples have to be taken. The latter should be performed preferably at home and after having taken a bath or a shower (Ni, Cd, Pb).
- *Level 3:* Normal conditions, i.e. healthy, nonexposed subjects. Here also rather small concentration differences due to nutritional or smoking habits sometimes have to be analyzed in epidemiologic or basic investigations on normal trace metal levels. This obviously constitutes the most difficult task and requires utmost care from the preparatory steps until final analysis. In this area, the last decade, particularly the period from 1979 onwards has been characterized by growing evidence that a considerable part of formerly reported data on trace metal contents in body fluids can be regarded as erroneous. Recent findings with improved methods demonstrate that the content of various elements in body fluids is at the µg/l level or even below for normal, nonexposed subjects. An overview based on the present state of knowledge for various important metals and arsenic in body fluids is given in Table 1. This information is taken from several very recent

papers based on extremely careful analysis including proper sampling and sample preparation techniques (Versieck and Cornelis, 1980; Brockhaus et al, 1980; Sunderman, 1980; Everson and Patterson, 1980; Hall and Fell, 1981; Brockhaus et al, 1981; Stoeppler, 1982; Frech et al, 1982; Vahter, 1982, Veillon et al, 1982, Ping et al, 1983; Jawaid et al, 1983; Stoeppler and Apel, 1983, Ostapczuk et al, 1983)

TABLE 1

Probable Median Endogenous Trace Metal Contents or Ranges in Body Fluids of Nonexposed Subjects.
(Data from various Recent Publications; Values in μg/l)

Element	Urine	Serum/Plasma	Whole Blood	Remarks
Al	<5	<2	≤ 2	
As	<10	<1	≤ 1	Inorganic Arsenic
Cd	≤ 0.5	<0.05	≤ 0.5	Nonsmokers
	<1	<0.1	≤ 1	Smokers
Co	≤ 0.5	<0.2	<0.2	
Cr	<1	<0.5	<1	
Hg	≤ 2	<2	≤ 2	
Ni	≤ 2	≤ 1	≤ 2	
Pb	<10	<1	$<50-300$	Females Lower
Tl	≤ 0.5	<5	<5	

3. SAMPLE COLLECTION - CONTAMINATION CONTROL

During sampling, losses may occur, but contamination from vessels used for material collection, from knives, pincers etc. for tissue preparation, from improper handling (urine voiding) and from extraneous sources (dust, analyst) is more likely. Thus, proper selection and preparation of all sampling gear is of paramount importance and all relevant operations have to be performed with utmost care in order to minimize contamination and losses (Behne, 1980; Frech et al, 1982, Heydorn and Damsgaard, 1982; Versieck et al, 1982, Nieboer, 1983; Stoeppler, 1983 a). Contamination is a severe source of error at this stage, particularly for elements with ubiquitous occurrence like aluminium, manganese and lead, but also for chromium, and cobalt (and especially for nickel) due to the wide use of highly resistant steels like the numerous stainless steels. Therefore it is evident that without rigid contamination control measures, considering the analyst himself, no sound results can be obtained, at least at normal or slightly elevated levels, if there is any significant risk of contamination at the level investigated. Therefore, an appropriate concept of clean sampling and laboratory conditions is necessary to ensure proper working conditions, ranging from rather simple to highly sophisticated approaches. The latter predominantly consist of different combined dust control systems with clean sampling and sample processing operations (Tschöpel et al, 1980, Harrison et al, 1981, Moody, 1982, Stoeppler et al, 1982).

4. SAMPLE WEIGHT AND ITS DEFINITION

For solid biological materials with significant moisture content the concentrations are often expressed in the term content per *fresh weight*. Fresh weight, however, has the disadvantage that, owing to the continuous loss of moisture, even during short-term storage in freezers, the fresh weight reported may decrease considerably depending on storage time and storage temperature. Hence, if it is not possible to perform sub-sampling and weighing of these sub-samples exactly at the time of sampling, the term *fresh weight* has to be replaced by the more practical terms *wet weight* or *actual weight*. This, however, needs proper definition which could be done in the following order: *fresh weight* ⩾ *wet weight* ⩾ *actual weight* ⩾ *dry weight*. Actual weight can also be used to define the condition of a material, e.g. previously freeze-dried but later again containing some moisture.

From this it is obvious that terms like *wet weight* and *actual weight* only mean that a certain material still has a considerable moisture content. In a large sample, for example a whole organ, the outer parts will contain less moisture than the inner. This has to be considered if subsampling is required. Another fact, already mentioned above, is the change in weight and elemental composition of autopsy materials if not taken immediately post mortem.

This may illustrate that, for analytical protocols, terms such as *wet weight* and *actual weight* are somewhat nebulous and should be replaced by a more certain weight definition. Therefore, numerous analytical chemists prefer, for materials with initially high moisture contents, to report the analytical data on the basis of *dry weight*. With a fresh weight determined as reliably as possible, this will be very often rather typical than an individual fresh weight, the data reduction to fresh weight being possible, in principle.

These points are not mentioned here to advocate the continuous use of dried materials for analysis. There are elements or compounds that may be, at least partially, lost or decomposed during drying. Also drying prior to analysis can enhance contamination risks at ultratrace levels and may lead to difficulties in decomposition. For such materials, subsamples should be taken for an independent drying procedure to obtain a normalized weight as well as for further treatment without any drying step.

The best means of *dry weight* determination, however, is still under discussion in order to arrive at a biologically sound basis:

- Oven-drying to constant weight around 100°C is doubtless the method of choice for robust materials. Biological materials, however, can be sensitive to higher temperatures and may decompose to some extent. Hence they may not reach a constant weight within a reasonable time.
- For sensitive materials, careful drying at temperatures far below 100°C, drying in a desiccator (e.g. with phosphorus pentoxide) or freeze drying seems to be more reliable.

In analytical practice, however, these methods can only be applied with success if they have been carefully checked using an appropriate material preferably with known contents or added radionuclides (Krivan, 1982). Such investigations commonly provide enough information for weight normalization and at least arbitrary content-weight relationships. Properly standardized drying procedures thus commonly lead to content-weight relationships with error limits well below the average error of the required analytical determinations. In the analysis of biological materials other normalization factors could be additionally used, for example internal biochemical standardization by determination of carbon, sulfur or nitrogen. This deserves to be investigated in some detail in further studies on biologically meaningful data normalization (Stoeppler and Nürnberg, 1983).

5. HOMOGENEITY AND HOMOGENIZATION

Analytical results significantly depend on the homogeneity of the specimen because a certain level of homogeneity is necessary for comparable data. Specimens that have a structure leading to an uneven distribution of the trace metal to be analyzed, for instance cadmium in the kidney and kidney cortex, can be of particular value in that micro-samples taken either for digestion or for direct analysis reflect such a structure. In general, however, the analyst would require a homogeneous sample, that means a sample for which inhomogeneity is significantly lower than the standard deviation of the method applied. Thus, the sample amount taken for analysis has to be chosen in such a way that these prerequisites can be fulfilled. If this is not the case, or if the tremendous detection power, for instance of Zeeman compensated AAS systems, is used for direct analysis of solids the materials have to be homogenized for subsequent determination.

Homogenization can be performed by different approaches, either wet or after deep freezing to obtain a final sample without loss of possibly volatile compounds (Stoeppler et al, 1982), or after a drying procedure. In the latter case, a better homogeneity down to the mg-level can be obtained at present with e.g. ball mills made from inorganic materials such as zirconium oxide, corundum etc. or PTFE. However, grinding procedures can introduce contamination from the materials of the mill and thus have to be carefully investigated if analysis at very low levels is required. Corundum components, for example, show significant contamination for chromium, manganese, nickel and lead (van Grieken et al, 1980). Thus for biological materials (tissue) PTFE mills using the brittle fracture technique are most suitable according to recent experience (Harrison et al, 1981).

6. SAMPLE STORAGE

During storage, contamination as well as losses are likely to occur requiring careful selection of storage conditions and storage containers. Short-term storage at about $+4^{\circ}C$ is feasible for body fluids and tissues. Anticoagulants (heparin, citrate, EDTA) are commonly added to blood, depending on the analytical requirements, in order to prevent clotting. For medium- and long-term storage, however, storage at lower temperatures (organs, blood, serum) or acidification (urine) is more appropriate. We have observed, however, that for medium- and long-term storage of urines, storage at lower temperatures (below $-16^{\circ}C$) without acidification appears to provide the best analytical results (Mohl et al, 1983). Losses due to the adsorption of trace elements from urine onto the container walls are negligible in the present author's experience.

If speciation measurements are necessary, acidification must always be avoided and the materials stored deep-frozen. The latter is also favourable if contamination and losses have to be minimized. In specimen banks storage temperatures can be as low as $-195^{\circ}C$ (liquid nitrogen). Satisfactory information on long-term behaviour of chemical species is not yet available but this is now being investigated in various research areas, particularly within projects on specimen banking and monitoring related to banking (Gills et al, 1979, Harrison et al, 1981, Stoeppler et al, 1982).

There is some indication in the literature that following storage recovery problems may exist for some metals, particularly for aluminium (Savory et al, 1983). This should be thoroughly investigated because of its general importance for any effort on retrospective analysis and quality control. From present knowledge, however, it can be generally stated that sample storage appears to be feasible for numerous elements in various biological materials.

7. SAMPLE TREATMENT

The extraordinary gain in detection power and the introduction of instrumentation capable of directly determining numerous elements in body fluids and also in solids have brought new dimensions into trace analytical chemistry as far as AAS is concerned (Slavin et al, 1983, Stoeppler, 1983 b).

Another method, however - voltammetry - has gained in importance and detection power as well during the last decade. It is evident that for a couple of important toxic metals the detection limits in solution are significantly lower than those of AAS (Nurnberg, 1982; Stoeppler and Nurnberg, 1983). Voltammetry, however, is at its best if the metals are in solution from a complete prior digestion.

Thus, this perspective to some extent has altered strategies in sample treatment. Besides the still necessary digestion for reference methods and of appropriate amounts of less homogeneous materials, the most important development in sample treatment is the improvement in digestion blanks in order to cope with instrumental progress. Table 2 gives a general overview of the most frequently applied digestion principles. There are two nearly comprehensive reviews on this matter, which are still valuable for general information (Bock, 1978, Tschöpel, 1980). Further information on very recent progress in digestion procedures, mainly for biological materials, is given below in the sections dealing with details of the techniques mentioned in Table 2.

7.1 *Dry ashing at higher temperatures*

If the volatility of a particular element or compound in a certain matrix is low, i.e. less than 800°K (527°C), dry-ashing can be the method of choice in that it provides a completely ashed residue. It can be performed in various types of muffle furnace offering temperature programming and, if necessary, with inner parts made from quartz for rather clean procedures. Dry ashing is undoubtedly useful for milligram per kilogram contents. Furthermore, oven ashing is inexpensive and easy to perform at these concentration levels. However, if analyses at lower levels are required oven ashing cannot be recommended for routine use. At this level errors may be introduced due to contamination from oven walls, dust or ashing aids such as sulfuric acid, nitric acid or magnesium nitrate. Moreover, losses can occur because of adsorption on the walls of the crucibles containing the material to be ashed. Despite the use of dry ashing in a couple of laboratories even for analyses at the microgram per kilogram level a warning should be extended that this requires particular skill and experience. A general use of dry ashing for sample pretreatment at micrograms per kilogram levels is therefore not to be recommended if there is no substantial experience of ultratrace analytical chemistry.

TABLE 2

Overview of the Most Commonly Applied Digestion Principles for Sample Treatment Prior to Analytical Determinations

Principle	Performance	Remarks
Dry ashing in various furnace types at temperatures up to 1000°C	Complete ashing, sometimes ashing aids required, throughput medium to high	Inexpensive instrumentation. Acceptable for higher contents, difficult (contamination) at lower levels and for volatile elements. Useful for all methods
Low-temerature ashing at about 150°C in excited oxygen plasmas. Use of cooling traps for volatile elements possible	Low blanks and complete ashing. For several materials extremely long ashing time, throughput low	Expensive instrumentation Improvements possible and already commercially available
Oxygen combustion with cooling traps for volatile elements	Low blanks Very effective for fatty materials, throughput low	Expensive instrumentation, less useful for routine analyses
Wet digestion in open systems (glass or quartz)	Complete ashing possible. Various acid mixtures applicable, automated systems also available, throughput medium to high	Low blanks require purified reagents. Method of choice prior to DPASV analysis. Difficult for volatile elements
Wet digestion in closed systems (PTFE, quartz etc) under pressure	Digestion not complete, multisample systems also available, throughput medium to high	Low blanks require purified reagents, excellent for volatile elements. Less useful for DPASV, but with exceptions. Instrumentation expensive, but price acceptable due to capacity and versatility
Tissue solubilization with quaternary ammonium compounds	Only partial digestion, thus not useful for all methods, throughput high	Blanks du to the reagents possible, very cheap, possibly new application fields with Zeeman-AAS

7.2 *Dry ashing at lower temperatures*

Trace metal losses and contamination also from the laboratory environment are minimal under proper working conditions with low temperature ashers (LTA). The ashing is performed in microwave-excited oxygen plasmas with partial pressures below 1 torr and at average temperatures below 423 K (150°C). This technique is often considered to be the most promising approach for trace and ultratrace analyses.

The majority of commercial instruments formerly available did not offer any possibility of trapping volatile elements such as mercury, arsenic, selenium and probably also chromium. In order to eliminate these sources of error, systems have been designed recently which avoid these disadvantages by cooling traps, so that low temperature ashing can be used successfully for all relevant metals (Tschöpel, 1980). Systems of this kind are commercially available now. Limitations of this method, however, are the expensive ashing devices and the usually time-consuming procedure. As the excited plasma only reacts at the sample surface, this can pose problems in the course of the ashing procedure if materials are treated that contain large amounts of inorganic constituents. Materials of that kind are, for example, bones if human specimens are concerned.

Because of these disadvantages modifications have been described very recently to improve this method. Introduction of fluorine into the oxygen plasma can drastically reduce ashing times but when added in gaseous form the use of fluorine leads to severe etching of all quartz surfaces. This now appears to be overcome by simply using PTFE dishes as both the source of fluorine and the sample container. This approach can accelerate ashing times up to about a factor of three (Williams, 1982).

A new low-temperature ashing system with a single vessel, cooling trap and stirrer to avoid incomplete ashing due to the formation of salt residues on the sample surface has also been developed very recently which appears to substantially improve the practicability of the method (Knapp, 1983).

7.3 *Combustion*

Combustion of organic material in a pure oxygen stream is a very effective ashing approach, particularly for fatty materials such as adipose tissue. As the reaction takes place at very high temperatures, trapping of volatile elements and compounds is necessary. This is best done by using appropriate absorbing solutions, with cooling "fingers", and cooled surfaces. A disadvantage of conventional and more recent systems is that high surface areas are involved which, despite an excellent ashing potential, prevents their extensive use in trace and ultratrace analysis due to contamination possibilities as discussed elsewhere (Stoeppler and Nurnberg, 1983).

Based on an earlier concept, an improved all-quartz combustion system with minimal surface area has been introduced very recently which appears to combine the advantages of combustion without the hitherto observed limitations. This device can be used in the absolute nanogram to microgram range for sample amounts up to 0.5 g dry weight as was demonstrated by the application of different analytical determination methods (Knapp et al, 1981; Raptis et al, 1982).

Methods of this kind, however, can only offer a comparatively low sample throughput but are indispensable as alternative procedures for difficult samples and to confirm trace analytical results with as many independent methods as possible.

7.4 *Wet ashing, open systems*

Wet digestion of biological materials, commonly with a single acid such as nitric acid with the aid of UV radiation or microwaves, but also with various acid mixtures has been extensively discussed in the literature. This technique still is the most frequently used at present prior to numerous and widely differing methods of determination. The reason is obvious: wet digestion is basically simple and cheap in its general approach and can be adapted to nearly any analytical task. It is furthermore possible, without extraordinary expense, to achieve a high sample throughput and to process samples up to several grams dry weight.

Series of samples can be digested in sand or quartz sand baths, on heating plates, in heating blocks, in microwave devices as well as in mechanized programmable systems which are commercially available. Digestion vessels consist of simple vessels of various forms or tubes made from laboratory glass, pyrex or quartz, frequently in connection with reflux systems to enhance the digestion potential and to minimize as far as possible reagent consumption as well as trace element losses and contamination. One of the most effective digestion mixtures is that of nitric and perchloric acid, sometimes with the addition of sulfuric acid. This mixture is capable of completely decomposing most biological materials for the interference-free subsequent voltammetric determination of numerous metals down to the lowest levels (Nürnberg, 1982, Ostapczuk et al, 1983). The effectiveness of this procedure could also be demonstrated in biological materials for metabolically incorporated radioactive selenium (Robberecht et al, 1982).

Limitations are the high consumption of acids if complete digestion is required and related contamination problems from the vessels and the reagents. Solutions from wet digestion may also pose problems for graphite furnace AAS if they contain too much acid, for example sulfuric or perchloric acid, which can destroy tubes and even attack platforms and thus may require evaporation prior to determination.

7.5 *Wet ashing, closed systems (pressurized digestion)*

In ultratrace analysis and particularly in the analysis of volatile elements, pressurized decomposition offers some advantages over most of the open digestion procedures. Nitric acid is mainly used as oxidant but the addition of other acids for particular purposes, for example hydrofluoric acid for silica containing samples, is of advantage. As the oxidation potential of nitric acid increases significantly at elevated temperature and pressure, the consumption of acid is usually lower than in open systems. This is also advantageous if blanks have to be extremely low.

Numerous devices, equipped with safety systems that allow blow-off in cases of overpressure to circumvent an explosion, are commercially available for sample weights up to several grams (dry weight). Systems are also offered for the simultaneous decomposition of several samples. Since a blow-off can be hazardous every material to be digested needs careful program evaluation and current control.

Very recently, the hitherto used decomposition vessels, mainly made from PTFE, have been replaced in new systems by others from glass or quartz. This allows the use of higher digestion temperatures (PTFE is limited to about $160^{o}C$) up to $300^{o}C$ or even higher (Knapp, 1983) and, in addition, can drastically reduce blanks as was shown recently for the ultratrace determination of mercury (May and Stoeppler, 1983).

These higher temperatures can now be expected to eliminate a severe disadvantage of hitherto used systems with PTFE vessels, i.e. that only partial ashing is achieved even for relatively simple materials which prevents, for example, direct DPASV analysis in digests from pressurized decomposition (Stoeppler and Nürnberg, 1983). This also applies to incomplete digestion of arsenobetaine compounds in urine after consumption of fish (Stoeppler, 1983 b, Stoeppler and Apel, 1983).

7.6 *Tissue solubilization*

For analytical methods that require only a homogeneous liquid sample (radiotracer studies, flame and furnace AAS, ICP-AES) simple tissue solubilization by quaternary ammonium compounds can be a rapid and reliable approach. Blood, tissue and hair samples can be solubilized at low temperatures, preferably over night, and the solution directly analyzed. Limitations are contamination from reagents and matrix interferences, mainly for GFAAS. However, the introduction of more effective compensation principles, such as Zeeman and Smith-Hieftje background compensation, may extend the application of this technique in the near future.

7.7 *Requirements for Reagents*

The steady increase in detection power particularly for DPASV and AAS also requires minimization of blanks for all digestion principles. Since commercially available reagents even of ultra-pure quality usually contain trace metals at a level of 0.1 to 1 µg/l additional purification for wet or pressurized digestion is frequently necessary. In conjunction with clean-room techniques for numerous elements blanks below 0.1 µg/l can now be achieved (Mitchell, 1982; Moody and Beary, 1982), so that the lowest natural levels of trace metals are very often made accessible even if a prior digestion is performed.

8. PRECONCENTRATION AND SEPARATION

Preconcentration and/or selective separation prior to determination is still required for numerous tasks. The reasons are either a lack of sensitivity or matrix interferences, sometimes both. This is generally achieved by the application of liquid extraction, extraction chromatography, evaporation and coprecipitation principles, amply discussed in the literature (e.g. Zolotov, 1978, Bächmann, 1981). Thus only a few aspects particularly relevant to biological materials will be mentioned here.

Separation of numerous metals directly from body fluids and appropriate digests of body fluids and tissues can be easily performed by liquid extraction with metal chelates e.g. ammonium pyrrolidine dithiocarbamate(APDC), sodium diethyldithiocarbamate, (NaDDTC; see also Fig. 1) hexamethylene ammonium hexamethylene dithiocarbamate (HMAHMDC), dimethylglyoxime, dithizone and many other compounds. Since the partition coefficients between the aqueous and the organic phase are of the order of about 50 to >1000 this will permit effective preconcentration and selective elemental or at least group separation. Particular conditions such as pH, addition of complexing agents, treatment with stripping reagents or selective chelators (e.g. dimethylglyoxime for nickel) also allow separation of certain metals or compounds.

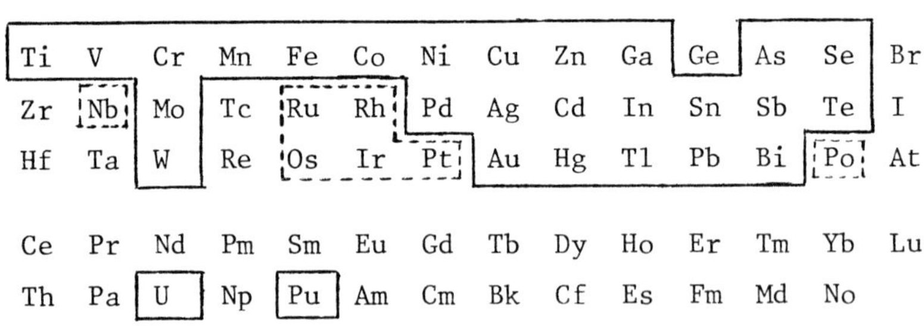

Fig. 1 Some elements which are completely extractable ———— or partially extractable --------- with sodium diethyldithiocarbamate and organic solvents.

For methods such as x-ray fluorescence spectroscopy, which require uniform thin-layer preparations, coprecipitation using carriers for example with carbamates, is very useful. The importance of contamination control and minimization for these procedures is obvious, at least if preconcentration is involved.

9. REFERENCES

K. Bächmann (1981) *CRC Crit. Rev. Anal. Chem.*, 12, 1-60.

D. Behne (1980) in *Trace Element - Analytical Chemistry in Medicine and Biology, Vol 1* (P. Brätter and P. Schramel Eds), Walter de Gruyter, Berlin pp 909-927.

R. Bock (1978) *A Handbook of Decomposition Methods in Analytical Chemistry*, International Textbook Company Glasgow.

A. Brockhaus, R. Dolgner, U. Ewers, U. Krämer, H. Soddemann and H. Wiegand (1981) *Int Arch Occup Environ Health* 48, 375-389.

A. Brockhaus, I. Freier, U. Ewers, B. Baginski, U. Krämer and R. Dolgner (1980) *Int Arch Occup Environ Health* 46, 59-70.

J. Everson and C.C. Patterson (1980) *Clin. Chem.*, 26, 1603-1607.

W. Frech, A. Cedergren, C. Cederberg and J. Vessman (1982) *Clin. Chem.*, 20, 2259-2263.

T.E. Gills, H.L. Rook, R.A. Durst, Eds (1979) *The National Environmental Specimen Bank Research Program*, EPA-600/1-79-017, United States Environmental Protection Agency.

R. van Grieken, R. van de Velde and H. Robberecht (1980) *Anal. Chim. Acta*, 118, 137-143.

D.J. Hall and G.S. Fell (1981) *Anal. Chim. Acta*, 129, 205-211.

S.H. Harrison, R. Zeisler, S.A. Wise Eds (1981) *Pilot Program for the National Environmental Specimen Bank - Phase I*, EPA-600/1-81-025, United States Environmental Protection Agency.

K. Heydorn and E. Damsgaard (1982), *Talanta*, 29, 1019-1024.

M. Jawaid, B. Lind and C.G. Elinder (1983) *Talanta* 30, 509-513.

G. Knapp (1983) in *Analytiktreffen* 1982 (K. Dittrich, Ed.) Karl-Marx-Universität Leipzig, pp. 87-101.

G. Knapp, S.E. Raptis, G. Kaiser, G. Tölg, P. Schramel and B. Schreiber, (1981) *Z. Anal. Chem*, 308, 97-103.

V. Krivan (1982) *Talanta*, 29, 1041-1080.

K. May and M. Stoeppler (1983) *Z. Anal. Chem.* (in press).

J.W. Mitchell (1982) *Talanta*, 29, 993-1002.

C. Mohl, M. Roth, P. Ostapczuk and M. Stoeppler (1983) in *Atomspektrometrische Spurenanalytik*, Band 2, (B. Welz, Ed), Verlag Chemie, Weinheim (in press).

J.R. Moody (1982) *Anal. Chem.* 54, 1258 A - 1376 A.
J.R. Moody and E.S. Beary (1982) *Talanta,* 29, 1003-1010.
E. Nieboer (1983), these Proceedings, pp.
H.W. Nürnberg (1982) *Pure Appl. Chem.,* 54, 853-878.
P. Ostapczuk, P. Valenta, M. Stoeppler and H.W. Nürnberg (1983), these Proceedings, pp.
L. Ping, K. Matsumoto and K. Fuwa (1983) *Anal. Chim. Acta* 147, 205-212.
S.E. Raptis, G. Kaiser, G. Tölg (1982) *Anal. Chim. Acta,* 138, 93-101.
H.J. Robberecht, R.E. van Grieken, P.A. van den Bosch, H. Deelstra and D. van den Berghe (1982) *Talanta* 29, 1025-1028.
J. Savory, A. Berlin, C. Cortoux, M.R. Willis and B. Yeoman, Eds (1983) *Proc. Int. Workshop on the Role of Biological Monitoring in the prevention of Aluminium Toxicity in Man,* Academic Press, in preparation.
W. Slavin, G.R. Carnrick, D.C. Manning and E. Pruszkowska (1983) *atomic spectroscopy,* 4, 69-86.
M. Stoeppler (1982) in *Edit. Proc. 3rd. Int Cd Conf. Miami 1981,* Cd. Assoc. London, Cd Council, New York, ILZRO, New York, pp. 95-102.
M. Stoeppler (1983 a) in *Trace Element Analytical Chemistry in Medicine and Biology,* Vol 2 (P. Brätter and P. Schramel, Eds), Walter de Gruyter, Berlin pp 909-927.
M. Stoeppler (1983 b), *Spectrochimica Acta, Part B,* (in press).
M. Stoeppler and M. Apel (1983), *Z. Anal. Chem.,* (in press).
M. Stoeppler, H.W. Dürbeck and H.W. Nürnberg (1982) *Talanta* 29, 963-972.
M. Stoeppler and H.W. Nürnberg (1983) in *Analysis of Heavy Metals in Human Toxicology* (A. Vercruysse, Ed), Elsevier, Amsterdam (in press).
F.W. Sundermann, Jr. (1980) *Pure Appl Chem.,* 52, 527-544.
P. Tschöpel (1980) in *Ullmanns Encyclopädie der Techn. Chemie,* 4. neubearb. und erw. Auflage, Band 5, Verlag Chemie, Weinheim, pp 27-40.
P. Tschöpel, L. Kotz, W. Schulz, M. Veber and G. Tölg (1980) *Z. Anal. Chem.* 302, 1-14.
M. Vahter, Ed (1982) *Assessment of Human Exposure to Lead and Cadmium Through Biological Monitoring,* Dept. of Environmental Hygiene, Karolinska Institute, Stockholm.
C. Veillon, K.Y. Patterson and N.A. Bryden (1982) *Clin. Chem.* 28, 2309-2311.
J. Versieck, F. Barbier, R. Cornelis and I. Hoste (1982) *Talanta,* 29, 973-984.
J. Versieck and R. Cornelis (1980) *Anal. Chim. Acta,* 110, 217-254.
E.V. Williams (1982) *Analyst,* 107, 1006-1012.
Yu. A. Zolotov (1978) *Pure Appl. Chem.,* 50, 129-148.

THE UTILIZATION OF SYNCHROTRON RADIATION FOR TRACE ELEMENT ANALYSES IN TOXICOLOGY OF METALS

A. L. Hanson, K. W. Jones, H. W. Kraner, B. M. Gordon, and J. R. Chen*

Brookhaven National Laboratory, Upton, NY 11973 U.S.A.
**State University College of Arts and Sciences, Geneseo, NY 14454 U.S.A.*

1. INTRODUCTION

The utilization of x-ray fluorescence (XRF) for trace element analysis has long been recognized in the biomedical community as an attractive technique for trace element analysis. Electron synchrotron-excited x-ray fluorescence (SXRF) provides several major improvements over conventional tube-excited fluorescence. Some of the improvements include: (1) higher brightness, several orders of magnitude improvement, resulting in the availability of monochromated photons with shorter analysis times; (2) high polarization of the photon beam so that scattering from the target can be reduced; (3) better sensitivities; calculations show sensitivities in the low ppb range for thin carbon matrices with short run times (Gordon (1982), Grodzins (1983)); (4) the photons have low divergence in both the horizontal and vertical direction so the beam can be focussed to a small spot, probably at least 1 μm. The proposed design for the BNL x-ray microprobe has been given previously by Hanson et al. (1983) ; (5) with monoergic beams, elements can be selectively fluoresced and chemical states can be determined.

The National Synchrotron Light Source (NSLS) at Brookhaven National Laboratory is a large facility dedicated only to the production of high intensity photon beams. When fully operational it will be able to support over 100 simultaneous experiments. In preparation for operation of the NSLS we have carried out several experiments at the CHESS facility at Cornell University. Initially, radiation damage to the structure of human blood and tissue samples was investigated with "white" (unfiltered) radiation. Subsequently, we studied sensitivities for trace element analyses and investigated several materials.

Chemical Toxicology and Clinical Chemistry of Metals
ISBN 0.12.137520.X
Copyright © 1983 by IUPAC
All rights of reproduction in any form reserved

2. RADIATION DAMAGE TO HUMAN BLOOD CELLS AND TISSUES

This experiment was to study the radiation damage resistance of certain types of cells on a gross basis, as it would relate to trace elemental analyses of entire cells and portions of tissue. Several samples of K_3EDTA-anticoagulated human blood were spun to a monolayer on a glass microscope slide and then air-dried (Slatkin, et al.(1983)). The slides were then irradiated in white radiation for 10-33000 seconds to fluences of 7×10^{15} to 2.4×10^{19} photons/cm^2. The average photon energy was 15 keV. Light microscope and SEM investigations showed no mass loss from either erythrocytes or leukocytes with fluences below 6×10^{18} photons/cm^2. Obvious mass loss from the leukocytes occurred at 1.6×10^{19} photons/cm^2, and destruction of the leukocytes occurred at 2.0×10^{19} photons/cm^2. Figure 1a shows a sample of cells outside the zone of radiation on one of the slides, and Figure 1b shows cells irradiated with 1.6×10^{19} photons/cm^2. This level of dose corresponds to 1.7×10^4 seconds radiation at the NSLS assuming a high bandpass (2%) monochromator at 10 keV with the machine operating at its full design specification of 2.5 GeV at 500 mA. Even though the samples irradiated with lower fluences showed no structural damage, they did show a resistance to staining. Samples of 8 μm human renal tissue were also irradiated for 1×10^4 to 2×10^4 seconds. These tissue samples also showed no apparent structural damage but stained poorly.

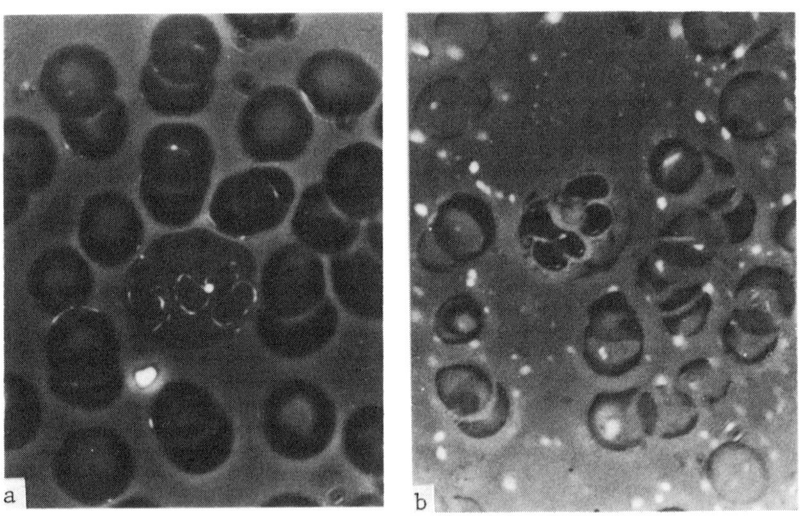

Figure 1. Air-dried K_3EDTA-anticoagulated human blood a) unirradiated and b) irradiated with 1.6×10^{19} photons/cm^2.

3. TRACE ELEMENT ANALYSES WITH SYNCHROTRON RADIATION

Several samples of NBS standard reference materials were analyzed for trace metals. In Hanson (1983) we showed a plot of minimum detectable limits (MDLs), as defined by Currie (1968), for trace elements in 1-mm thick samples of pressed NBS Orchard Leaves at several incident photon energies. All of the analyses were performed with a 30-mm^2 Si(Li) detector 15 cm from the sample. In Hanson et al. (1983) we presented the best MDL for each element analyzed. Between nickel and strontium we were able to obtain 160-290 ppb. The MDLs for elements lighter than nickel were considerably worse; 0.67 ppm for iron, 0.84 ppm for manganese, 14 ppm for calcium, and 23 ppm for potassium. One of the problems with analysis of the lighter elements was poorer transmission of photons through the beam line components at lower, more optimum energies.

Figure 2 shows a fluorescence spectrum of yeast fluoresced with an 8.4-keV monochromated photon beam, the monochromator having a bandpass of 5-10 eV. The sample was several layers of yeast spores sandwiched between two 10 $\mu g/cm^2$ formvar films. The spectrum was accumulated for 600 seconds with a 30-mm^2 Si(Li) detector. The interest in the yeast was in relation to ingestion of chromium and nickel at trace levels. The chromium content was less than 20 $\mu g/cm^2$ for a total of 3 μg of chromium in the beam, and the nickel concentration was 14 $\pm$ 2 $\mu g/cm^2$ for a total of 2 μg. This spectrum is notably

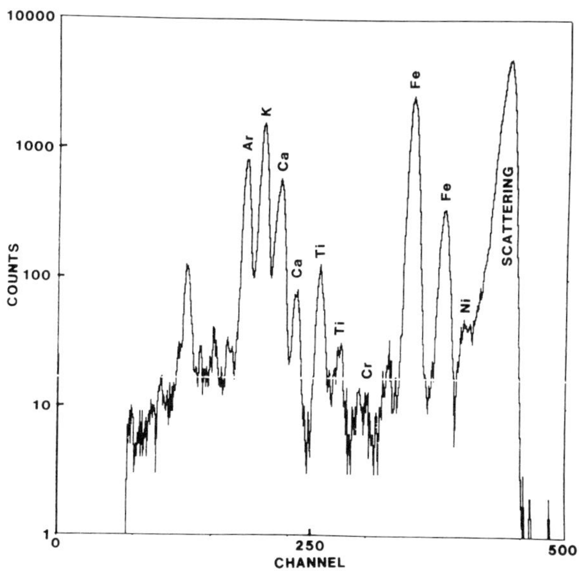

Figure 2. X-ray fluorescence spectrum of yeast fluoresced with an 8.4-keV monochromated photon beam.

cleaner (under the photo peaks) from background radiation than spectra from tube-excited XRF. Since monochromated radiation was used, the scattered radiation is confined to a fairly small energy region just below the incident photon energy. Since energy dispersive (Si(Li)) detectors have problems with incomplete charge collection, 3-5% of the photons at a particular energy will be distributed throughout the spectum at lower energies. Therefore, reducing the scattered radiation will reduce the background. Since the photons are highly polarized, proper placement of the detector can greatly reduce the scattered radiation. It is estimated that less than 10% of the photons were able to scatter in the direction of our detector at 90°.

4. CONCLUSIONS

The use of SXRF will nicely complement other more widely used analytical techniques for trace elements. The experiments at CHESS showed minimum detectable limits for 1-mm thick organic matrices with monochromated photon beams to be of the order of 160-300 ppb for Ni to Sr with minimal structural damage to the material being irradiated. Extrapolations to operating conditions at the NSLS, with a facility designed for XRF, indicate that MDL limits of 10-100 ppb should be achievable. The utilization of wavelength dispersive detectors should gain an order of magnitude in sensitivity, but with trade-off of some flexibility in multielemental analyses.

5. ACKNOWLEDGMENTS

This work was sponsored by the Processes and Techniques Branch, Division of Chemical Sciences, Office of Basic Energy Sciences, U. S. Department of Energy under Contract No. DE-AC02-76CH00016.

6. REFERENCES

L. A. Currie (1968) Anal. Chem, 40, 586.

B. M. Gordon (1982) Nucl. Instrum. Methods, 204, 223-229.

L. Grodzins (1983) Proceedings of Workshop on Uses of Nuclear Analytical Methods in Metal Toxicology, Brookhaven National Laboratory, October 11-12, 1983, Neurotoxicology (to be published).

A. L. Hanson, H. W. Kraner, K. W. Jones, B. M. Gordon, R. E. Mills and J. R. Chen (1983) IEEE Trans. Nucl. Sci. NS-30, 1339-1342.

D. N. Slatkin, G. C. Finkel, A. L. Hanson, K. W. Jones, H. W. Kraner and J. B. Warren (1983) Damage to Air-Dried Human Blood Cells and Tissue Sections by Synchrotron Radiation (in preparation).

DETERMINATION OF METAL CHEMOTHERAPEUTIC AGENTS (Al, Au, Li, and Pt) IN HUMAN BODY FLUIDS USING ELECTROTHERMAL VAPORIZATION WITH INDUCTIVELY COUPLED PLASMA ATOMIC EMISSION SPECTROSCOPY

Henryk Matusiewicz* and Ramon M. Barnes

Department of Chemistry, GRC Towers
University of Massachusetts, Amherst, MA 01003-0035, USA
**On leave from Technical University of Poznan,*
Department of Analytical Chemistry
60-965 Poznan, Poland

1. INTRODUCTION

The identification and quantitation of trace metals in biological systems is of ever increasing importance, particularly in the areas of clinical and forensic pathology. During treatment of rheumatoid arthritis with gold compounds described by Schattenkirchner and Muller (1983), manic-depressive psychosis with lithium carbonate reviewed in Palladino et al. (1983), certain human malignancies with anti-tumor agent cis-platin surveyed in Lippard (1983), and administration of aluminum-containing phosphate binding agents discussed by Alfrey et al. (1980), the development of analytical methodology capable of monitoring total metal levels in biological fluids is necessary.

The inductively coupled plasma (ICP) has over the past decade developed into a powerful tool for element analysis by atomic emission spectrometry (AES). Despite its widespread use, ICP-AES remains limited in routine applications by sample introduction methods that require large volumes of solutions. Considerable impetus has existed and continues for the development of ICP sample introduction techniques capable of accepting very small quantities of solutions.

An atomic emission spectrometric method is described for the direct determination of Al, Au, Li, and Pt at elevated levels in microvolume samples of human body fluids such as serum and urine. Simple modifications of two commercial electrothermal vaporizers (ETV) are described to convert them into sampling units for an ICP discharge.

Chemical Toxicology and Clinical Chemistry of Metals
ISBN 0.12.137520.X
Copyright © 1983 by IUPAC
All rights of reproduction in any form reserved

2. EXPERIMENTAL

(a) Apparatus

Experimental facilities and operating conditions are listed in Table 1.

TABLE 1

ICP-AES Facilities and Operating Conditions

ICP Generator	Plasma-Therm, 40.68 MHz with 3-turn load coil
Nebulizer	Babington with double glass spray chamber
Power	0.5 kW
Plasma Torch	Conventional, 18mm inside diameter, 1.0 mm aerosol orifice
Observation height	16 mm above induction coil
Argon flow rates	Outer 16 l/min, intermediate 0.2 l/min.
Spectrometer	1 m Czerny-Turner monochromator, 1:1 image, slit widths 50 μm, slit height 5 mm.
Detector	RCA 1P28 (700 V) and 4832 (1100 V) photomultipliers
Readout	log/linear recording picoammeter or computer (PDP 11/23)

The electrothermal vaporization arrangement is an adaption of a commercial instrument (Varian-Techtron Carbon Rod Atomizer CRA-90), in which a single graphite electrode replaced the standard electrode cup. A number of significant modifications were made since the original design was described by Barnes and Fodor (1983). Also, a modified controlled-temperature furnace atomizer (CFA) (Instrumentation Laboratory IL655) is used in both systems; the sample is vaporized from the modified graphite electrode or microboat, which is confined in a special double-walled argon-purged chamber or in an atomizer cell panel, into the ICP discharge, followed by atomization and excitation in the plasma.

(b) Procedure

Sample Preparation. Commercially available (or pooled normal human serum) control serum, Miles Lab., and pooled urine were used. Two parallel experiments were conducted with serum and urine for each element under the element's individual optimum analysis condition. In the direct ETV-ICP analysis, 5-μL standard additions were made to aliquots of human body fluids, to a "synthetic" standard (3300 ppm Na, 1000 ppm K, 400 ppm Ca, 100 ppm Mg, 1 ppm Cu, 1 ppm Fe and 1 ppm Zn) and to distilled water, and 5-μL aliquots of these solutions were measured directly by ETV-ICP. In the digestion study, after insertion of the 0.2 mL sample into a multi-vessel Teflon bomb described by Matusiewicz (1983), 0.6 mL of concentrated nitric acid and 0.1 mL of concentrated perchloric acid were added, and the bomb sealed and placed in an oven at 130°C for 4 h. The contents were transferred to a volumetric flask, and diluted with water to 2 mL. In addition, metal-spiked serum or urine were used, and desired analyte element concentration were measured. Calibration curves were based on distilled water, "synthetic" standard, and actual samples.

Spectroscopic Analysis. After the plasma was ignited, the outer and intermediate gas flows, power, and viewing height were adjusted to the compromise conditions listed in Table 1. The dry, ash and atomized stages of the vaporizer supply were then set as listed in Table 2. The appropriate wavelength for each element was selected by aspirating a solution containing the analyte. The pneumatic nebulizer was then disconnected from the plasma torch and replaced by the ETV system. Then 5 μL sample solution was deposited into the electrode or microboat and the firing sequence was initiated. The desolvated, vaporized sample was carried into the plasma by the argon carrier gas. The emission intensity for the selected line of the element of interest was then recorded.

TABLE 2

Operating Conditions for ETV-ICP

Element	Wavelength (nm)	Flow rates (L/min) Carrier	Outer	Temperature[1] (°C) Ashing	Vaporization[2]	C_L[3] Limit of detection[3] (mg/mL)	(pg)	CV[4] (%)
Al	308.2	0.6	0.9	1500	2500	1.5	7.5	10
		0.75	5.0	1500	2500	1.0	5.0	7
Au	242.8	0.75	5.0	1000	2700	3.5	17.5	5
Li	670.8	1.0	5.0	800	2200	1.0	5.0	6
Pt	214.4	0.6	5.0	1100	2800	4.0	20.0	6

[1] Dry at 120°C for 40 s; ash for 40 s; vaporize for 1 s or 5 s.

[2] Heating rate 800°C/s

[3] Corresponding to the concentration giving a signal equivolent to 3-times standard deviation of the background.

[4] At a concentration 100-times limit of detection.

3. RESULTS AND CONCLUSION

The recovery for Pt and Li added at a concentration of 1 µg/mL to serum and urine samples is illustrated in Table 3.

TABLE 3

Analysis of Spiked Biological Samples

Element	Matrix	% Recovery
Li	H_2O	100
	Synthetic	138
	Urine	135
	Digested Urine	134
Pt	H_2O	100
	Synthetic	97
	Serum	103
	Digested Serum	96

The results obtained for Al, Au, and Pt by the dilution with water, "synthetic" standard, real sample, and Teflon bomb digestion method are consistent. On the other hand, the high electrolyte concentrations present in serum and urine tend to enhance the Li signal. Accordingly, it is recommended that lithium in human serum or urine samples be determined by the method of standard additions or by calibration from artificial serum standard ("synthetic"). In addition, the organic constituents must be completely decomposed prior to aerosol sampling to avoid irreproducible aerosol sampling and double peaks. This was achieved by a gradual increase of the furnace temperature up to 1500°C.

4. REFERENCES

A.C. Alfrey, A. Hegg, and P. Craswell (1980) *Am. J. Clin. Nutr.* 33, 1509-1516.

S. J. Lippard (Ed.) 1983, Platinum, Gold and other Metal Chromotherapeutic Agents, *American Chemical Society*, Washington DC (1983)

A. Palladino, R.G. Longenecker, and L. J. Lesko (1983) *J. Clin. Psychiatry*, 44, 7-9.

M. Schattenkirchner and W. Muller, Eds, Modern Aspects of Gold Therapy (Rheumatology: An Annual Review, Vol. 8, S. Karger, AG, Basel (1983) Switzerland.

THE DETERMINATION OF ALUMINIUM IN BONE

Philip J. Metcalfe, J. Philip Day, Fiona M. Garstang, Kenneth C. Hodge and Peter Ackrill*

Department of Chemistry, University of Manchester Manchester M13 9PL, U.K.
**Artificial Kidney Unit, Withington Hospital, Manchester M20 8LR, U.K.*

1. INTRODUCTION

Knowledge of aluminium concentrations in bone is of importance in the management of renal dialysis patients, particularly in the diagnosis and treatment of fracturing dialysis osteodystrophy (osteomalacia).

Aluminium has been determined separately in marrow and trabecular bone, which provides a more reproducible analysis and hence more reliable chemical parameters than unspecific "whole bone" analysis. The technique has been applied extensively to studies of aluminium-related encephalopathy and bone disease in renal patients. In particular, the progress of treatment of osteomalacia by chelation therapy to remove aluminium has been monitored by bone histology and analysis for aluminium (Ackrill et al., 1982; Day et al., 1983).

2. METHOD

Bone specimens are obtained from renal patients by transiliac biopsy and sub-divided for histological examination and chemical analysis. Samples (ca 10mg) of cancellous bone are vacuum dried (20°C), weighed and then subjected to controlled gas phase oxygen plasma degradation (oxygen at ca 1 torr with 100W radiofrequency excitation). Initially, a first order decrease in mass is observed, with a half-life of about one hour. The process is halted when the softer marrow has been effectively ashed to completion (usually 8 hours), but when the trabecular fraction has remained essentially unaffected, confirmed by microscopic examination of the sample (continued ashing effects a much slower mass change, which we interpret as being due to oxidation of the protein component of the trabecular bone). Removal of the ashed marrow with 10^{-4}M nitric acid allows the subsequent dissolution of the trabeculae

Chemical Toxicology and Clinical Chemistry of Metals
ISBN 0.12.137520.X
Copyright © 1983 by IUPAC
All rights of reproduction in any form reserved

in 10M nitric acid. Aluminium in both fractions is determined by graphite furnace atomic absorption spectrometry using direct injection of the analyte solution. At the concentrations at which the other bone constituents appear in the final solutions, the aluminium determination is effectively free from chemical interferences.

3. RESULTS

Bone samples have been analysed from over 65 renal dialysis patients and 25 controls. In some cases patient biopsies have been repeated annually for up to 5 years.

The relative proportions of marrow and trabeculae vary widely, from 10:90 to 90:10 parts by (dry) weight, respectively (Table 1). It is also clear that the marrow and trabecular aluminium concentrations in each sample are not well correlated. Taken together, these facts suggest that "whole bone" analyses must present a less reproducible chemical measurement than separate aluminium determinations for the marrow and trabeculae.

Confirmation of the analytical results has been provided by specific staining for aluminium in thin sections of cancellous bone. There is a close correlation between the histochemical grading and the analytically determined aluminium (Table 2).

TABLE 1

Proportion of Trabecular Bone in Specimens Obtained from Renal Dialysis Patients and Controls

Mass* of trabecular bone as a percentage of the whole bone mass (in ranges of 20%)	Percentage of samples within each range
0 - 20	2.5
20 - 40	11.3
40 - 60	47.5
60 - 80	36.2
80 - 100	2.5

* Mass determinations refer to vacuum dried material.

TABLE 2

Mean Trabecular Aluminium Concentrations Related to Aluminium Staining Intensities

Number of samples	Mean aluminium concentration and range (mg/kg)	Intensity of stain*
19	8 (0 - 24)	0
10	28 (2 - 57)	1
18	102 (44 - 139)	2
21	183 (89 - 213)	3
15	302 (177 - 631)	4

* Stain intensity is estimated visually on the scale 0 - 4, where 0 indicates no aluminium and 4 indicates a strong aluminium stain throughout the trabeculae.

Both trabecular and marrow aluminium levels are raised in renal dialysis patients (range from 20 to 600mg/kg, compared with "normal" levels below 20mg/kg). In general, trabecular aluminium relates more closely to aluminium exposure from dialysis water supplies, summed over a long period (e.g. 5 years), whilst marrow aluminium is more variable and, like serum aluminium, relates more closely to recent aluminium exposure.

In renal patients with fracturing dialysis osteodystrophy, characterised by the bone condition osteomalacia, trabecular aluminium levels are usually raised well above 50mg/kg, with some correlation between aluminium concentration and severity of osteomalacia (OM), characterised histologically (Table 3). Elevated bone aluminium concentrations are also associated with the condition osteitis fibrosa (OF).

An effective treatment for fracturing dialysis osteodystrophy has proved to be by chelation therapy using desferrioxamine and subsequent removal of the aluminium complex by normal dialysis (Ackrill, et al.,1982; Day et al., 1983). The removal of aluminium is associated with the elimination of osteomalacia and the restoration of bone turnover, sometimes accompanied by osteitis fibrosa, and with a dramatic improvement in the patient's mobility and freedom from pain. We have concluded that aluminium is an important causative agent in this form of osteomalacia. Although the mechanism of operation remains to be determined, we speculate that aluminium at low concentrations in the osteoid, perhaps by preferential binding to the calcium carrier, may hinder the normal processes of calcification.

TABLE 3

Trabecular Aluminium Concentrations in Controls and Various Groups of Renal Dialysis Patients

Group*	C	R	OM1	OM2	OF
Al (mg/kg) (std dev)	9 (8)	45 (30)	192 (134)	282 (148)	125 (85)
Number in Group	25	9	17	27	12

* Groups: C = control group of non-renal patients; R = renal dialysis patients (rdp) with no adverse bone conditions; OM1 = rdp with less severe osteomalacia; OM2 = rdp with more severe osteomalacia; OF = rdp with osteitis fibrosa.

4. CONCLUSIONS

Separate trabecular and marrow aluminium determinations are possible, and provide inherently more reliable measurements than whole bone determinations.

Aluminium levels in trabecular bone in renal patients are related to levels of aluminium exposure integrated over a period of years.

Trabecular aluminium is a useful parameter in the diagnosis and treatment of fracturing dialysis osteodystrophy.

5. ACKNOWLEDGEMENTS

We wish to thank Professor J. Ball and Mr J. Denton, of the Department of Rheumatology, University of Manchester, for communicating to us the results of their aluminium staining on the bone specimens mentioned. Grants from the Medical and Science Research Councils, and the Kidney Unit Endowment Fund, Withington Hospital, are gratefully acknowledged.

6. REFERENCES

P. Ackrill, J.P. Day, F.M. Garstang, K.C. Hodge, P.J. Metcalfe, Z. Benzo, K. Hill, A.J. Ralston, J. Ball and J. Denton (1982) Proc. European Dialysis and Transplant Assocn, 19; 203-207 and Kidney International, 23, 552.

J.P. Day, P. Ackrill, F.M. Garstang, K.C. Hodge, P.J. Metcalfe, M. O'Hara, Z. Benzo and R.A. Romero-Martinez (1983), These Proceedings.

Determination of Cadmium and Lead in Whole Blood and Urine by Zeeman Flameless Atomic Absorption Spectroscopy

Stephen T. Wang, George Strunc and Frank Peter

Biochemistry Laboratory, Laboratory Services Branch, Ontario Ministry of Health, Toronto, Ontario, M9W 5K9 Canada

1. INTRODUCTION

The determination of lead and cadmium in blood and urine using graphite furnace atomic absorption spectrophotometer has become widely used in clinical laboratories. Several techniques such as matrix modification, Zeeman background correction and L'vov platform have been applied for the determination of lead in blood and urine. Fernandez and Hilligoss (1982) published a simple direct determination of lead in blood using the graphite furnace. Subramanian and Meranger (1981) reported a rapid graphite furnace procedure for determining cadmium and lead in whole blood using $(NH_4)_2HPO_4$ as a matrix modifier. Pleban and Pearson (1979) described a method for lead determination in whole blood and urine using Zeeman background correction. They used a standard curve from diluted whole blood matrix for blood lead determination and the standard addition method for urine lead. The paper presented here describes a direct methodology with simple dilution for the determination of cadmium and lead in blood and urine using a Hitachi Zeeman Atomic Absorption Spectrophotometer equipped with an automatic sampling system.

2. METHODS AND RESULTS

(a) *Apparatus and Instrument Setting*

A Hitachi Zeeman Graphite Furnace Atomic Absorption Spectro photometer, Model 180-80, equipped with a Hitachi data processing unit, Model 180-0205 and an automatic sampler, No.180-0127 were used. A standard graphite cup-type cuvette from Hitachi

Chemical Toxicology and Clinical Chemistry of Metals
ISBN 0.12.137520.X
Copyright © 1983 by IUPAC
All rights of reproduction in any form reserved

was employed. The instrumental parameter settings are:

Wavelength:	283.3 nm for Lead; 228.8 nm for Cadmium
Lamp Current:	5 mA for Lead; 7.5 mA for Cadmium
Slit:	1.3 nm
Analytical Mode:	Concentration
Measurement Mode:	Peak Area
Equation Type:	Linear Fit
Background Corrector:	Zeeman AA

To improve the precision and accuracy of the pipetting system, a MICRO LAB M pipettor from Hamilton, Switzerland was used to deliver the matrix solution.

(b) *Analytical Procedure*

One volume of the blood specimen is diluted with 9 volumes of 12.5 g/L $(NH_4)_2HPO_4$ - 5 g/L Triton-X-100. The stock solution of 50 g/L $(NH_4)_2HPO_4$ is purified by the Reagent Cleaning System of the Environmental Science Associates (ESA) Model 2014P, to remove the contamination of the metals. The urine specimens are diluted 1 + 1 with 25 g/L $(NH_4)_2HPO_4$ - 2 g/L Triton-X-100. The calibration curve is constructed by preparing the standards in the same matrix. Ten uL of the diluted sample is transferred to a graphite cup-type cuvette by the autosampler. Typical graphite furnace conditions for lead and cadmium are as follows:

Drying:	100-120°C, 60 s
Ashing:	700°C, 30 s
Atomization:	2000°C, 7 s
Cleaning:	2400°C, 3 s
Carrier Gas:	Argon, 200 mL/min

The optimum amount of $(NH_4)_2HPO_4$ was obtained by investigating the concentration at 0, 2.5, 7.5 and 12.5 g/L while keeping the Triton-X-100 at 5 g/L. We selected the matrix solution of 12.5 g $(NH_4)_2HPO_4$ - 5 g Triton-X-100 per litre for lead and cadmium in blood and 25 g $(NH_4)_2HPO_4$ - 2 g Triton-X-100 per litre in urine. The optimum ashing temperature was obtained by studying the temperature range from 500-1000°C. In the presence of the matrix modifier $(NH_4)_2HPO_4$, no loss of lead up to 1000°C was found and as high as 800°C can be used for cadmium. For routine analysis 700°C was the choice for both lead and cadmium in blood and urine. The slope of the calibration curve for lead and cadmium in aqueous matrix is parallel to the curve prepared from diluted blood matrix by the standard addition method over the linear analytical range. The method was linear to 0.97 µmol/L in diluted specimen both for blood and urine lead and to 89 nmol/L for blood and urine cadmium.

(c) *Precision*

The within-run and between-run precisions for lead and cad- in blood and urine are shown in Table 1. The between-run precision was estimated on the same specimens over a period of two weeks to one month. The coefficient of variation (CV) of within-run and between-run precision was 2.4-7.4% for blood lead; 3.1-8.1% for urine lead; 2.9-6.7% for blood cadmium and 3.9-7.2% for urine cadmium. The reproducibility obtained for lead and cadmium appears to be satisfactory.

TABLE 1

Precision of ZAAS Measurement of Lead and Cadmium in Blood and Urine

	Within-Run Precision			Between-Run Precision		
Sample	N	Mean	CV%	N	Mean	CV%
Blood Pb	20	1.62	3.5	21	0.94	7.4
(μmol/L)	20	3.80	2.4	22	3.05	4.7
	20	5.76	2.4			
Urine Pb	20	0.24	4.8			
(μmol/L)	20	0.60	3.1	11	0.60	8.1
Blood Cd	25	53	6.7	10	54	4.8
(nmol/L)	22	249	3.2	10	293	3.3
	22	463	2.9			
Urine Cd (nmol/L)	20	32	3.9	10	30	7.2

(d) *Analytical Recovery*

The recovery was determined by assaying a blood or urine sample after the addition of a known amount of lead or cadmium. The average analytical recoveries in % (standard added, μmol/L) are as follows:

Blood Lead: 99.6% (1.21); 101.6% (2.41) and 101.0% (4.83)
Urine Lead: 102.0% (0.24); 96.0% (0.48) and 98.0% (0.97)
Blood Cadmium: 107.6% (0.22) and 101.2% (0.45)
Urine Cadmium: 106.0% (44 nmol/L) and 96.0% (89 nmol/L)

The overall recovery for lead and cadmium at various concentrations was in the range of 96 to 106%.

(e) Comparison Studies

We found good correlation between the ZAAS (Y) and the Anodic Stripping Voltammetry (X) methods, when using an ESA 3010 for the determination of blood lead. The linear regression equation was: $Y=0.88\ X + 0.27$ µmol/L; correlation coefficient 0.9811, N=21; Mean $\pm$ S.D. (µmol/L), by Y method, 1.73 ± 1.0; by X method, 1.66 ± 1.13. Also, we analyzed blood lead proficiency testing samples supplied by the Centre for Disease Control and control blood for cadmium obtained from the Behring Institute to verify the accuracy of the method. Our results versus the assigned values are as follows: Our CDC blood lead result, 1.53 (assigned 1.53 µmol/L), 3.30 (3.34) and 5.14 (5.11); Behring blood cadmium 54 (assigned 49 nmol/L) and 293 (268). Our results are all within the acceptable ranges.

3. CONCLUSIONS

The Zeeman AAS method is being used routinely for the determination of lead and cadmium in blood and urine. Since there is a significant deterioration of the graphite cup after about 300 firings, frequent measurements of quality control specimens (Control Blood and Control Urine for Metal from the Behring Institute, Tox-EL from A.R. Smith and Ortho Urine Control II) are desirable. The precision of this method showed a better result than the method used by Pleban and Pearson (1979). The accuracy of the method is satisfactory as indicated by our results obtained in the CDC blood lead Proficiency Testing Program.

4. REFERENCES

F.J. Fernandez and D. Hilligoss (1982) At. Spectrosc., 3, 130-131.

P.A. Pleban and K.H. Pearson (1979) Anal.Lett., 12, 935-950.

K.S. Subramanian and J.C. Meranger (1981) Clin.Chem., 27, 1866-1871.

VOLTAMMETRIC DETERMINATION OF NICKEL AND COBALT IN BODY FLUIDS AND OTHER BIOLOGICAL MATERIALS

Peter Ostapczuk, Pavel Valenta, Markus Stoeppler and Hans Wolfgang Nürnberg

Institute of Applied Physical Chemistry,
Nuclear Research Center (KFA) Juelich
P.O.Box 1913, D-517o Juelich
Federal Republic of Germany

1. INTRODUCTION

A new extremely sensitive voltammetric method, based on application of differential pulse voltammetry after prior interfacial accumulation by an adsorption layer of suitable heavy metal chelates at the hanging mercury drop electrode, was used for the determination of Ni and Co in the lower µg/kg range (Pihlar et al., 1981; Ostapczuk et al., 1983).

The goal of this work was to develop a reliable procedure based on adsorption differential pulse voltammetry (ADPV) for the routine voltammetric determination of nickel and cobalt in a variety of biological materials.

2. EXPERIMENTAL

A Polarographic Analyzer System consisting of a Model 384 Console, RE0082 Digital Plotter, Model 3o3 SDME and a Model 3o5 Stirrer all from EG&E, Princeton, N.J. was used. The Model 384 is a second generation microprocessor-based polarographic analyzer. By controlling each step the microprocessor automates the polarographic or voltammetric determination.

The analyte medium was o.o5 M ammonia buffer of pH 9.2. It was prepared by weighing appropriate amounts of 25 % NH_3, 3o % HCl and deionized water into a 25 ml flask. Using Suprapur grade chemicals (E.Merck) and deionized water from a Milli-Q-Water-Purification System, blank levels for Ni and Co remained lower than 1o ng/l. The optimal concentration of the selected chelator, dimethylglyoxime (DMG), in the cell was $1.1o^{-4}$ M.

a) Sample pretreatment

For various biological materials a practical and universally applicable procedure prior to voltammetric determination is the following wet digestion: o.5 ml of a liquid or 5 to 2oo mg of a solid sample are mixed with o.1 to o.5 ml 7o % $HClO_4$ (E.Merck,

Chemical Toxicology and Clinical Chemistry of Metals
ISBN 0.12.137520.X
Copyright © 1983 by IUPAC
All rights of reproduction in any form reserved

Suprapur) and o.2 to 1 ml 65 % HNO_3 (E.Merck, Suprapur) in a quartz cup of 1o ml volume covered with a quartz watch glass. The temperature is maintained at 1oo°C until the evolution of nitric oxides almost ceases. The temperature is then increased to 2oo°C, the mixture is heated until a light yellow or colorless liquid results and then evaporated to dryness. The resulting white crystalline residue is dissolved in 2 ml of deionized water and 5µl of 7o % $HClO_4$.

b) Determination by adsorption differential pulse voltammetry

At first 5 ml of the blank analyte medium (o.o5 M ammonia buffer with $1x1o^{-4}$ M DMG) are put into the voltammetric cell and are deaerated for 4 min with pure Ar (99.999 %). Adjusting an adsorption potential of -o.7 V vs. the Ag/AgCl reference electrode for an adsorption time of 1 min adsorption of the $Ni(DMG)_2$ and $Co(DMG)_2$ traces in the blank analyte occurs. To speed up mass transfer the solution is stirred (8oo rpm) with a magnetic bar. Subsequently the stirring is turned off and after a rest period of 3o s the background voltammogram is recorded in the differential pulse mode (pulse height 5o mV, pulse duration 57 ms, clock

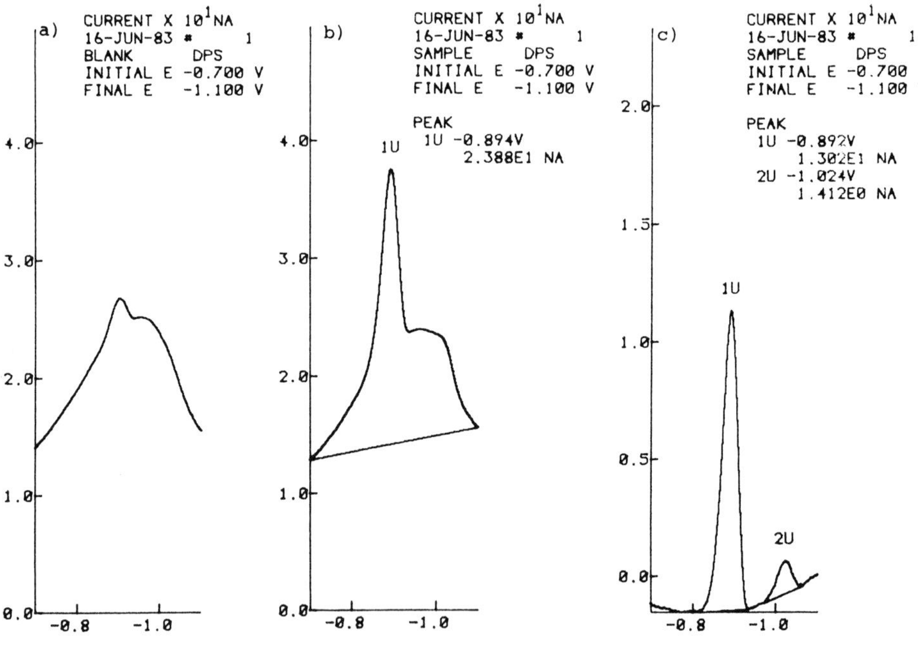

Fig.1 Ni and Co determination with Model 384 EG&G Polarographic Analyzer. a) Blank, b) Sample: o.1 µg/1 Ni and o.o1 µg/1 Co, c) Blank corrected signal.

time of pulses 0.5 s) by scanning with 4 mV/s into the negative potential direction. The blank curve is stored automatically. Then up to 1 ml of the sample solution resulting from the wet digestion is added. After new deaeration for 90 s the whole procedure with the adsorption stage and the subsequent recording of the voltammogram as before for the blank analyte is repeated. The stored background voltammogram is subtracted from the sample (Fig.1). This step improves significantly the determination power. The concentration of Ni and Co determined is automatically evaluated after three standard additions.

3. RESULTS and DISCUSSION

The precise determination of very low concentrations of Ni and Co is very important to establish normal levels in specimens from healthy adults. Voltammetric detection limits of 1 ng/l for Ni and Co in the analyte solution are attainable with good precision and accuracy; however, the real limits are blank levels and contamination. In the case of extremely low Ni contents contamination can be reduced significantly by laminar air flow clean bench operation during the digestion and voltammetric determination stage of the described procedure. The concentration of Ni and Co in Suprapur grade chemicals (E.Merck) is low and the blank values are about 0.1 ng for Ni and 5 pg for Co per digestion mixture.

TABLE 1

Ni and Co in Body Fluids and Tissues

Sample	n	Nickel Mean ± SD (µg/kg)	Nickel Range (µg/kg)	Cobalt Mean ± SD (µg/kg)	Cobalt Range (µg/kg)
Whole blood (nonexpos.)	9	1.41±0.37	0.7-1.8	0.09±0.02	0.08-0.12
Whole blood (exposed)	4	5.65±1.18	4.5-5.7	1.42±0.43	0.65-1.27
Urine (nonexpos.)	9	2.30±0.58	1.6-3.3	0.25±0.11	0.18-0.47
Urine (exposed)	4	5.27±1.22	3.9-6.7	0.88±0.51	0.50-1.75
Saliva	5	2.51±1.12	0.4-6.9	0.28±0.14	0.08-0.63
Human liver	15	7.87±6.34	3.0-37	42.3 ±15.2	20.7-75.6
Fingernail	3	1360 ±420	440 -3120	83.2 ±18.1	43.2-151
Toenail	2	240 ± 20	230 -289	29.7 ± 5.2	24.8-39.1

In Table 1 are listed recent results for Ni and Co contents in body fluids and biopsy tissues from living, nonexposed and exposed adults.

The normal level for nickel in whole blood samples is less than 2 µg/l which is lower than reported in the literature (Sunderman, 198o). Among persons who have been exposed to Ni powder, higher levels of Co and Ni were found in blood and urine.

In contrast to results with other heavy metals, such as Cd, Pb and Co, no enhanced concentration of Ni in human liver was observed.

The use of ordinary scissors to prepare fingernail samples for Ni and Co determination can lead to significant Ni contamination amounting to 3 - 2o mg/kg. When Ni and Co free scissors are used, differing concentrations in various fingernails can be detected. For right-handed persons the right thumb nail shows the highest Ni concentration. The results are reverse for left-handed persons. These differences are probably attributed to environmental influences. Thus, fingernails appear not to be useful indicator specimens for establishing normal levels of Ni and Co.

In contrast to fingernails no significant variation among toenails was found. Thus the toenails might be useful for the determination of normal levels in nails, although further studies are needed.

4. ACKNOWLEDGEMENT

This work has been financially supported by Deutsche Forschungsgemeinschaft under contract No Sto 149/2-1.

5. REFERENCES

B. Pihlar, P.Valenta and H.W.Nürnberg (1981) *Z. Anal. Chem.*, 3o7, 337-346

P.Ostapczuk, P.Valenta, M.Stoeppler and H.W.Nürnberg (1983) *Z.Anal.Chem.*, (in press)

F.W.Sunderman (198o) *Pure Appl.Chem.*, 52, 527-544

A DIRECT DETERMINATION OF COBALT IN BLOOD AND URINE BY ZEEMAN ATOMIC ABSORPTION SPECTROPHOTOMETRY

J. Molin Christensen, S. Mikkelsen* and A. Skov

Danish National Institute of Occupational Health,
DK-2900 Hellerup, Denmark
* *Department of Occupational Medicine, Rigshospitalet*
DK-2100 Copenhagen Ø, Denmark

1. INTRODUCTION

Cobalt is an essential trace nutrient as a component of vitamin B_{12}. It is widely used in industry as metal, catalyst and cobalt-blue colour. In animal cobalt has been proven to be a carcinogenic substance. Thus, there is much interest in monitoring this metal in biological fluids.

However, cobalt is a difficult element to determine reliably in biological material. Earlier published values of cobalt in blood and plasma of healthy individuals illustrate this: range 0.5 - 190 ng/ml (Barfoot and Pritchard, 1980).

Most of the previously published methods require extraction from the matrix (Wittkopf and Angerer, 1981), or digestion with acid followed by ion exchange separation (Lidums, 1979). Direct analysis is preferable because it is faster and minimizes possibilities for contamination during handling.

We describe a rapid, simple and sensitive method for direct determination of cobalt in whole blood and urine using ZEEMAN effect for background correction.

2. EXPERIMENTAL METHODS AND RESULTS

(a) *Apparatus*

A Perkin-Elmer Model 5000 atomic absorption spectrophotometer equipped with a ZEEMAN graphite furnace HGA-500 programmer, AS-40 autosampler and pyrolytically coated furnace tubes was used. The instrumental parameters were: Wavelength, 240.7 nm; Slit "Low", 0.7 nm; Scale expansion x5; Maximum power and Argon gasstop mode; drying at 90 °C to 250 °C in 105 sec.; charring at 1100 °C for 20 sec. and atomization at 2400 °C for 6 sec. A Nicolet digital oscilloscope was used to collect high speed data.

Chemical Toxicology and Clinical Chemistry of Metals
ISBN 0.12.137520.X
Copyright © 1983 by IUPAC
All rights of reproduction in any form reserved

(b) *Materials*

Cobalt standard solution 1.00 g/l (BDH Chemicals, England) was diluted in 1 M HNO_3 to 0.5 - 12 µg/l: Minisorb^R polypropylene tubes from Nunc (Denmark): Sure-Sep^R (plasma/serum separators) from General Diagnostics (Morris Plains, NJ, USA).

(c) *Procedure for Whole Blood*

Four 0.5 ml aliquots of vortex mixed blood were transferred to Minisorb^R tubes containing 100 µl of Triton x-100/K_2-EDTA (2.5 %/1 %). Each aliquot was then spiked with 1500 µl of solutions containing 0, 0.5, 1.0 and 2.0 ng/ml cobalt in 1 M HNO_3 respectively. Samples were vortex mixed for 5 sec., plugged up and allowed to stand at 60 °C for 30 min. (to prevent contamination packed in plastic bags). The samples were cooled and after adding Sure-Sep^R the solids were separated by centrifugation at 3000 g for 15 min. The supernatant was recentrifugated using Sure-Sep^R.

80 µl were pipetted into the graphite tube. Results were plotted and the initial cobalt concentration calculated from peak heights using linear regression on the four samples.

(d) *Procedure for Urine*

The well mixed urine specimen was filtrated through filter paper. Four 0.5 ml aliquots were placed in sample cups. Each aliquot was then spiked with 0.5 ml of solutions containing 0, 3.0, 6.0, 12.0 ng/ml cobalt in 5 % HNO_3. 10 µl were pipetted into the graphite tube. Calculation as for whole blood.

(e) *Precision*

Because there is no certified reference materials for blood and urine cobalt it was found useful to collect pool samples of urine and blood, which were run every day of analysis. That serves as a long-term quality control measure of the method. At a concentration of 2.6 ng/ml and 11.2 ng/ml the coefficients of variation for 10 blood and 10 urine samples were 6.2 % and 6.7 % respectively.

(f) *Interferences*

We also studied the effect of other metals as Fe, Ni, Mn, Zn and Cu in blood and calcium phosphate in urine. Fig. 1 shows the spectral interferences when determining cobalt in a matrix containing Fe recpectively Cu and Ni. Fig. 2 illustrates the cobalt and calcium phosphate background absorbance in urine during atomatization. In both cases spectral interference does not exist using Zeeman background correction.

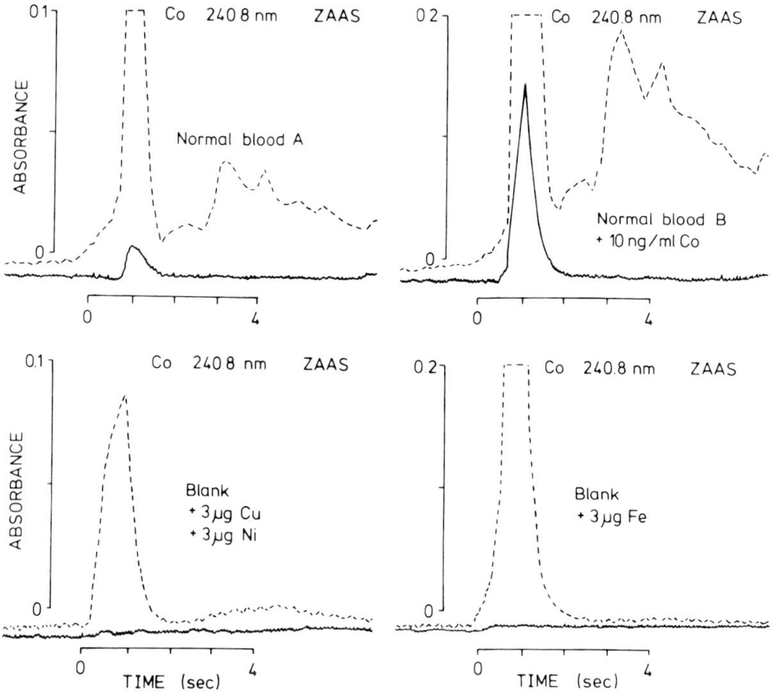

Fig. 1. The effect of metal ions on the determination of Co in blood.
Dashed lines: Background and Co absorbance
Solid lines: Zeeman background corrected absorbance

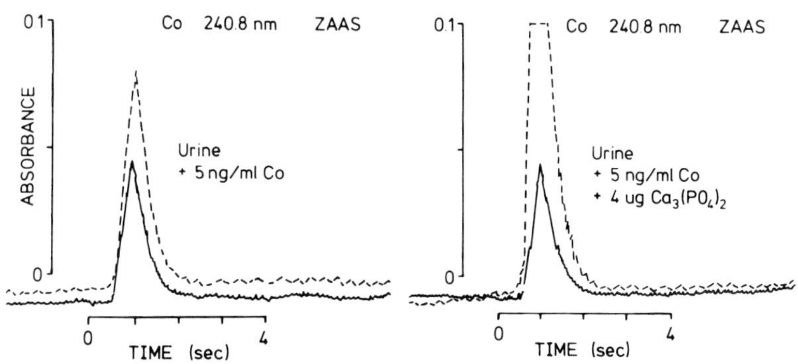

Fig. 2. The effect of calcium phosphate on the determination of Co in urine
Dashed lines: Background and Co absorbance
Solid lines: Zeeman background corrected absorbance

(g) *Inter-method Comparison*

Blood and urine samples from unexposed and exposed workers were analysed independently by Anodic Stripping Voltametry (DPASV) after wet digestion and by the present method. The results shown in Table 1 indicate good agreement.

Table 1

Comparison of Results of Analysis of Human Blood and Urine by AAS-ZEEMAN (ZAAS) and Anodic Stripping Voltametry (DPASV)

Sam-ple no.	Blood cobalt (ng/ml)			Urine cobalt (ng/ml)		
	ZAAS	DPASV	Deviation (%)	ZAAS	DPASV	Deviation (%)
63	1.6	1.56	+ 2.6			
67	7.0	7.18	- 2.5			
144	0.1	0.104	- 3.8			
8				126	120.8	+ 4.3
74				44.5	46.8	- 4.9
80				2.7	2.73	- 1.1
207				22.2	23.5	- 5.5

DPASV analyses performed at the Institute of Applied Physical Chemistry of the Nuclear Research Centre (KFA) Juelich, F.R.G.

(h) *Reference Values*

The sensitivity for cobalt in blood and urine was 0.1 ng/ml. The normal values for blood and urine cobalt were between < 0.1 - 0.6 ng/ml (median 0.15 ng/ml), < 0.1 - 7.7 ng/ml (median 0.3 ng/ml) respectively. Analysis of blood and urine from 50 exposed workers employed in a Danish porcelain factory gave concentrations between 0.2 - 24 ng/ml (median 1.0 ng/ml) 0.4 - 848 ng/ml (median 25.6 ng/ml) respectively.

3. CONCLUSIONS

A sensitive, precise and accurate analysis for cobalt in blood and urine was developed using Zeeman AAS. The results were similar to those obtained by anodic stripping voltametry (DPASV). The method is simple which minimizes contamination and it is applicable to determine Ni in whole blood and urine.

4. REFERENCES

R.A. Barfoot and J.C. Pritchard (1980), *Analyst*, 105, 551-556.

V.V. Lidums (1979) *At. Absorp. Letters*, 18, 71-72.

E. Schumacher-Wittkopf and J. Angerer (1981) *Int. Arch. Occup. Envir. Health* 49, 77-81.

ASSESSMENT OF QUALITY CONTROL SERA SUITABLE FOR ALUMINUM DETERMINATION BY ATOMIC ABSORPTION SPECTROMETRY

Fred Y. Leung and A. Ralph Henderson

Department of Clinical Biochemistry,
University Hospital,
London, Ontario, Canada N6A 5A5

1. INTRODUCTION

The determination of serum aluminum by atomic absorption spectrometry is important for the diagnosis and treatment of aluminium intoxication in renal dialysis patients (Parkinson *et al.*, 1981). The lack of assayed quality control sera for the performance of this test prompted us to prepare "in-house" sera containing added aluminum in known amounts. This material at various concentration levels is used for estimating precision and detecting systematic deviations in the analytic system. We also randomly selected and determined the content of aluminum in 13 commercially manufactured quality control sera to establish their suitability for use as control material of this test.

2. MATERIALS AND METHODS

The protocol for the preparation of the "in-house" quality control sera at three aluminum concentrations, and for the determination of aluminum in serum by atomic absorption spectrometry using a stabilized temperature platform furnace and matrix modifiers is as previously described (Leung and Henderson, 1982).

Commercially manufactured quality control sera used in this study are listed in Table 1. Each product was prepared as described by the Vendor except that purified water (Leung and Henderson, 1982) was used to dissolve the lyophilized sera. Freshly reconstituted sera were tested for up to 10 replicates to establish a mean aluminum concentration. A portion of each specimen was aliquoted into 1 ml fractions in washed polypropylene containers and stored frozen at -20°C. The remainder of the serum was stored in its original glass vial at 4°C, and assayed approximately every two to three days for up to four weeks. Liquid sera in ethylene-glycol base were stored at 4°C in their supplied plastic bottles during the study.

Chemical Toxicology and Clinical Chemistry of Metals
ISBN 0.12.137520.X
Copyright © 1983 by IUPAC
All rights of reproduction in any form reserved

TABLE 1

Identity of Control Sera Studied

DESCRIPTION		LOT NO.
Beckman Instruments Inc., Fullerton, CA 92634	Decision 1	C104031
	Decision 2	C104032
	Decision 3	C104033
American Dade Miami, FL 33152	Monitrol Ix	XLT-414
	Monitrol IIx	XPT-9610
General Diagnostics Morris Plains, NJ. 07950	Versatol Lo	4J718/4D726
	Versatol Hi	4H853/4D765
Fisher Scientific Co. Orangebury, N.Y. 10962	Sera-Chem N	203-081
	Sera-Chem A	434-111
Hyland Diagnostics Deerfield Ill. 60015	Omega I	4811W003C
	Omega II	4822Y004B
Ortho Diagnostics Raritan, NJ. 08869	WPQAS-I	X39X06
	WPQAS-II	X40X03

Standard statistical methods were used for the single classification anova (Sokal and Rohlf, 1969).

3. RESULTS

(a) *Assayed Values*

The commercially manufactured quality control sera were found to contain aluminum which spanned a wide range from 4 to over 1000 µg/l (37,000 nmol/l) as listed in Table 2. The within-run and between-run CV ranged between 0.7-7.5% and 3.1-11.7%, respectively.

(b) *Vial Variations*

Representative samples of four lots of lyophilized control sera with mean concentrations for aluminum of 18,77,124 and 276 µg/l were tested for vial-to-vial variation. Six vials of each lot were tested and analyzed by anova. The F ratios of between 1.1429 and 1.6302 indicated no significant difference between vials ($P>0.05$).

TABLE 2

Variance of Al in Quality Control Sera*

	Within-run			Between-run		
Serum	n	$\bar{x}$±SD	CV%	n	$\bar{x}$±SD	CV%
Decision 1	8	4±0.3	7.5	4	4±0.5	11
Decision 2	8	6±0.4	6.7	5	6±0.7	11.7
Decision 3	8	15±0.8	5.3	5	15±1.5	10
Monitrol I	5	381±6	1.4	9	385±12	3.1
Monitrol II	8	121±1.5	1.2	8	119±8	6.7
Versatol Lo	5	59±.9	3.2	6	59±5.5	9.3
Versatol Hi	6	94±3.5	3.7	7	98±10	10.2
Sera-Chem N	9	72±3	4.2	8	69±5	7.4
Sera-Chem A	9	17±1.5	8.8	8	16±1.8	11.2
Omega I	8	1248±9	0.7	7	1251±44	3.5
Omega II	8	1179±21	1.8	6	1149±35	3.0
WPQAS I	7	123±2.8	2.3	7	117±11	9.4
WPQAS II	8	271±5.2	1.9	8	261±19	7.3
In-house pool:						
Level 1	10	25±1.2	4.8	13	26±2	7.7
Level 2	10	54±2.0	3.7	14	53±4	7.5
Level 3	10	126±4.2	3.3	14	129±7	5.4

*Triplicate Readings; units µg/l (to convert to nmol/l multiply by 37.06).

(c) *Storage Effects*

The effects of storage at 4°C and -20°C on the concentration of aluminum are shown for four control sera (Figure 1). A significant increase ($P<0.05$) in aluminum concentration after 48 h is noted for sera stored at 4°C in their original glass vials when compared to samples stored at -20°C in micropolypropylene tubes. This effect becomes more marked after four weeks at 4°C. Liquid sera stabilized by ethylene-glycol were reproducible during storage at 4°C over a three week period of the study.

Pooled human sera prepared "in-house" were found to be stable during storage at -20°C over a 6-month period. No significant difference ($P>0.05$) was detected for these samples which had been aliquoted into 1 ml fractions and stored frozen.

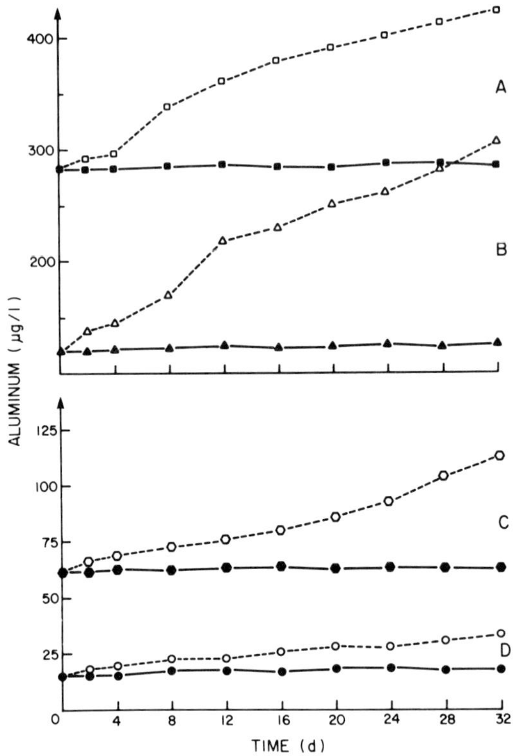

FIG.1 Storage effects on serum Al. A=WPQAS II, B=WPQAS I, C= SeraChem-N, D=SeraChem-A at -20°C(lines) and 4°C(dashed lines).

4. CONCLUSION

We demonstrated that quality control sera prepared "in-house" can be used effectively to monitor the determination of aluminum by atomic absorption spectrometry. Commercially manufactured material designed for routine chemistry application contains appropriate amounts of aluminum which can also be used for this purpose. With either source of supply, one would need to establish the residual mean amount of aluminum content by repeated analysis. Reconstituted or fresh serum samples should be aliquoted into (Al-free) clean polypropylene containers for storage at-20°C.

5. REFERENCES

F.Y. Leung and A.R. Henderson (1982) *Clin. Chem.* 28, 2139-2143.

I.S. Parkinson, M.K. Ward and D.N.S. Kerr (1981) *J. Clin. Path.* 34, 1285-1294.

P.R. Sokol and F.J. Rohlf (1969) in *Biometry*, Freeman and Company, San Francisco pp. 204-252.

BLOOD AND URINE SAMPLES FOR QUALITY CONTROL OF HEAVY METALS

B.Müller-Wiegand, D. Steuer and P. Heessels*

Research Laboratories, Behringwerke AG, 3550 Marburg, F.R.G.
**Diagnostics Division, Hoechst Canada Inc., Montreal-Quebec-H4R 1R6, 4045 Cote Vertu, Canada*

1. INTRODUCTION

Quality control of heavy metals lacked until recently commercially available samples. New special control bloods and urines have been produced for the quality control of heavy metals in industrial medicine and toxicology. The assigned values were determined by reference laboratories of the "Deutsche Gesellschaft für Arbeitsmedizin" with the most common methods according to the design of Hansert and Stamm (1980).

2. CONTROL BLOOD

(a) Composition

Control blood for metals is produced from bovine blood, collected by venipuncture: 1 mg EDTA/ml is added as an anticoagulant. It is heamolysed by ultrasonication prior to the addition of the metal components.

Control blood for metals contains the constituents at 2 different levels:

Concentration 1 contains the metals at the upper limit of the normal range or at the marginally affected range. Metals in concentration 2 are in a strong affected or toxic range.

The constituents, the concentration levels, and the methods of analysis are summarized in table 1.

Chemical Toxicology and Clinical Chemistry of Metals
ISBN 0.12.137520.X
Copyright © 1983 by IUPAC
All rights of reproduction in any form reserved

TABLE 1

Characteristics of Blood Preparations for Quality Control of Heavy Metals

Constituent Method of Analysis	Concentrations 1	 2	
Lead a. Extraction and flame atomic absorption b. Standard addition and electrothermal atomic absorption	200-400	700-800	µg/l
Cadmium Standard addition and Electrothermal atomic absorption	5- 10	25- 35	µg/l
Chrome Atomic absorption	8- 12	---	µg/l
Nickel Anodic stripping voltametry	5- 10	---	µg/l
Mercury Cold vapor atomic absorption	10- 15	90-100	µg/l

In order to avoid microbial growth, gentamycin and PHB ester have been added as preservatives.

(b) Stability

The control blood is lyophilized for ensuring long term stability (serveral years). It has to be reconstituted with 5 ml water, which naturally must be free of contaminant heavy metals.

Except for mercury, stability after reconstitution is 4 weeks at refrigerator temperatures (+2 to +8° C) or 2 weeks at room temperature (+20 to +25° C). The stability for mercury is 4 weeks at -20° C if frozen only once, or 2 days at +2 to +8° C, or 1 day at +20 to +25° C (Angerer et al., 1981).

TABLE 2

Characteristics of Urine Preparations for Quality Control of Heavy Metals

Constituent Method of Analysis	Concentrations Levels 1	 2	 3
5-Aminolevulinic acid (mg/l) ionexchange chromatography and photometry	4- 5	15-20	45- 65
Arsenic (μg/l) atomic absorption hydride method	50-100	200-200	600-700
Lead (μg/l) extraction and flame atomic absorption	20- 30	100-200	500-600
Cadmium (μg/l) a. extraction and flame atomic absorption b. standard addition and electrothermal atomic absorption c. anodic stripping voltametry	2- 3	10- 20	40- 50
Chrome (μg/l) standard addition and electro-thermal atomic absorption	3- 5	20- 25	50- 65
Fluoride (mg/l) ion selective electrode	1-1.5	4- 5	20- 25
Nickel (μg/l) extraction and electrothermal atomic absorption	3- 4	25- 30	70- 80
Mercury (μg/l) cold vapor atomic absorption	5- 15	100-200	800-1000
Trichloroacetic acid (mg/l) a. Fuijwara reaction and photometry b. gas chromatography	20- 30	150-200	---

(c) Package Forms

The two concentration levels of the control bloods are offered as Control Blood for Metals, for each concentration a pack with 4 vials to be reconstituted with 5 ml water.

3. CONTROL URINE

(a) Composition

Because of the variability of urine a synthetic matrix has been elaborated. It is composed of electrolytes and organic substances resembling native human urine.

Three control urines are produced with different concentration levels. Concentration 1 contains the metals at the upper limit of the normal range, concentration 2 in the elevated range, and concentration 3 in the toxic range.

Table 2 shows the constituents, the methods of analysis and the concentration ranges. Fluoride, 5-aminolevulinic acid and trichloroacetic acid have been added as diagnostically important constituents besides the well known heavy metals for industrial medicine.

(b) Stability

The stability of the lyophilized control urines is at least 2 years.

After reconstitution the stability for mercury and 5-aminolevulinic acid is 4 weeks at -20° C if frozen only once, 2 days at +2 to +8° C and 1 day at +20 to +25° C. All other constituents are stable after reconstitution for 4 weeks at +2 to +8° C and 2 weeks at +20 to +25° C.

(c) Package Forms

The three concentration levels of the control urines are offered under the trade mark *Lanonorm Metals* for each concentration;a pack with 12 vials to be reconstituted with 50 ml water.

4. CONCLUSIONS

Both the control bloods and the control urines for metals have been successfully applied in the internal and external quality control of heavy metal determinations (Stoeppler, 1982).

5. REFERENCES

J.Angerer, K.-H.Schaller and R.Heinrich (1981) Arbeitsmed., Sozialmed., Praventivmed. 5/81, 125-127.

E.Hansert and D.Stamm (1980) J. Clin. Chem. Clin. Biochem. 18, 461-490.

M.Stoeppler (1982) Proc. Third Int. Cadmium Conf., Miami, 1981, ILZRO, New York, pp. 95-101.

EVALUATING LABORATORY PERFORMANCE VIA AN INTERLABORATORY COMPARISON PROGRAM FOR TOXIC SUBSTANCES IN BLOOD AND URINE

J.Y. Savoie and J.P. Weber *

IRSST, 1199, de Bleury, Montreal, Qc, Canada H3B 3J1
** Centre de Toxicologie du Quebec*
2705, Boul. Laurier, Sainte-Foy, Qc, Canada G1V 4G2

1. INTRODUCTION

The determination of exposure to toxic metals in the workplace is often best achieved by measuring the levels of the metals in the biological fluids of the affected workers. This type of analysis is difficult to perform with accuracy due to the low concentration of metals involved, the complexity of the matrix and the lack of standardised methodology. Consequently, results from different laboratories can vary significantly. This situation tends to cast doubt upon the reliability of results in general thus hampering the use of biological monitoring both for regulatory and medical purposes.

Interlaboratory exchange programs using representative biological samples allow participants to assess the accuracy of their analytical work. However, the only ongoing programs extant appeared to be for blood lead e.g.West Allis Memorial Hospital (Maher *et al.* 1979) and the Center for Disease Control's blood lead proficiency testing (Boone *et al.* 1979). We consequently decided to initiate, in 1979, a program which would also include other substances of interest to us, e.g. mercury, arsenic, cadmium.

2. OBJECTIVES

The main objective of this program was to allow participants to mutually compare the results obtained after analysing identical samples of representative biological material. Each participant could then assess the accuracy of his determinations by examining the overall distribution of results. Using the data collected during the course of the program we wished to evaluate the reliability of laboratory results for each substance as well as to bring to light causes of error in the analytical methods used.

3. PROCEDURE

Participants were solicited from Canadian and U.S. laboratories involved in toxicological analysis. Control samples were prepared by pooling specimens from exposed workers except in the case of blood cadmium (spiked human blood) and some

Chemical Toxicology and Clinical Chemistry of Metals
ISBN 0.12.137520.X
Copyright © 1983 by IUPAC
All rights of reproduction in any form reserved

urine arsenic samples (spiked with specific arsenicals). Aliquots were distributed to participants at regular intervals (5 times/year). Results were compiled and elementary statistical calculations were performed (mean, median, standard deviation, histogram). The accumulated data were compared to the target values to evaluate performance during the course of the program.

4. RESULTS AND DISCUSSION

a) *BLOOD LEAD*

Number of participants : 34
Number of control samples : 45 (15 runs)
Methods used: APDC/MIBK extraction/flame AA (12 labs)
Delves cups (5 labs)
Graphite furnace AA (11 labs)
Anodic stripping voltammetry (6 labs)

The overall results for blood lead are quite good with the average error not exceeding ± 5% for most samples. However, individual results can show much greater fluctuations. Participation in the program did not greatly influence performance. Many laboratories were already obtaining very acceptable results and kept on doing so. Unfortunately, most of the labs who could not achieve good results at the start of the program were in the same position at the end.

The best method, after a somewhat inauspicious beginning, turned out to be the APDC/MIBK extraction. The two other AA methods also gave good, albeit more variable results. However, the ASV method systematically underestimated the lead content by some 15%. Only at the very end of the program did this situation improve somewhat.

b) *URINE MERCURY*

Number of participants : 31
Number of control samples : 44 (15 runs)
Methods used: Cold vapour AA with $SnCl_2$ as reductant (29)
Cold vapour AA with $NaBH_4$ as reductant (1)
APDC/MIBK extraction (1)

Results obtained for urine mercury varied greatly among the participating laboratories. The average deviation from the reference value was usually within ± 20%. However, individual participants were frequently in error by more than ± 50%. The two laboratories using non-standard methods reported consistently unacceptable results. It appears that the standard cold vapor is the best although the dispersion of results is still unacceptable.

c) *URINE ARSENIC*

Number of participants : 28
Number of control samples : 42 (14 runs)
Methods used:

- colorimetric (Ag/DEDC) (10 laboratories)
- arsine generation/AA (7 laboratories)

- graphite furnace AA (3 laboratories)
- other (8 laboratories)

The determination of urine arsenic is very difficult to achieve with accuracy, the average deviation being of the order of ± 30%. Of the three methods compared, the conventional colorimetric method yielded results closest to the reference value whereas the flame AA technique tended to underestimate the arsenic content. However, the wide scatter of results precluded more detailed analysis. Upon examining the individual results it was observed that some laboratories always reported acceptable results, regardless of the method used. It would appear that the care taken in applying a method was more important than the choice of the method itself.

The large scatter of data points also made it impossible to quantify performance according to the type of arsenic present. It was observed that the dispersion of results was greater for urines spiked with DMAA and urines containing arsenobetaine which probably indicates that some laboratories were unable to oxidise these species completely.

Inspection of the individual results showed some improvement during the third year of the program.

d) *BLOOD CADMIUM*

Number of participants : 11
Number of control samples : 12 (4 runs)
Methods used:

Deproteinisation, GFAA	(4 labs)
Dilution with Triton-X GFAA	(2 labs)
Acid digestion, GFAA	(2 labs)
Other	(3 labs)

For samples containing more than 6 μg/L of Cd accuracy and precision were good. In the case of normal samples (Cd levels < 0.5 μg/L) the reported results were often too high. This systematic error is probably due to one or more of the following factors: insufficiently sensitive methods, high reagent blanks, contamination in the laboratories.

Laboratories using the deproteinisation technique Stoeppler *et al.* 1978) obtained the best results. They were closer to the target value and also showed less scatter, especially for low cadmium concentrations.

The dilution method led to poor recovery and low sensitivity. Wet digestion methods tended to yield high results, probably due to contamination. The one participant who used APDC/MIBK extraction followed by flame AA obtained excellent results for concentration above 5 μg/L. Only the deproteinisation method yielded acceptable results throughout the range of interest.

5. LONG-TERM RELIABILITY OF ANALYTICAL WORK

The long term performance (over three years) for each laboratory was estimated by compiling the number of results within a set percentage of the target value. For blood lead, a deviation of ± 15% was considered acceptable, this being in keeping with the CDC's standards. However, for urine mercury or urine arsenic such a stringent requirement could have been met by only a very small fraction of participants. Therefore deviations of 25% and 50% respectively were accepted. For blood cadmium the program had not been in operation for a sufficiently long time to allow meaningful comparisons. The results at two reliability levels (75% and 50% of a laboratory's results being within acceptable limits of the target value) are shown below:

SUBSTANCE	CRITERION	% OF PARTICIPANTS
Blood lead	75% of results within 15% of t.v.	53
	50% of results within 15% of t.v.	82
Urine mercury	75% of results within 25% of t.v.	47
	50% of results within 25% of t.v.	83
Urine arsenic	75% of results within 50% of t.v.	39
	50% of results within 50% of t.v.	68

6. CONCLUSION

This program was designed to allow participants to mutually compare results. We expected that laboratories whose results differed significantly and repeatedly from the mean would take appropriate steps to improve their performance. Although this was true in some cases, many laboratories showed no improvement during the course of the program. On the other hand laboratories who performed adequately at the outset continued to do so until the end.

Analytical performance for blood lead and blood cadmium was on the whole quite acceptable. For urine arsenic and to a lesser degree urine mercury, reported results were scattered widely about the target value.

Several methods were identified as having the potential for generating accurate results while others were shown to be a cause of systematic error. However, in the proper hands, most methods could be made to yield acceptable results.

7. REFERENCES

Boone, J., Hearn, T. and Lewis, S.
Comparison of interlaboratory results for blood lead with results from a definitive method.
Clinical Chemistry 1979, 25(3), 389.

Maher, C.C., Roettgers, M.S. and Conlon, H.J.
Interlaboratory comparison blood lead determinations.
AIHA Journal 1979, 40, 230.

Stoeppler, M., Brandt, K. and Rains, T.C.
Analyst 1978, 103, 714.

CLINICAL REFERENCE MATERIALS CERTIFIED FOR ULTRA-TRACE ELEMENT LEVELS : UTOPIA OR FAR-OFF REALITY ?

Rita Cornelis and Jacques Versieck*

Laboratory for Analytical Chemistry, Rijksuniversiteit Gent, Proefteuinstraat 86, B-9000 Gent, Belgium

**Department of Internal Medicine, Division of Gastroenterology, Rijksuniversiteit Gent, De Pintelaan 185, B-9000 Gent, Belgium*

1. INTRODUCTION

Biomedical sciences exhibit a growing interest in the ultra-trace elemental composition of human body fluids, organs, tissues and other materials (such as dialysate or parenteral fluids). The term ultra-trace, in this paper, covers the concentration range below a few tenths of a ng/g. Whatever the objective of the ultra-trace elemental pursuit - be it epidemiological, occupational or pathological - the first step consists in producing accurate and precise analytical data. In the light of the impressive number of publications and of the many contradictory statements in them, it can be said that the time has now come to assess certain methods and to improve the expertise. The proliferating number of interlaboratory comparisons aiming at low levels and yielding results covering several orders or magnitude, clearly bears out this statement.

2. WHAT ARE THE PROBLEMS AND HOW CAN THEY BE SOLVED ?

Contamination is one of the main problems. It already turns up at the sample collecting and handling stages, and it looms large throughout the whole analytical procedure. This often results in a blank, which exceeds the true level of the element being investigated. In addition, the analytical requirements for the determination of the ultra-trace levels are very taxing. In many instances the researcher has to work close to the detection limits, a situation which jeopardizes accuracy and precision.

The relevance of very accurate and precise trace elemental analyses can be exemplified with the values registered e.g. for serum Mo in health and disease. The normal mean concentration (± S.D.) is 0.55 ± 0.21 ng/ml (n=50), whereas in case of hepato-biliary disease, e.g. during the initial phase of acute viral hepatitis, it amounts to 3.10 ± 1.46 ng/ml (n=14) (Versieck et al., 1981). It is clear that the actual differences between these mean values might have gone by unnoticed, if there had been a substantial contamination problem.

Chemical Toxicology and Clinical Chemistry of Metals
ISBN 0.12.137520.X

Copyright © 1983 by IUPAC
All rights of reproduction in any form reserved

Clinical conclusions are only valid when supported by accurate data. The sole remedy resides in a strict quality control, but this is, so to speak, missing in almost every publication, through lack of clinical reference materials certified for ng/g or subng/g amounts. The fact that a publication is short of data on quality control, does not necessarily imply that the results are wrong, but it is nevertheless impossible to evaluate their accuracy.

The definition of a clinical reference material for ultra-trace elements implies a similar matrix as the sample to be analysed and certified for physiological concentrations of the ultra-trace elements. However, the available biological standard reference materials (SRM's) are certified for much higher contents and are therefore not really to the point, because their analyses do not proceed along the same lines. Table 1 compares the certified and non-certified contents of 15 elements present in the well known and very valuable NBS SRM 1577 bovine liver, to those thought to be normally present in the serum of healthy individuals. The certified values for Mn, Pb, Cd, Cr, As and Hg exceed those for serum with a factor 2.10^4 to 10. The other, non-certified contents are similarly several orders of magnitude higher. The same criticism applies to other reference materials (issued e.g. by the Bureau of Reference Materials of the Commission of the European Communities, the International Atomic Energy Agency, ...). What is really needed are materials, certified for values identical to those present in the clinical specimens under study.

3. REALIZATION OF A CLINICAL SRM FOR ULTRA-TRACE ELEMENTS

The production of such an SRM will be an excessively difficult task. There is first of all a need to set up technologies for the semi-industrial production of a homogeneous and stable material. Collection, drying, homogenization, subsampling, sterilization and storage should be performed without the slightest contamination. This requires a very special infrastructure, such as class 100 clean room conditions and high purity containers, rigorously cleaned with ultra-pure reagents and ultra-pure water.

Next comes the certification of the ultra-trace elements : a difficult and tedious task. Basically the development of a high-quality certified reference material requires so called absolute or definitive methods. Usually collaborative certification tests are performed with the aid of two or more chemically independent methods with proven accuracy and precision, versus a primary standard. This procedure results in a certification by means of consensus.

Last but not least comes the cost of such a SRM. Through the courtesy of Dr. Rook and Dr. Alvarez of the U.S.National Bureau of Standards (NBS), it was possible to learn the breakdown of costs of the NBS bovine liver SRM 1577a (as at 1981).

TABLE 1

Comparison of a Selection of Trace Constituents in NBS SRM 1577 Bovine Liver, within the Ultra-trace Levels in Plasma or Serum of Healthy Individuals

	NBS SRM 1577 Bovine liver ng/g		Normal concentration in human plasma or serum ng/ml	reference
	Certified values (± 95% conf.limits)			
Mn	10 300 ± 1 000		0.58	Halls and Fell, 1981
Pb	340 ± 80		0.02	Everson and Patterson, 1980
Cd	270 ± 40		0.11	Jagner et al., 1981
Cr	88 ± 12		0.14	Kayne et al., 1978
As	55 ± 5		1.07	Damsgaard et al., 1973
Hg	16 ± 2		1.8	Wester, 1973
	Non-certified values (NBS)			
Ag	60		0.68	Kasperek et al., 1979
Co	180		0.108	Versieck et al., 1978
Mo	3 400		0.58	Versieck et al., 1981
Sb	5		0.52	Kasperek et al., 1979
	Literature values			
Al	30 000	Gladney, 1980	2.1	Alderman and Gitelman, 1980
Cs	17	id.	0.74	Versieck et al., 1977
Ni	400	id.	2.0	Spruit and Bongaarts, 1977
V	61.5	Cornelis et al., 1981	0.031	Cornelis et al., 1981

The total cost for 2000 bottles each of 50 g amounted to circa $ 159,200.- : circa $ 4,200.- for purchase and freeze-drying of the livers, circa $ 5,000.- for homogenisation and subsampling and circa $ 150,000.- for certification. As far as a SRM for ultra-trace elements is concerned, it is no overstatement to claim that the know-how to manufacture a good product virtually exists already. The costs for the development and for the building of the infrastructure to produce it, on a scale sufficient to fulfill the demand of the scientific community, is expected to be prohibitively high. After a long and tedious preparation will come an even longer and more tedious certification. Therefore the chances are that a rough estimate of a real cost amounts to a figure exceeding that of the NBS bovine liver, cited above, by 10 times.

As a consequence, SRM's certified for ultra-trace elements may very well remain a utopia. They are, however, so vital in our search for a better understanding of the biochemical role of trace elements in the human body, that they must become reality.

4. REFERENCES

F.R.Alderman and H.J.Gitelman (1980) Clin. Chem., 26, 258-260.

R.Cornelis, J.Versieck, L.Mees, J.Hoste and F.Barbier (1981) Biol. Trace Element Res., 3, 257-263.

E.Damsgaard, K.Heydorn, N.A.Larsen and B.Nielsen (1973) Risø Report (Denmark) n° 271, 1-32.

J.Everson and C.C.Patterson (1980) Clin. Chem., 26, 1603-1607.

E.S.Gladney (1980) Anal. Chim. Acta, 118, 385-396.

D.J.Halls and G.S.Fell (1981) Anal. Chim. Acta, 129, 205-211.

D.Jagner, M. Josefson, S.Westerlund and K.Årén (1981) Anal. Chem., 53, 1406-1410.

K.Kasperek, G.V.Iyengar, J.Kiem, H.Borberg and L.E.Feinendegen, (1979) Clin. Chem., 25, 711-715.

F.J.Kayne, G.Komar, H.Laboda and R.E.Vanderlinde (1978) Clin. Chem., 24, 2151-2154.

D.Spruit and P.J.M.Bongaarts (1977) Dermatologica, 154, 291-300.

J.Versieck, J.Hoste, F.Barbier, H.Michels and J.De Rudder (1977) Clin. Chem., 23, 1301-1305.

J.Versieck, J.Hoste, F.Barbier, H.Steyaert, J.De Rudder and H.Michels (1978) Clin. Chem., 24, 303-308.

J.Versieck, J.Hoste, L.Van Ballenberghe, F.Barbier, R.Cornelis and I.Waelput (1981) J. Lab. Clin. Med., 97, 535-544.

P.O.Wester (1973) Acta Med. Scand., 194, 505-512.

OCCUPATIONAL AND ENVIRONMENTAL EXPOSURES

THE APPLICATION OF ATOMIC SPECTROSCOPY IN THE CONTROL OF INDUSTRIAL EXPOSURE TO TOXIC ELEMENTS

Ivar Martinsen and Yngvar Thomassen

Institute of Occupational Health
P.O. Box 8149 Dep, OSLO 1, Norway

1. INTRODUCTION

Analytical toxicology has undergone dramatic changes in the last ten years. The development and improvement of atomic spectroscopic methods is due in large measure to recent advances in electronic technology and analytical understanding. The basic principle underlying spectroscopic elemental analysis was already demonstrated in the classic experiments of Kirchhoff and Bunsen (1859). Almost one hundred and twentyfive years later the analytical techniques based on their discovery are recognized for their high specificity, selectivity, precision and accuracy. These atomic spectroscopic techniques have become the most extensively employed for the determination of elements in a wide variety of materials. This paper will discuss the application of the two most common methods (inductively coupled plasma atomic emission spectroscopy (ICP-AES), and atomic absorption spectroscopy (AAS)) in the analysis of toxic elements in body fluids and workroom atmosphere for the evaluation of industrial exposure.

2. ANALYTICAL AND PRACTICAL CRITERIA

The most important criteria considered in making the choice of suitable methods for elemental analysis in atmospheric and biological monitoring programs are listed in Table 1.

Chemical Toxicology and Clinical Chemistry of Metals
ISBN 0.12.137520.X
Copyright © 1983 by IUPAC
All rights of reproduction in any form reserved

TABLE 1

Important analytical criteria

Accuracy	Element coverage
Precision	Single or multielement
Sensitivity	Possibility for speciation studies
Speed	
Cost	Direct analysis of gases, particulates and body fluids
Technique availability	Interferences

Despite the claims of instrument manufactures and supporters of special techniques, no method is likely to satisfy all criteria. The final choice for the analyst is therefore a compromise between the listed criteria. In occupational and environmental hygiene, there exists the known concept of environmental monitoring; which is measurement of airborne concentrations of chemical substances. In addition there is also biological monitoring which is measurement of internal exposure through analysis of biological fluids as whole blood and/or urine specimen. The use of biomonitoring is in general limited to few elements (Hg, Cd, Pb) for which health risk evaluation of exposure to workroom atmosphere is accepted.

The task for an analyst in occupational hygiene is to determine ultra trace concentrations of elements in body fluids and the elemental composition of the workroom atmosphere. The concentrations, dependent on element, range from less than 0.1 to 500 μg/l in body fluids and from ng to mg/m^3 of the elements in air. A factor of considerable importance is the identification and determination of a particular chemical form of an element (speciation) in the atmosphere (e.g. $Ni(CO)_4$, AsH_3).

(a) Body fluids

Adequate analytical sensitivity is crucial in the determination of the elemental composition of body fluids. During the sampling operation of biological fluids from workers numerous possibilities of contamination of the samples excist. The reported concentrations of trace elements in body fluids vary depending upon a number of sampling and analytical factors. As new knowledge is obtained the quality of the analytical data is considerably improved. Table 2 illustrates the concentration of some selected elements in non-exposed Norwegian adults.

TABLE 2

Typical blood serum and urine concentration of some selected elements in non-exposed Norwegian adults

Element	Concentration in μg/l	
	Blood plasma	Urine
Al	3	-
As	2	10
Bi	<0.5	<0.5
Cd	<0.5	0.2
Co	<0.5	<0.5
Cr	<0.3	0.2
Hg	2	5
Mn	<0.5	<0.5
Ni	2	5
Pb	1	10
Sb	<0.5	<0.5
Se	110	50
V	<0.2	<0.2

TABLE 3

Typical detection limits for ICP-AES, flame AAS and EAAS

Element	Detection limits (μg/l)		
	ICP/AES[a)]	Flame AAS[b)]	EAAS[c)]
Al	50	30	0.01
As	60	400	0.2
Bi	80	30	0.1
Cd	4	2	0.003
Co	10	15	0.02
Cr	6	3	0.003
Hg	40	400	2
Mn	1	3	0.002
Ni	18	10	0.06
Pb	60	14	0.05
Sb	80	40	0.15
Se	12	100	0.1
V	8	50	0.02

a) ARL 34000 ICP
b) Perkin-Elmer Model 560
c) Perkin-Elmer HGA 500

Despite the fact that a number of factors control the sensitivity and detection limit of a particular analytical method we have nevertheless listed some typical detection limits in table 3. As a minimum criterion the American Chemical Society's Committee on Environmental Improvement has recommended that the region for quantitation should be clearly above the detection limit. In order to illustrate the applicability of conventional commercially available equipment in the determination of trace elements in body fluids, the ratios between analyte concentration and detection limit for ICP-AES, flame AAS and EAAS are given in Table 4. In considering these ratios we may conclude that only EAAS has in general the required sensitivity for direct analysis of most of the listed elements. However, several ways of producing an aerosol or a vapor have been shown to increase the sensitivity in ICP-AES and AAS considerably, namely

ultrasonic nebulization
hydride generation
electrothermal volatilization
direct solid sampling
atom trapping system
cold vapor generation
Delves cup

Pretreatment of samples by acid digestion and/or preconcentration by solvent extraction to obtain lower detection limits is also widely used in AAS and ICP-AES. Thus, stricter control is required on reagent purity and analytical conditions.

TABLE 4

Applicability of some atomic spectroscopic methods for the determination of selected elements in blood serum

Element	Reference value/detection limit		
	ICP/AES	Flame AAS	EAAS
Al	0.06	0.1	300
As	0.03	0.005	50
Bi	<0.006	<0.02	< 5
Cd	<0.1	<0.2	<150
Co	<0.05	<0.03	< 25
Cr	<0.05	<0.1	<100

(Table 4 cont.)

Hg	0.05	0.005	1
Mn	<0.5	<0.2	<250
Ni	0.1	0.2	33
Pb	0.02	0.07	20
Sb	<0.006	<0.02	< 3
Se	9	1	1100
V	<0.03	<0.004	< 10

(b) Workroom atmosphere

The applicability of the atomic spectroscopic techniques in the determination of elements in body fluids is considered briefly because the monitoring programs for worker exposure to toxic chemicals are mainly based on air sampling. Allowable or recommended worker exposure limits to chemical substances in the workplace are based on the best available information from industrial experience, experimental human and animal studies. However, the majority of epidemiological studies investigating occupationally related causes of disease falter when it comes to establishing the exposure histories of populations of workers under study. This is an area where analysts can make a major contribution to future occupational health research. The evaluation of workplace atmospheres that contain a variety of elements, often in specific chemical forms, requires sampling and analytical techniques which are capable of determining a large number of elements over a broad concentration range in a variety of materials. Atomic absorption spectroscopy is a well established analytical technique for the industrial hygiene evaluation of worker exposure to airborne particulates. Since the 60ies elemental analysis has been dominated by flame atomic absorption with sequential measurement in solution. However, flames are limited by relatively low temperatures so that it is difficult to analyze refractory elements or elements with high excitation energies. With the higher temperatures of the ICP compared to combustion flames, the high linear concentration range and with the greatly reduced inter-element interferences ICP-AES meets the above requirements and satisfy most of the criteria listed in Table 1. The capabilities of ICP-AES satisfy the analytical chemists demands for the multi-element characterization of workroom atmospheres. The analytical excellence of ICP-AES in the characterization of the workroom environment portends increasing application of this technique.

Sampling of workroom atmosphere and analysis by ICP-AES

The collection of particulates by filters is used extensively for air monitoring. The air particulates on the filter media must be quantitatively brought into solution. To obtain best optimum multi-elemental dissolution conditions of workroom particulates we recommend pressure decomposition in teflon autoclaves of the sample with a mixture of aqua regia and hydrofluoric acid at 180° C.

However, it is crucially important that not only the decomposition method but also the sampling method is suitable in monitoring metal species. In Table 6 we have listed some sampling procedures for element speciation of workroom atmospheres.

TABLE 6

Some sampling techniques for speciation studies of workroom atmospheres

Component	Filter media	Backup-pad	Adsorbent
Alkyl-lead			Charcoal
$Ni(CO)_4$			Charcoal
$Fe(CO)_5$			Charcoal
AsH_3		*$AgNO_3$ or	Charcoal
SbH_3		*$AgNO_3$	
PH_3		*$AgNO_3$	
Cr(VI)	PVC		
As_2O_3	*Glycerol/Na_2CO_3		
HF	PVC	*KOH	

* Impregnation reagent

Although it is important to use recommended sampling and sample preparation methods, it still must be recognized that the sample matrix is never the same, it is at the mercy of time and space. The occurrence of spectral interferences is common in ICP-AES and the selection of appropriate analytical lines may require some compromise between detecting power and least spectral interference from major concomitants. Multielement analysis by ICP-AES may be accomplished in two different ways; sequential or simultaneous determinations. While polychromators do offer advantages of speed and simultanety which are useful for the routine determination of elements in large number of samples of similar composition, the sequential scanning monochromator offers greater flexibility, allowing the laboratory to easily change the choice of wavelength or to scan a spectral region of interest. This is an obvious advantage for the determination of workroom atmospheres where the particulates are of varying composition. The choice between a sequential or simultaneous spectrometer is thus dependent on the need of versatility and sample throughput. In our laboratory we have given versatility the highest priority in the application of ICP-AES to occupational analysis. Nevertheless, a user of a sequential system try to analyze most of the samples under the same analytical conditions. Table 7 contains a partial listing of our multi-element analytical conditions applying a Perkin-Elmer ICP/5500 sequential scanning spectrometer.

TABLE 7

Analytical conditions in the multi-element determination of air particulates

Element	Analytical wavelength nm	Background correction	Spectral interference	Administrative norm (mg/m^3)
Al (I)	396.15	-0.09	Ca (/)	
As (I)	189.04			0.05
Ba (II)	455.40			0.5
Be (I)	249.47			0.002
Ca (II)	393.37			
Cd (II)	214.44			0.02/0.05
Co (II)	228.62			0.05
Cr (II)	205.55	+0.07	Fe (∧)	0.02/0.5
Cu (I)	324.75			0.1/1
Fe (II)	259.94			3.5
Hg (II)	194.23			0.05
Mg (II)	279.55			
Mn (II)	257.61			2.5
Mo (II)	202.03	+0.13	Fe (/)	5
Nb (II)	212.65			
Ni (II)	231.60			0.1
P (I)	214.91		Cu (∧), Large	
Pb (II)	220.35	-0.07	Cu (−), Ni (−), Al (/)	0.05
Sb (I)	206.83			0.5
Si (I)	251.61			
Sn (II)	189.98			2
Ta (II)	223.95	+0.06	Cu (\), Ni (−)	5

Ti (II)	307.86			
V (II)	310.23			0.05/0.5
W (II)	207.91	-0.07	Al (-)	1/5
Y (II)	371.03			1
Zn (II)	206.20			1/5
Zr (II)	343.82			5

(-) indicates simple background shift
(\), (/) indicates sloping background shift
(∧) indicates direct spectral overlap

The sample introduction system is corrosion-resistant, and allows the direct aspiration of the digested filter samples with no need for complexing excess hydrofluoric acid. It must be recognised that extremely care must be exercised in avoiding incompatible combinations of the elements in the multi-element standard solutions which could give losses from solution by precipitation. Cross-contamination of one element at low concentrations by impurities in an element present at high concentrations must similarly be avoided. The contamination of the major components in glassware is a general problem in preparation of the multi-element standard solutions (Si, Ca, Al). The range of the analytical precision of using a single set of compromised experimental values for forward power, observation height, and carrier gas flow is 2-5% R.S.D. for sample concentrations around Administrative Norms (Norwegian guidelines for airborne concentrations of chemical compounds in workroom atmospheres). Applying the above spectral conditions, only copper interferes seriously in the determination of the listed elements (P) on the assumption that the airborne concentrations do not exceed the administrative norms more than 10 times. This spectral interference is illustrated in figure 1. Despite of this direct spectral overlap the 214.91 nm phosphorus line is still the most suitable emission line in the caracterization of workroom atmosphere. The use of background correction eliminates the baseline shift from solutions containing copper and nickel on the 223.95 tantalum and 220.35 lead lines, respectively. High aluminium concentrations affect the accuracy in the determination of low levels of lead as seen in figure 1. Table 8 shows an analytic report applying our report software in the elemental determination of air filter samples. Lowest reported values of airborne concentrations when filtering 1 m^3 of air are 0.002 mg/m^3 (decomposed filter sample diluted to 15 mL). Beryllium and cadmium are reported down to 0.0003 mg/m^3.

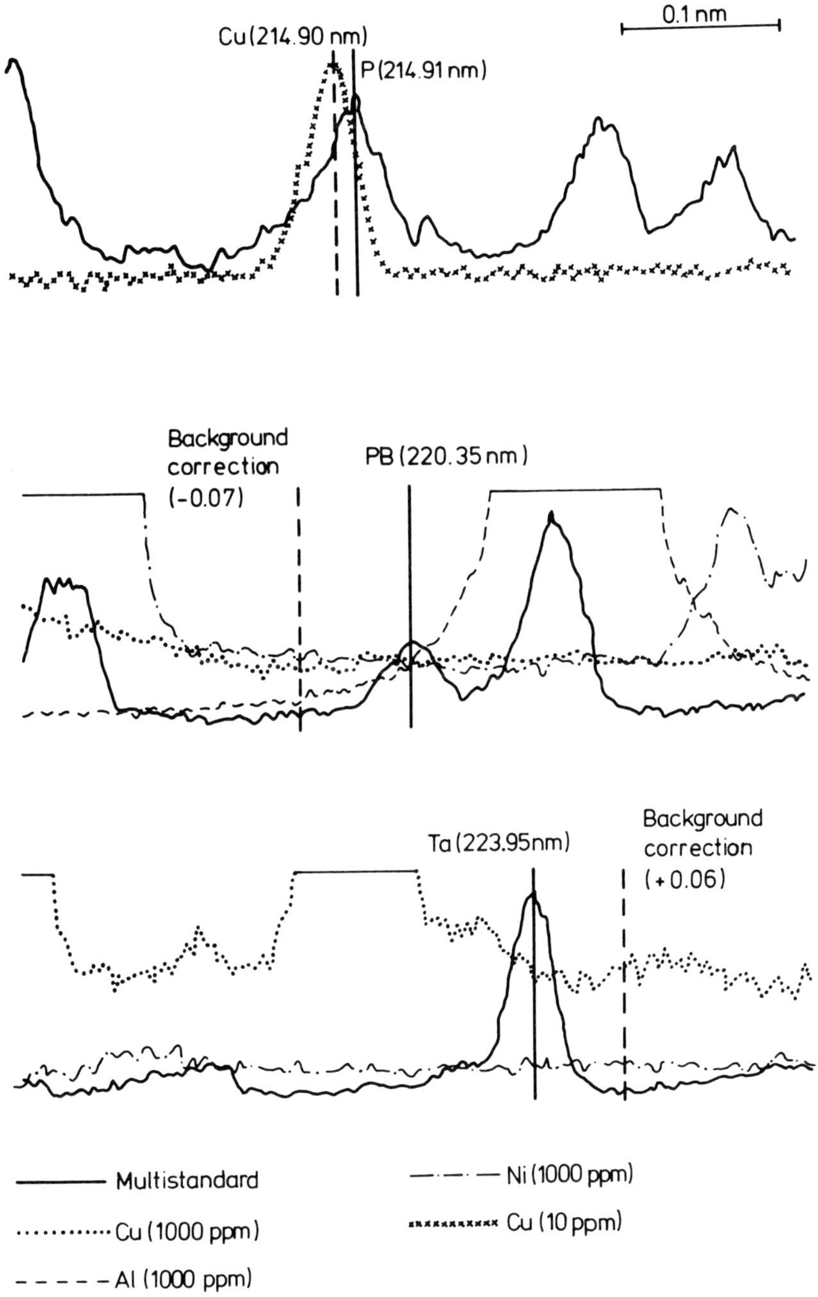

FIG. 1 Examples of spectral line-overlap, line-broadening and background shift interferences. The concentrations of P, Pb, Al and Ta in the multistandard are 5, 2, 5 and 10 μg/ml, respectively.

TABLE 8

Analysis of a filter sample from workroom atmosphere during welding operation by ICP-AES: an example of our report software

Sample no: 555

Date		070283
Sampling time (min)	:	127
Working operation	:	Welding, MIG Ar/He, Steel
Sample code	:	NIUIA2 T20
Personalia		
Sex	:	Male
Name	:	-
Date of birth	:	050846
Air Volume (m^3)	:	0,253

Concentration in mg/m^3

As:0.006	Sn:0.014	Hg:<0.006	Mo:<0.006	Cr:<0.006	Zn:0.54
Sb:<0.006	W :<0.006	Nb:<0.006	Cd:<0.0008	P :0.10	Pb:0.012
Co:<0.006	Ni:<0.006	Be:<0.0008	Si:0.63	Mn:0.65	Fe:5.22
Mg:0.022	Ta:<0.006	Ti:0.014	V :<0.006	Cu:0.18	Zr:<0.006
Ca:0.026	Al:0.056				

Total dust: 12.69

The accuracy of the method is not readily measured and generally involves analysis of reference materials or comparison of results obtained by independent methods. In U.S. National Institute of Occupationl Safety and Health provides laboratories with a quality assurance service called the Proficiency Analytical Testing (PAT) Program. Our ICP-AES analytical results of these metal spiked distributed filter samples are acceptable (recovery 100 $\pm$ 5% for Cd, Pb and Zn).

The recovery of an analyte from syntetic samples will only prove the accuracy of a method. The accuracy of the results can only be proved by obtaining agreement with certified standard reference materials. Standard filter materials which approximate the general matrix composition and the analyte concentration ranges of interests are not available. Interlaboratory comparisons of results obtained on real samples by the different methods in use with regard to the decomposition and the quantitative steps offer the simplest and most reliable means of evaluating accuracy.

To improve the analytical performance among nordic industrial hygiene laboratories an inter-laboratory testing program based on real workroom atmospheric filter samples was initiated in 1979. A field sampling unit capable of sampling 83 parallell filters has been developed at the Institute of Occupational Health in Oslo. The homogeneity of the exposed filters is adequate for round-robin programs (< 2% R.S.D.). The latest inter-laboratory quality program (elements) organized by our institute is listed in Table 9.

TABLE 9

Nordic Analytical Testing Program on Workroom Atmosphere (1983)

Filter exposed during	Component
Stainless steel welding	Cr, Ni, Fe, Mn, Cr (VI)
Crystal glass production	As, Pb
Smelting/steel production	Pb, Zn, Mn, Fe

All values for Cr (welding) and Pb (glass production) reported in the latest round (1983) have been arranged in order from the lowest to highest and plottet sequentially against the actual value of each result to form S-distribution curves (fig. 1). On each graph are shown the aritmetic mean ($\bar{x}$) and the limits set by one standard deviation ($\bar{x} \pm s$). Most of the participating laboratories have analyzed the samples by AAS. Of fourteen participating laboratories three are presently applying ICP-AES in their industrial hygiene monitoring program with good accuracy. In the absence of available certified standards for element determination in workroom atmospheres, the collection of parallell filters is promising in the evaluation of the accuracy of multi-element methods.

3. CONCLUSION

The applicability of atomic spectroscopy in the direct determination of toxic elements in biological fluids is limited and only EAAS has the required sensitivity. Neither flame AAS nor ICP-AES is comprehensive for direct determination of the listed elements and each technique must be supplemented by preconcentration procedures. In order to conduct a satisfactory analysis of airborne pollutants it is advisable to obtain as much information as possible.

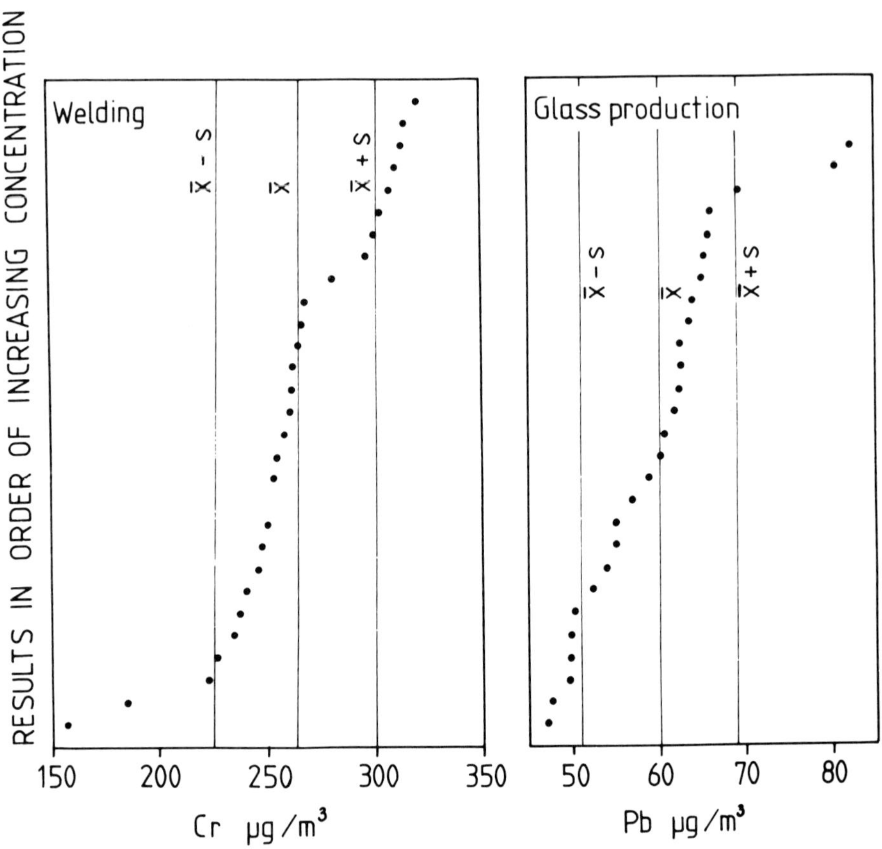

FIG. 2 S-distribution curves of the reported results on analysis of real parallell exposed workroom filters. On each graph are shown the aritmetic mean and the limits set by one standard deviation.

With appropriate sampling and decomposition methods ICP-AES exhibits excellent conditions for multi-element analysis of workroom atmospheres.

4. REFERENCE

ACS Committee on Environmental Improvement (1980). Anal. Chem., 52, 2242-2249.

MONITORING OF ENVIRONMENTAL EXPOSURES TO TOXIC METALS

Philippe Grandjean

Institute of Community Health, Odense University, 5000 Odense C, Denmark

1. INTRODUCTION

Prevention against environmental exposures to toxic metals should be directed primarily towards minimizing emissions from the most important sources. As a consequence, the ideal monitoring of such exposures should furnish appropriate information on the current necessity to institute additional emission control. Unfortunately, toxic metals originate from a complicated spectrum of natural and anthropogenic sources, and they reach humans through multiple pathways. The toxic metals have useful applications, but harmful exposures may be caused by production, use and disposal of a vast number of metal-containing products. Under these circumstances, an efficient surveillance of environmental exposures becomes extremely complicated, and an early warning system becomes difficult to design. Further, concentration limits may be impossible to define in the absence of detailed information on the relevance to human absorption levels, dose-response relationships and risk assessments. Thus, environmental monitoring is, to some extent, useful to continuously evaluate emission levels from a small number of primary sources, but such monitoring is rarely a sufficient safeguard against toxic exposures to metals.

The control of environmental exposures to toxic metals is the means to achieve the goal of preventing hazardous absorption levels. However, individual variation with regard to degree of exposure, intake/inhalation, fractional absorption and other factors makes it impossible to predict individual human absorption levels from environmental concentrations.

Biological monitoring has therefore become an important supplement with regard to assessing the significance of metal exposures. Biological, or health-related, monitoring usually means repeated or continuous measurement of metals in human blood or other samples as an indication of individual levels of metal absorption. Occasionally, measurement of enzyme activities or metabolite levels may be substituted when they provide a specific indication of the metal concentrations in the body. Biological monitoring results may in addition provide information on individual risk of intoxication, and a close relationship to health surveillance and diagnostic procedures is therefore obvious.

Chemical Toxicology and Clinical Chemistry of Metals
ISBN 0.12.137520.X
Copyright © 1983 by IUPAC
All rights of reproduction in any form reserved

2. SELECTION OF BIOLOGICAL INDICATORS

Biological monitoring of environmental exposures shares important features with methods used in screening for disease. Criteria for the proper identification and use of screening parameters have been discussed in a WHO publication by Wilson and Jungner (1968). These criteria may be discussed under three subheadings: a valid test should be available, the results should be applicable, and the monitoring program should be socioculturally acceptable (Table 1).

TABLE 1

Criteria for Selection of Biological (health-related) Monitoring Procedures

1. Test validity
   - Analytical validity
   - Availability of instrumentation, skills
   - Diagnostic validity
2. Relevance to prevention
   - Guidance for preventive actions
   - Reversibility of changes monitored
   - Availability of treatment (if needed)
3. Acceptability
   - Risk-benefit considerations
   - Confidentiality, medical ethics
   - Cost-benefit considerations

(a) Test validity

The methods used for biological monitoring should, in principle, satisfy the following three requirements. They should show a high degree of analytical validity, i.e. being both accurate and precise. Ideally they should be relatively inexpensive and only require simple instrumentation and a minimum of complex skills. Finally, the components measured should preferably have a high degree of diagnostic validity which will allow straightforward conclusions concerning exposure levels and individual risk. Such conclusions will only be possible when a metabolic model for the individual metal has been developed and detailed exposure-dose and dose-response relationships under chronic exposure conditions have been described. These requirements, if applied stringently, would certainly exclude the majority of current possibilities for biological monitoring.

(b) Relevance to prevention

Biological monitoring is only one aspect of the preventive approach to metal exposures. The monitoring methods should produce results which may elicit appropriate responses, and just like diagnostic methods they should be performed only when the results may cause or guide a proper therapeutic action. Most commonly, the biological monitoring is performed as a preventive surveillance, where the average level indicates the general effect of exposure control measures. If a certain limit (below the threshold for toxicity) is exceeded, additional preventive actions should go into force. From the viewpoint of preventive medicine, changes in biological indicators should be reversible, so that any early abnormality disappears when the exposure is decreased or removed. This requirement obviously will cause difficulties with regard to carcinogenic metals. Alternatively, under certain conditions, a pathological change may be acceptable (and actually useful as a disease marker) when successful treatment is available. In that case, the biological monitoring becomes a screening program for metal toxicity.

(c) Acceptability

The collection of blood and other specimens must be performed with a minimum of risk and inconvenience to the population group involved in a biological monitoring program. Even though a test may only entail a small risk, this disadvantage may not be counterbalanced by the minimal benefit to the individual from participating in the program. The spectrum ranges from collecting hair from the floor at a hairdresser's shop to stimulating X-ray fluorescence of metals in vivo by neutron or X-ray activation. These questions also need to be settled with regard to the practice of medical ethics. From a societal viewpoint, cost/risk-benefit considerations should be taken into regard before biological monitoring is instituted.

3. DEVELOPMENTAL TRENDS

Although the necessity of biological monitoring is well recognized, some current programs may proceed in a less-than-ideal fashion, and progress is only achieved in a piecemeal fashion. New developments in analytical chemistry, such as sophisticated multi-element analysis of minute hair segments, are useful research tools, but their applications in biological monitoring may be premature. In fact, commercially available multi-element analysis of hair samples has frightened many subjects into seeking medical assistance for allegedly abnormal trace metal levels or ratios between metal concentrations in hair. Thus, new techniques must produce results which are interpretable before these methods are introduced in biological monitoring. Apart from the rapid improvements in analytical chemistry, important developmental trends hold significant promises for biological monitoring in the future.

4. IMPROVEMENT OF VALIDITY

(a) Analytical performance

In a recent WHO document on laboratory quality control, Aitio (1981) recommended that future development should focus on international quality assessment, production of certified reference materials, elaboration of reference methods and establishment of reference laboratories. While these recommendations still hold true, some progress has been recently made.

With regard to interlaboratory comparison studies and external quality control, the continuing CEC program for quality assessment of blood lead determinations has shown favorable results. Perhaps even more optimistic are the results from the lead and cadmium analyses performed in connection with the UNEP/WHO Pilot Project on Assessment of Human Exposure to Pollutants through Biological Monitoring (Vahter, 1982). The average bias for 30 quality control samples varied mostly between 10 and 20 µg Pb/l blood and between 0.4 and 0.8 µg Cd/l blood in ten laboratories in industrialized and developing countries. This fine result was achieved after considerable improvement during a training phase. Intercomparison studies of nickel levels in biological fluids have also suggested an improved quality of laboratory performance (Sunderman et al., 1982).

The analytical problems related to metal determinations in body fluid may be less extensive than the problems associated with analysis of tissue samples. One study has been performed to assess the variability of lead level determinations in powdered premolar teeth (Stack and Delves, 1982). The results varied between 4.3 and 15.5 µg Pb/g with an overall average of 8.1 µg Pb/g. We have subsequently analyzed aliquots of a sample kindly supplied by Dr. M.V. Stack. In our method of analysis, a few milligrams of calcified tissue is dissolved in nitric acid in a sampling cup for electrothermal atomic absorption. Our results varied between 4.9 and 12.3 with a mean of 9.1 µg/g, and this variation was probably related to the fact that the tooth material primarily contained particle sizes of 0.1 - 0.5 mm, thereby making it virtually impossible to obtain a representative aliquot of a few milligrams. These experiences will be utilized in a proposed quality assurance study for tooth lead measurements organized by the Institute of Community Health, Odense University, in cooperation with the Institute of Medical Hygiene, Düsseldorf University.

Zinc protoporphyrin determination in erythrocytes has become a popular biological indicator of lead levels in the body, in particular because a portable front-face fluorometer (the hematofluorometer) enables an extremely fast analysis (Blumberg et al., 1977). In Danish screening studies, we have examined up to 180 employees at a workplace for zinc protoporphyrin in one day. Unfortunately, the instrument calibration is not stable, and interferences from bilirubin and other fluorescent substances may occur (Grandjean and Lintrup, 1981). Two individual instruments may each yield very precise results, but the unsatisfactory agreement when aliquots of the same blood sample are analyzed at the same time suggests that the accuracy could be variable even though precision is excellent. This widely used instrument needs to be frequently calibrated by comparison with a reference method. In addition, external quality assurance programs are badly needed, and one will hopefully be organized in connection with the implementation of the CEC directive on occupational lead exposure.

Reference materials with low trace metal levels and based on relevant tissues have not been available until recently. The International Atomic Energy Agency has developed reference materials of animal bone and freeze-dried animal blood, the assigned values being based on intercomparison studies. In addition, a sample of homogenized human hair has been tested in a number of laboratories. We have examined the day-to-day variation of lead levels in 1 mg aliquots of the animal bone material. Our results for 11 runs were (mean $\pm$ S.D. in μg Pb/g) 3.41 $\pm$ 0.46, while the assigned value was 3.1 $\pm$ 0.28 and the "accepted" intercomparison results were 3.0 $\pm$ 1.08. These results indicate a high degree of homogeneity in the material. In addition, the CEC Community Bureau of Reference is now developing reference materials of lyophilized blood with certified levels of lead and cadmium. Such materials with low metal concentrations are in much demand.

The development of reference methods is an important task which has been considered by IUPAC subcommittees. A reference method for nickel in serum/plasma and urine has been carefully tested and verified (Brown et al., 1981), and work is in progress concerning nickel in whole blood, cadmium (and lead) in blood, and aluminium in serum/plasma. Isotope dilution mass spectrometry is frequently referred to as a definitive method, but even such a method may give different results in different laboratories, as illustrated be a recent study of low blood lead values (Vahter, 1982). We may have to accept that we will never know the true concentrations, but close approximations may be achieved with the help of reference methods.

(b) Metal speciation

Current convention is to express metal concentrations in biological fluids as if each metal only appeared as such in only one relevant form. Unfortunately, each metal may occur as compounds with entirely different kinetic and toxic properties. This is true for chromium which may occur as Cr(III) or Cr(VI) in compounds which may be easily or sparingly soluble. The resulting spectrum of toxicokinetic properties is a main reason for the limited success of biological monitoring for chromium exposures so far (Langård, 1982). Additional experience from prospective monitoring of exposures and the resulting levels in erythrocytes, serum/plasma and urine may provide sufficient insight to allow a cautious interpretation of the levels found. A similar situation occurs in case of nickel. In case of pulmonary exposures to sparingly soluble nickel compounds, small and relatively stable increases in plasma nickel levels are found (Høgetveit et al., 1980). On the other hand, easily soluble compounds result in rapid and large increases in the nickel levels urine and plasma (Tola et al., 1979; Bernacki et al., 1980). In shipyard workers with exposures from various sources, plasma and urine nickel levels suggest mixed toxicokinetic patterns (Grandjean et al., 1980). Thus, for both chromium and nickel, plasma and urine levels are difficult to interpret without detailed information on the metal compounds of the exposure.

Mercury speciation has definite toxicological importance and, in addition, significant implications for biological monitoring. These aspects have been properly addressed, e.g. by Clarkson (1977), who in particular stressed the difference between mercury vapor and methyl mercury. Other metals may occur in methylated forms as well. This fact is of special relevance in the case of arsenic which undergoes methylation in the body. Inorganic arsenic compounds, such as arsenates, are excreted in the urine partly as monomethylarsonic and dimethylarsinic acids, while the remainder has been found to be inorganic arsenic (Buchet et al., 1981). Organic arsenic compounds from seafood interfere with analyses of total arsenic, while the selective detection of inorganic arsenic and the two methylated compounds appears to be specific for inorganic arsenic exposure.

Due to the continued production and use of tetraalkyllead, human exposures to organolead compounds may occur. Currently, biological monitoring of occupational exposures to these compounds is based on the urinary lead excretion level. However, the major excretion route for organolead compounds is through the bile (Jensen, 1983), and inorganic lead exposures may contribute to urinary lead levels. Animals experiments have suggested that diethyllead is the major lead species excreted in urine shortly after exposure to tetraethyllead, and this finding has been confirmed in a case of human poisoning (Yamamura et al., 1981). Thus, the determination of diethyllead levels in urine may become a more specific biological indicator of exposure to organolead compounds. Analysis of blood levels may be useful too, and the results published by Beattie et al. (1972) suggest that increased blood lead levels in case of tetraethyllead poisoning are mainly, if not entirely, due to high lipid-extractable lead levels in the blood.

(c) Sampling procedures

Much attention has duly been paid to problems concerning the replicability of sampling procedures, the avoidance of contamination, and sample transportation and storage (Aitio, 1981). Additional problems occur when specimens other than biological fluids are used for biological monitoring. Thus, both hair and teeth concentrate and store toxic metals and may be used as indicators of long-term exposures or absorption levels. However, questions arise concerning the homogeneity of these tissues and the procedure to obtain representative samples.

One of the questions in relation to hair is the significance of hair color. Early analyses have suggested that hairs with different hair colors contain different metal concentrations (Schroeder and Nason,

1969). However, such differences could be related to variations in exposures. In order to examine one aspect of this problem more closely, we obtained samples of single hairs with their roots from 18 grey-haired individuals. There is no such thing as grey hair, and each grey-haired individual had a mixture of white and pigmented hairs. We analyzed the first 1-cm segment of several adjacent white and pigmented hairs from each individual. In none of them was the lead concentration higher in white hairs higher than in pigmented, and in ten individuals this difference was statistically significant (Grandjean, 1979). Hair color is obviously important.

Concentrations of lead and other metals may increase along the hair shaft as an indication of external contamination. Thus, the first 1-cm segment of the hair adjacent to the root is to be preferred for analysis. In a cross-sectional study of 87 males with occupational lead exposures, a blood lead level of 60 μg/100 ml was found to correspond to approximately 3 ng of lead in the first centimeter of hair (Grandjean, 1979). More recently, new insight has been offered by a severe case of lead poisoning due to oral intake of lead nitrate during several months. This patient ultimately developed a blood lead level of approximately 200 μg/100 ml at the time of diagnosis. Lead levels in 1-cm segments of hair collected subsequently at two points of time were up to about 1.3 ng (Fig. 1), i.e. significantly less than the workers with exposure to airborne lead particles. This experience suggests that even the first 1-cm segment of hair may be contaminated if exposure to lead dusts has occurred.

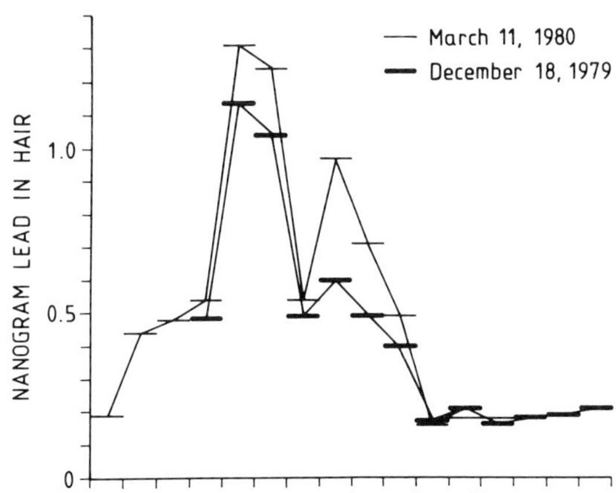

FIG. 1. Lead content of 1-cm segments of two hair samples collected three months apart from a lead-poisoned patient with a history of possible long-term ingestion of lead nitrate, except during a one-month vacation corresponding to the dip in lead levels 3 cm and 6 cm from the root (the growth rate is approximately 1 cm/month)

Shed deciduous teeth are regarded as important indicators of past metal exposures, because sparingly soluble metal phosphates are stored and may accumulate in the dental tissues, particularly in the secondary dentin. Different types of teeth vary in size and in the content of enamel and secondary dentin, and metal storage may also vary. Thus, a tendency has been found towards higher lead levels in incisors, particularly at increased lead exposures, but large variations in lead concentrations in both central and lateral incisors have been documented when comparing teeth obtained from the same children (Delves et al., 1982). On the other hand, circumpulpal dentine levels in a complete set of teeth from an adult individual were found to be independent of tooth type with a total coefficient of variation of 35 percent (Grandjean et al., 1979). Preliminary experience with circumpulpal lead levels in shed teeth from Danish children suggests that variation with tooth type is of only minor significance (Table 2). Such problems need careful evaluation with regard to the sampling procedures adopted.

TABLE 2

Lead Levels (μg/g) in Secondary Dentin Adjacent to the Pulpal Chamber of 185 Shed Deciduous Incisors Sampled in 1982 from First Grade School Children in Aarhus, Denmark

Tooth type	Number		Mean	
Medial, upper right	37	97	10.8	10.7
- - left	24		10.3	
- lower right	18		10.4	
- - left	18		11.5	
Lateral, upper right	32	88	10.3	11.5
- - left	21		13.3	
- lower right	17		11.8	
- - left	18		11.2	

(d) Metabolic models and thresholds for toxicity

A thorough understanding of the toxicokinetics of the metal or metal compound of concern is necessary for a satisfactory interpretation of biological monitoring results. Rarely does the biological fluid analyzed constitute the target for a toxic metal, and interpretation must therefore consider the possible metal levels in the target, or critical, organ. In addition, approximate thresholds for early toxicity should be known so that action limits well below the thresholds can be defined. These thresholds should be identified in the proper units of measurement for the relevant biological sample used for monitoring.

Acute metal intoxications have fortunately become rare, but the widespread and continued occurrence of chronic exposures and subtle or delayed toxicity have actually made prevention more difficult. The metabolic models have to deal with more or less steady-state conditions under long-term exposures, and the toxicity concerned may occur as a result of metal levels accumulated over long time periods, or it may occur after a considerable latency period, as in the case of metal-related carcinogenicity.

Unfortunately, these problems may cause increasing problems with regard to the use of biological monitoring. Serious research efforts will be necessary to define the role, if any, of biological monitoring under conditions of long-term, low-level exposures from a multitude of sources.

5. TESTS FOR INTEGRATED RESPONSES

The complex dissemination and uptake of several metals in different compounds may render environmental monitoring rather tedious and, perhaps, inefficient. Also, analysis of biological fluids for current or recent absorption of several metals may not be a sufficient safeguard against the combined effect of metal exposures. Such questions could be addressed by means of epidemiological surveillance of morbidity and mortality rates, but these effects may only occur after a considerable latency period, and this kind of surveillance cannot function as an early warning system for the prevention of toxic metal exposures. The question then arises whether or not biological indicators of early or subclinical responses to combined effects of toxic metals may be available for possible use in biological monitoring (see e.g. Goldstein, 1981).

(a) Genotoxicity

Among the several possibilities for assessing early genotoxic effects, two tests may be carried out on human lymphocytes as an indication of the combined effect on the DNA of these cells (Vainio et al., 1981). Most lymphocytes are rather long-lived with a half-life of a couple of years, and, mainly being in the presynthetic G_0 stage of the cell cycle, these cells may accumulate mutations over a long period of time (Natarajan and Obe, 1982). Chromosomal aberration has until recently been a preferred indicator which has shown clear responses to exposures to known mutagens. However, the technique requires considerable expertise and the laborious examination of at least 100 cells. The new "fluorescent plus Giemsa" method for assessment of sister chromatid exchange (SCE) has resulted in important advances, because of the speed and simplicity of this technique. Although the basic cytogenetic mechanism for SCE induction is incompletely understood and the long-term significance of increased SCE rates is unknown, SCE assessment provides a measure of genotoxic effects with potential relevance to health (Brewen and Stetka, 1982).

Increased frequencies of chromosomal aberrations have been found in relation to lead levels in the complex exposures in a copper smelter (Nordenson et al., 1978), but "clean" lead exposure appears not to induce significant aberrations (Nordenson et al., 1982). However, increased SCE rates in battery factory workers have been found to correlate closely with augmented levels of zinc protoporphyrin in the blood, thus indicating a relationship to long-term lead exposures (Grandjean et al., 1983). Long-lasting effects may be caused by arsenic (Burgdorf et al., 1977; Nordenson et al., 1979). With regard to nickel, chromatid gaps may be a sensitive indicator of nickel-related genotoxicity, while SCE has not shown any significant response (Waksvik and Boysen, 1982). In case of Cr(VI), exposed plating workers have exhibited increased SCE rates (Stella et al., 1982). In addition, considerable evidence from in vitro studies suggests that SCE induction may be caused by several metal compounds (Andersen, 1983). Thus, the possibility exists that the SCE test may provide an integrated genotoxicity response assessment under complex metal exposure conditions. Although this technique is not yet ready for biological monitoring programs, future research will hopefully help to define the conditions for the use of this test and the detailed interpretation of the results.

(b) Other tests of integrated response

A large proportion of absorbed metals is invariably excreted in the urine, and toxic effects on kidney function are therefore obviously possible. In fact, the kidneys comprise the critical organ in case of chronic exposures to cadmium and, most probably, inorganic mercury compounds (Nordberg, 1976). In addition, other metals may exert nephrotoxic effects, notably lead (Wedeen et al., 1979) and, in animal experiments, nickel (Sunderman and Horak, 1981). As reviewed by Sunderman and Horak (1981), studies of nephrotoxicity usually entail a battery of different tests. With regard to cadmium, however, most emphasis has been put on tubular proteins as an indication of toxicity in the proximal tubuli (Kjellström et al., 1977). Cadmium nephrotoxicity is of particular relevance because of its possible occurrence in elderly people with high cadmium burdens due to environmental exposures (Roels et al., 1981). However, since mixed metal exposures frequently occur, combined effects on kidney function could be expected. So far, however, little evidence is available to suggest the most sensitive indicator of such nephrotoxicity.

Serum levels of various enzymes are routinely used in medical diagnostics and occasionally for occupational surveillance programs. Intracellular enzymes may be released as a result of cellular toxicity and may provide a sensitive indicator of an integrated response. However, insufficient information is available on the release and fate of intracellular enzymes (Kristensen and Hørder, 1983), and the specific relevance to low-level metal exposures has not been documented. The availability of sophisticated enzyme assays should encourage studies in this area.

(c) Health surveillance programs

In the occupational setting, health surveillance programs are usually offered, in part, as an extra safeguard against adverse effects of occupational exposures, should other preventive efforts for some reason fail. Unnecessary health examinations have often been instituted or continued, perhaps due to the belief that such examinations promote health. Abnormal findings at a physical examination frequently occur at a late stage in the progression of an intoxication. Other findings may be entirely nonspecific, and specific preventive actions may be difficult to identify. Thus, biological monitoring and, perhaps, supplementary tests for integrated responses would comprise a superior safeguard for workers exposed to toxic metals than would a rather general health surveillance.

6. PROBLEMS IN APPLICATION

(a) Confidentiality

At the present time, biological monitoring is recommended and practiced in several countries. However, only monitoring for blood lead levels seems to be a common legal requirement. In the U.S., the blood lead test is part of the medical surveillance offered by the employer. The OSHA standard requires the employer to inform, within five working days, an employee of the blood lead result if it exceeds 40 μg/100 g. In the recently approved EEC directive on occupational lead exposure, the decision is left to the responsible physician to decide whether or not to inform an employee of the result and its interpretation. Detailed rules on actions required on the basis of blood lead determinations would conceivably necessitate that the employer receives the results of the biological monitoring.

In the implementation of the EEC directive, national regulations will govern the practice of confidentiality. In Denmark, for example, biological monitoring results are regarded as confidential health information. Results and their interpretation are given to the individual employee by the responsible physician. Information required by regulations is made available to the employer, e.g. appropriate intervals for blood lead levels, but not necessarily the exact results.

Different traditions prevail with regard to confidentiality. This question will probably receive increased attention as biological monitoring becomes more common and, in particular, if tests for integrated response are applied. Information on individual health should be regarded a matter of privacy between the individual and the physician. Any dissemination of biological monitoring results could potentially lead to misuse.

(b) Relevance to responses

An important disadvantage of biological monitoring is that the specific preventive action needed is not indicated by a biological test result which exceeds the limit. Critics of biological monitoring have claimed that such procedures use human beings as samplers or monitors while leaving alone the patient, i.e., the environment with the exposure sources. This statement is not true, as long as biological monitoring is only practised as a supplement to surveillance of environmental exposure sources.

Without information on the exposure sources, biological monitoring data may be misleading. For example, several epidemiological studies have suggested that smokers have higher blood lead levels than do nonsmokers. Scattered data indicated that tobacco indeed contained lead, some of which could be transferred to the mainstream smoke. Only recently was a population group interviewed concerning both smoking and alcohol habits, and the statistical evaluation showed that most of the excess blood lead in smokers could be explained by their alcohol comsumption habits (Grandjean et al., 1981). Smokers drink more than nonsmokers, and, probably due to the high lead content of wine, this difference results in a difference in blood lead levels. Had somebody intended to decrease the lead levels by discouraging people from smoking, this preventive action would have shown minimal results in terms of blood lead levels.

7. REFERENCES

A. Aitio (1981) Laboratory Quality Control, WHO Regional Office for Europe, Copenhagen (European Cooperation on the Health Aspects of the Control of Chemicals - Interim Document 4).

O. Andersen (1983) Environmental Health Perspectives, 47, 239-253.

A.D. Beattie, M.R. Moore and A. Goldberg (1972) Lancet, 2, 12-15.

E.J. Bernacki, E. Zygowicz and F.W. Sunderman, Jr. (1980) Annals of Clinical and Laboratory Sciences, 10, 33-39.

W.E. Blumberg, J. Eisinger and A.A. Lamola (1977) Clinical Chemistry, 23, 270-274.

J.G. Brewen and D.G. Stetka (1982) in Mutagenicity, New Horizons in Genetic Toxicology (Ed., J.A. Heddle), Academic Press, New York, pp.351-384.

S.S. Brown, S. Nomoto, M. Stoeppler and F.W. Sunderman, Jr. (1981) Pure and Applied Chemistry, 53, 773-781.

J.P. Buchet, R. Lauwerys and H. Roels (1981) International Archives of Occupational and Environmental Health, 48, 111-118.

W. Burgdorf, K. Kurvink and J. Cervenka (1977) Human Genetics, 36, 69-72.

T. Clarkson (1977) in Clinical Chemistry and Chemical Toxicology of Metals (Ed., S.S. Brown), Elsevier/North-Holland, Amsterdam, pp. 189-200.

H.T. Delves, B.E. Clayton, A. Carmichael, M. Bubear and M. Smith (1982) Annals of Clinical Biochemistry, 19, 329-337.
I.F. Goldstein (1981) Environmental Research, 25, 236-240.
P. Grandjean (1979) in Toxicology and Occupational Medicine (ed.,W. B.Deichmann), Elsevier/North-Holland, Amsterdam, pp. 311-318.
P. Grandjean and J. Lintrup (1981) Scandinavian Journal of Work, Environment & Health, 7, 311-312.
P. Grandjean, O.V. Nielsen and I.M. Shapiro (1979) Journal of Environmental Pathology and Toxicology, 2, 781-787.
P. Grandjean, E.J. Selikoff, S.K. Shen and F.W. Sunderman, Jr. (1980) American Journal of Industrial Medicine, 1, 181-189.
P. Grandjean, N.B. Olsen and H. Hollnagel (1981) International Archives of Occupational and Environmental Health, 48, 391-397.
P. Grandjean, H.C. Wulf and E. Niebuhr (1983) Environmental Research, in press.
A.C. Høgetveit, R.T. Barton and I. Andersen (1980) Journal of Occupational Medicine, 22, 597-600.
A.A. Jensen (1983) in Biological Effects of Organolead Compounds (Ed., P. Grandjean), CRC Press, Boca Raton, Florida, in press.
T. Kjellström, P.E. Evrin and B. Rahnster (1977) Environmental Research, 13, 303-317.
S.R. Kristensen and M. Hørder (1983) in Clinical and Analytical Concepts in Enzymology (Eds, S. Kristensen and M. Hørder), College of American Pathologists, Skokie, Ill., pp. 1-13.
S. Langård (1982) in Biological and Environmental Aspects of Chromium (Ed., S. Langård), Elsevier Biomedical, Amsterdam, pp. 149-169.
A.T. Natarajan and G. Obe (1982) in Mutagenicity, New Horizons in Genetic Toxicology, (Ed., J.A. Heddle), Academic Press, New York, pp. 171-213.
G.F. Nordberg, Ed. (1976) Effects and Dose-Response Relationships of Toxic Metals, Elsevier, Amsterdam.
I. Nordenson, G. Beckman, L. Beckman and S. Nordström (1978) Hereditas, 88, 263-267.
I. Nordenson, S. Nordström, A. Sweins and L. Beckman (1982) Hereditas, 96, 265-268.
I. Nordenson, S. Salmonsson, E. Brun and G. Beckman (1979) Human Genetics, 48, 1-6.
H. Roels, R. Lauwerys, J.P. Buchet and A. Bernard (1981) Environmental Research, 24, 117-130.
H.A. Schroeder and A.P. Nason (1969) Journal of Investigative Dermatology, 53, 71-78.
M.V. Stack and H.T. Delves (1982) in IUPAC Collaborative Interlaboratory Studies in Chemical Analysis (Eds, H. Egan and T.S. West), Pergamon, Oxford, pp. 115-118.
M. Stella, A. Montaldi, R. Rossi, G. Rossi and A.G. Levis (1982) Mutation Research, 101, 151-164.
F.W. Sunderman, Jr., S.S. Brown, M. Stoeppler and D.B. Tonks (1982) in IUPAC Collaborative Interlaboratory Studies in Chemical

Analysis (Eds, H. Egan and T.S. West), Pergamon, Oxford, pp. 25-35.
F.W. Sunderman, Jr., and E. Horak (1981) in Organ-Directed Toxicity (Eds, S.S. Brown and D.S. Davies), Pergamon, Oxford, pp. 55-67.
S. Tola, J. Kilpiö and M. Virtamo (1979) Journal of Occupational Medicine, 21, 184-188.
M. Vahter (1982) Assessment of Human Exposure to Lead and Cadmium through Biological Monitoring, National Swedish Institute of Environmental Medicine, Stockholm.
H. Vainio, M. Sorsa, J. Rantanen, K. Hemminki and A. Aitio (1981) Scandinavian Journal of Work, Environment & Health, 7, 241-251.
H. Waksvik and M. Boysen (1982) Mutation Research, 103, 183-190.
R.P. Wedeen, D.K. Mallik and V. Batuman (1979) Archives of Internal Medicine, 139, 53-57.
J.M.G. Wilson and G. Jungner (1968) Principles and Practice of Screening for Disease, Public Health Papers No. 34, World Health Organization, Geneva.
Y. Yamamura, F. Arai and H. Yamauchi (1981) Industrial Health, 19, 125-131.

IN VIVO TESTS TO MONITOR BODY BURDENS OF TOXIC METALS IN MAN

Robert R. Lauwerys

Medical and Industrial Toxicology Unit, Department of Occupational Medicine and Hygiene, Catholic University of Louvain, 30.54 Clos Chapelle-aux-Champs 1200 Brussels Belgium

1. INTRODUCTION

The evaluation of human exposure to toxic metals can be carried out by monitoring their concentrations in the environment or by determining their levels in biological tissues and fluids. Depending on the metal and the analyzed biological material, the result may reflect mainly either the amount of metal recently absorbed or a fraction of the total body burden. For several metals (e.g. arsenic, chromium, fluoride, lead, mercury, nickel), the levels in blood and in urine are an indicator of recent exposure and therefore depending on the biological half-life of the metal, the time of blood and urine sampling may be critical (Lauwerys, 1983). In this paper, I shall focus only on three metals (lead, cadmium, mercury) and on one organometallic substance (methylmercury) for which some in vivo tests are applicable in humans in order to evaluate their body burdens.

2. LEAD

Although a multicompartment model of lead distribution has been proposed to take into account all the observed kinetic data for lead (Baloh, 1974; Rabinowitz et al., 1976) a simple two compartment model may, however, be adequate to characterize roughly its distribution in the human organism. The first compartment contains the rapidly exchangeable pool of lead. It includes the blood, the soft tissues and possibly also a mobile compartment within bone. The biological half-life of lead in this compartment is about 40 days. The skeleton represents the second compartment which contains the slowly exchangeable pool of lead. In a non-occupationally exposed adult, there is about 110 mg lead in the skeleton and 6.2 mg in the blood, liver and muscle of which 1.4 mg is in the blood (I.C.R.P., 1975).

Chemical Toxicology and Clinical Chemistry of Metals
ISBN 0.12.137520.X
Copyright © 1983 by IUPAC
All rights of reproduction in any form reserved

(a) Chelatable lead

The urinary excretion of lead after the administration of a chelating agent ($CaNa_2EDTA$; penicillamine) is considered to reflect the mobilizable pool of lead located in soft tissues, in blood and possibly also in a mobile compartment within bone. Under steady state conditions, the amount of lead excreted in urine in the 24-hour period following the administration of a chelating agent correlates with the lead level in blood (Alessio and Foa, 1980). Alessio et al. (1979) and Hansen et al. (1981) found, in workers who had been exposed to lead for at least one year, a correlation coefficient of 0.85-0.87 between lead in blood and the amount of lead excreted in urine within 24 hours after the intravenous infusion of $CaNa_2EDTA$ (12.5-25 mg/kg). In persons who have been occupationally exposed to lead in the past, the relationship between chelatable lead and lead in blood is different from that found in currently exposed subjects (Alessio et al., 1981). In adults non-occupationally exposed to lead, the urinary excretion of lead within 24 hours after administration of 1 g EDTA does not exceed usually 0.6-0.7 mg (Roche et al., 1960). Araki and Ushio (1982) have estimated that on average, 14 % and 19 % of the body burden of lead is mobilised into the urine during the 24-hour after an injection of 20 and 40 mg $CaNa_2EDTA$ per kg bodyweight respectively.

(b) Lead in bone

Ninety to ninety-five per cent of whole body lead burden is stored in the skeleton, mainly in the cortical bone. Measurement of skeletal lead concentration provides therefore a valid estimate of total body burden of lead. The levels of lead in bone increases with age in both sexes, more specifically in male subjects and in dense bone, varying between mean values of 2.16 ppm in the ribs of children to over 50 ppm in the petrous temporal lobes of elderly male adults (Barry, 1975). Ahlgren and Mattsson (1979) have first developed an X-ray fluorescence technique for in vivo determination of lead concentration using ^{57}Co as the source for stimulating lead X-ray production. Because of its relatively high limit of detection (20 µg/g or 20 ppm), the Ahlgren and Mattsson's method can only be used to monitor persons with elevated body burden of lead (e.g. occupationally exposed persons). Ahlgren et al. (1980) have determined the lead concentrations in the phalanges and in the blood of 22 subjects who had formely been exposed to lead in a storage battery plant. Of the 22 men studied, 15 (70 %) had lead concentrations in the phalange above 20 µg/g. The lead concentrations of the phalanges increase with the length of employment in the battery factory. The results of this study indicate that seven years after exposure had ended, the blood lead concentration is more dependent on the daily intake of lead than on the release of lead from the skeleton. The use of a ^{109}Cd gamma-ray source to excite lead X-rays in place of a ^{57}Co source has facilitated a decrease in the limit of detection of lead in bone (tibia) to 12 ppm (Somervaille et al., 1983).

(c) Lead in teeth

The determination of lead concentration in circumpulpal dentine of deciduous teeth has been proposed to estimate exposure to lead during early childhood (Altshuller et al., 1962; Shapiro et al., 1972, 1973). The secondary dentine is a zone of dentine that does not commence to form until after tooth eruption takes place. Lead in the secondary dentine zone, therefore, is a reflection of the amount of lead accumulated in the organism during a fixed and usually known time period i.e. from eruption to exfoliation. Frequently, published data on tooth-lead concentrations cannot be compared because different types of tooth and different components of teeth have been analysed. Secondary dentine has a much higher lead content than primary dentine. Furthermore, the lead content of secondary dentine varies considerably from tooth to tooth. Delves et al. (1982) have also found that the concentrations of lead in pairs of central and lateral incisors from children show large differences for some individuals. Nevertheless, the gross concentrations of lead in crowns, roots or teeth in general may be used to indicate excessive accumulation of lead in childhood. Differences between the lead concentration of dentine from children living in urban, surburban or rural areas has been reported by several authors (Needleman et al., 1972, 1979; Delves et al., 1982; de la Burdé and Shapiro, 1975). For example, Needleman et al. (1972) reported that the mean tooth lead levels of 40 suburban American children was 11.1 ppm and for urban Philadelphia children 51.1 ppm. It is currently possible to measure the lead content of teeth in situ by X-ray fluorescence (Block et al., 1976). The technique measures the characteristic radiation from lead contained within the substance of the tooth during irradiation with a pencil beam of gamma-rays from a ^{57}Co source. In 30 children who had been exposed to lead, Shapiro et al. (1978) found a statistically significant correlation between the tooth lead concentration divided by the child's age and the free erythrocyte protoporphyrin ($r = 0.51$). Fremlin and Edmonds (1980) using alpha-particle excitation and micro-autoradiography have shown dentine zones of lead enrichment related to abrupt changes in exposure.

3. CADMIUM

(a) Cadmium in liver and in kidney

Cadmium is a markedly cumulative toxin with a biological half-life of over 10 years in man. The newborn is practically free from cadmium but the cadmium concentrations in the organs increase with age. In persons non-occupationally exposed to cadmium, the concentration in the renal cortex at age 50 may range from 10 up to 100 ppm. The concentration in liver ranges from less than 1 up to 7 ppm (Kjellström, 1979). The highest values have been found in Japan. Smoking increases the body burden of cadmium. It is possible to measure directly by neutron activation the amount of the metal that has accumulated in the two main sites of deposition, i.e. the liver and the kidney (Al-Haddad et al., 1981, McLellan et al., 1975; Vartsky et al., 1977). We have attempted to determine the critical level of cadmium in kidney cortex i.e. the level at which the first signs of renal disturbances may occur in a certain percentage of the population. With the portable neutron activation system developed by Al-Haddad et al. (1981), we have measured the cadmium level in kidney and in liver in 309 workers employed in two cadmium smelters. The detection limits were about 30 mg/kg in kidney and 3 mg/kg in liver. At the same time, the renal status of the workers was evaluated by determining the urinary concentrations of total protein, β_2-microglobulin and albumin. Kidney function was considered to be abnormal if the urinary concentration of β_2-microglobulin exceeded 0.2 mg/g creatinine, or of albumin, 12 mg/g creatinine, or of total protein, 250 mg/g creatinine. In workers who fulfilled one of these criteria, the cumulative frequency distribution of kidney cadmium concentrations was not different from that found in those with normal renal function. Liver cadmium was significantly higher in workers with renal dysfunction than in those with normal renal function. The fact that the cumulative frequency distribution of kidney cadmium in people with normal kidney function was indistinguishable from that of people with abnormal kidney function suggests either that renal dysfunction is not related to the absolute cadmium concentration in the kidney or that renal cortical cadmium decreases progressively after the onset of kidney damage. The former hypothesis can be rejected in view of the relationship between cadmium accumulation in kidney and disturbance of renal function in animals exposed to cadmium. The latter appears reasonable since it is known that cadmium is lost from the kidney and is excreted in larger amounts in urine when renal damage develops. It thus appears that, following exposure to cadmium, kidney cadmium increases progressively up to a critical level and then kidney dysfunction develops. Hence it is reasonable that the cumulative frequency distribution of cadmium in kidney is the same in those with normal renal function as in those with abnormal renal function. It might be suggested that the critical level of cadmium in renal cortex lies in the range of the highest concentration found in those workers without kidney damage, i.e., around 285-385 ppm. This approach, of course, implies that the renal cortical cadmium concentration invariably increases to this level prior to the onset of renal dysfunction. However, due to the existence of individual biological variability, it may be assumed that the renal cortical cadmium level does not rise up to this level in all the cadmium workers before renal dysfunction occurs. Indeed, another type of analysis based on the study of the correlation between cadmium in liver and kidney in groups of workers with or without renal disturbances leads us to suggest that the minimal critical level of cadmium in the renal cortex could be as low as 215 ppm. In summary, the data collected on workers suggest that, depending on their individual susceptiblity, the critical level of cadmium in renal cortex may range from 215 to 390 ppm. The lowest value corresponds to a concentration of cadmium in liver of 30 ppm (Roels et al., 1981, 1983a).

Using a similar neutron activation instrument but with a lower detection limit (15 mg/kg for the kidney and 1.5 mg/kg for the liver), Ellis et al., (1979) were able to measure the cadmium content of the left kidney and the concentration of cadmium in the liver in 20 non-occupationally exposed adults. Data from the group under study demonstrate significant differences for organ cadmium levels between smokers and non-smokers. Their data suggest that an adult American male non-smoker (age 52) has a total body burden of approximately 19 mg of cadmium (assuming that kidneys and liver represent 50 % of the total body burden). A smoker (age 50, with a smoking history of 38.7 packyears) may have an average total body burden of 35 mg of cadmium. Their data confirm that in non-occupationally exposed persons, cigarette smoking may double the body burden of cadmium. Ahlgren and Mattsson (1981) have suggested the use of X-ray fluorescence analysis as an alternative method of cadmium detection in vivo. The X-ray fluorescence method is more advantageous than the alternative neutron capture gamma ray analysis techniques as far as the total irradiation of the patient is concerned. Unfortunately, the present detection limit for the determination of cadmium in kidney is higher than that obtained by neutron activation. It amounts to 20 mg/kg when the kidney surface is 3 cm below the skin but reaches 100 mg/kg when the distance is 5 cm.

(b) Cadmium in urine

It is now generally accepted that under conditions of moderate cadmium exposure (i.e. general environmental exposure or moderate occupational exposure), when the total amount of cadmium absorbed has not yet saturated all the available cadmium binding sites in the organism (in particular the kidney), the cadmium concentration in urine reflects mainly the cadmium level in the body and hence in the kidney. When integrated exposure has been so high as to cause a saturation of the binding sites, cadmium in urine may then be related partly to the body burden and partly to the recent exposure.

The relative importance of each factor depends on the intensity of exposure. It is also known that when renal dysfunction occurs, the loss of cadmium from the kidney increases. In non-occupationally exposed persons, cadmium in urine rarely exceeds a level of 2 µg/g creatinine. We have attempted to determine the critical level of cadmium in urine. Two approches have been used 1) the study of correlation between cadmium in urine and cadmium in kidney cortex measured by neutron activation 2) the examination of the prevalence of biological signs of renal disturbance as a function of cadmium in urine. Both approaches are in agreement; it seems that to prevent renal dysfunction in a certain percentage of workers exposed to cadmium, the concentration of cadmium in urine should not exceed 10 µg/g creatinine (Buchet et al., 1980; Roels et al., 1981).

(c) Metallothionein in urine

In urine cadmium is present mainly bound to metallothionein. In cadmium workers with normal renal function as well as in those with abnormal renal function, there is an excellent correlation between metallothionein levels in urine and those of cadmium in urine (Roels et al., 1983b). Hence, the determination of metallothionein in urine may provide the same information as the determination of cadmium, but metallothionein analysis presents an advantage over cadmium analysis since the former is not influenced by external contamination of the urine sample, in contrast to the determination of cadmium in urine.

4. METHYLMERCURY

Under steady-state conditions, mercury in whole blood and in hair correlates with methylmercury body burden and with the risk of signs of methylmercury poisoning (Bakir et al., 1973; Berlin, 1979). In humans exposed to methylmercury, the ratio between hair and blood mercury is on the average 250 (Amin-Zaki et al., 1976). It has been estimated that in persons chronically exposed to alkylmercury, the earliest signs of intoxication (paresthesia, sensory disturbances) may occur when the levels of mercury in blood and in hair exceeds 20 μg/100 ml and 50 μg/g respectively (WHO, 1976).

5. METALLIC MERCURY

The kidney and the brain are the two main sites of deposition of mercury following exposure to metallic mercury vapour. The mercury content of blood and urine is only a reflection of recent exposure to metallic mercury. Block and Shapiro (1981) have developed an X-ray fluorescence technique to measure the mercury concentration in situ of the superficial tissues in the head. A sealed 10 mCi ^{57}Co point source was used to excite the K-shell electrons of mercury. Dentists are slightly exposed to metallic mercury used for the preparation of amalgam fillings and by applying the above technique to 298 male dentists, Shapiro et al. (1982) found in more than two-third of them that the levels of mercury in the temporal region and the wrist were below the detectable level (20 μg/g tissue). But more than 13 % of the dentists had temporal mercury values greater than 40 μg/g. In almost all subjects in whom mercury deposits were detected by X-ray fluorescence, the head mercury concentrations were greater than those at the wrist. Normal kidney levels of mercury are 0.1 to 3 μg/g (wet weight). Smith et al. (1982, 1983) are attempting to monitor kidney levels of mercury in vivo by two different techniques 1) detection of prompt gamma-rays emitted following thermal neutron capture and 2) excitation of fluorescence X-rays using external gamma-rays sources (^{57}Co or ^{109}Cd). Presently, the techniques have only been applied to kidney phantom. Their limits of detection are still inadequate to detect the "normal" concentration of mercury in kidney.

6. REFERENCES

L. Ahlgren and S. Mattsson (1979) Phys. Biol., 24, 136-145.

L. Ahlgren, B. Haeger-Aronsen, S. Mattsson and A. Schütz (1980), Brit. J. Ind. Med., 37, 109-113.

L. Ahlgren and S. Mattsson (1981) Physics Med. Biol., 26, 19-26.

L. Alessio, M.R. Castoldi, O. Monelli, F. Toffoletto and C. Zocchetti (1979) Int. Arch. Occup. Env. Health, 44, 127-132.
L. Alessio and V. Foa (1980) in Human Biological Monitoring of Industrial Chemicals. 4. Inorganic lead. Commission of the European Communities, Luxembourg.
L. Alessio, M.R. Castoldi, P. Odone and I. Franchini (1981), Brit. J. Ind. Med., 38, 262-267.
I.K. Al-Haddad, D.R. Chettle, J.G. Fletcher and J.H. Fremlin (1981), Int. J. Appl. Rad. Isot., 32, 109-112.
L.F. Altshuller, D.B. Halah and B.J. Landing (1962) J. Pediat., 60, 224-229.
L. Amin-Zaki, S. Elhassani, M.A. Majeed, T.W. Clarkson, R.A. Doherty, M.R. Greenwood and T. Giovanoli-Jakubczak (1976) Am. J. Dis. Child, 130, 1070-1076.
S. Araki and K. Ushio (1982) Brit. J. Ind. Med., 39, 157-160,
F. Bakir, S.F. Damluji, M. Amin-Zaki, M. Murtadha, A. Khalidi, N.Y. Al-Rawi, S. Tikriti, H.I. Dhahir, T.W. Clarkson, J.C. Smith and R.A. Doherty (1973) Science, 181, 230-241.
R.W. Baloh (1974) Arch. Environ. Health, 28, 198-208.
P.S.I. Barry (1975) Brit. J. Ind. Med. 32, 119-139.
M. Berlin (1979) in Handbook on the Toxicology of Metals (Eds, L. Friberg, G.F. Nordberg and V.B. Vouk) Elsevier North-Holland, Amsterdam pp. 503-530.
P. Block, G. Garavaglia, G. Mitchell and I.M. Shapiro (1976) Phys. Med. Biol., 20, 4140-4144.
P. Block and I.M. Shapiro (1981) Med. Phys., 8, 308-311.
J.P. Buchet, H. Roels, A. Bernard and R. Lauwerys (1980) J. Occup. Med., 22, 741-750.
B. de la Burdé and I.M. Shapiro (1975) Arch. Environ. Health 30, 281-284.
H.T. Delves, B.E.Clayton, A. Carmichael, M. Bubear and M. Smith (1982) Ann. Clin. Biochem., 19, 329-337.
K.J. Ellis, D. Vartsky, I. Zanzi, S.H. Cohn and S. Yasumura (1979) Sciences, 205, 323-325.
J.M. Fremlin and M.I. Edmonds (1980) Nucl. Inst. Meth., 173, 211-215.
J.P.B. Hansen, M. Dossing and P.E. Paulev (1981) J. Occup. Med. 23, 39-43.
ICRP (International Commission on Radiological Protection) (1975) in Report of the Task Group on Reference Man. Pergamon Press, Oxford.
T. Kjellström (1979) Environ. Health Perspect. 28, 169-197.
R.R. Lauwerys (1983) in Industrial Chemical Exposure :

Guidelines for Biological Monitoring. Biomedical Publications, Davis, California.

J.S. McLellan, B.J. Thomas, J.H. Fremlin and T.C. Harvey (1975) Phys. Med. Biol. 20, 88-95.

H.L. Needleman, O.C. Tuncay and I.M. Shapiro (1972) Nature London, 235, 111-112.

H.L. Needleman, C. Gunnoe, A. Leviton, R. Reed, H. Peresie, C. Maher and P. Barrett (1979) New Engl. J. Med. 300, 689-694.

M.B. Rabinowitz, G.W. Wetherhill and J.D. Kopple (1976) J. Clin. Invest., 58, 260-270.

L. Roche, A. Badinand and E. Lejeune (1960) Arch. Mal.Prof., 21, 1-12.

H. Roels, R. Lauwerys, J.P. Buchet, A. Bernard, D.R. Chettle, T.C. Harvey and I.K. Al Haddad (1981) Environ. Res, 26, 217-240.

H. Roels, R. Lauwerys and A.N. Dardenne (1983a) Toxicology Letters, 15, 357-360.

H. Roels, R. Lauwerys, J.P. Buchet, A. Bernard, J.S. Garvey and H.J. Linton (1983b) Int. Arch. Occup. Environ. Health (in press)

I.M. Shapiro, H.L. Needleman and O.C. Tuncay (1972) Environ. Res., 5, 467-470.

I.M. Shapiro, B. Dokkin, O.C. Tuncay and H.L. Needleman (1973) Clin. Chim. Acta, 46, 119-123.

I.M. Shapiro, A. Burke, G. Mitchell and P. Bloch (1978) Environ. Res., 17, 46-52.

I.M. Shapiro, A. Summer, L.K. Spitz, D.R. Cornblath, B. Uzzel, I.I. Ship and P. Bloch (1982) Lancet, I, 1147-1150.

J.R.H. Smith, S.S. Athwal, D.R. Chettle and M.C. Scott (1982) Int. J. Appl. Radiat. Isot., 33, 557-561.

J.R.H. Smith, D.R. Chettle and M.C. Scott (1983) in Proceedings International Conference - Heavy Metals in the Environment. CEP Consultants Ltd, Edinburgh (in press).

L.J. Somervaille, E.E. Laird, D.R. Chettle and M.C. Scott (1983) in Proceedings International Conference in Heavy Metals in the Environment. CEP Consultants Ltd, Edinburgh (in press).

D. Vartsky, K.J. Ellis, N.S. Chen, S.H. Cohn (1977) Phys. Med. Biol., 22, 1085-1096.

WHO (1976) in Environmental Health Criteria. 1. Mercury. World Health Organization, Geneva, pp. 135.

URINARY AND BLOOD MANGANESE AS INDICATORS OF MANGANESE EXPOSURE IN MANUAL METAL ARC WELDING OF MILD STEEL

Jorma Järvisalo, Maritta Olkinuora, Antti Tossavainen, Matti Virtamo, Pentti Ristola* and Antero Aitio

Institute of Occupational Health, Helsinki 29, Finland, and
**Helsinki Shipyards, Oy Wärtsilä AB, Finland*

1. INTRODUCTION

Manganese is an essential element, but in excessive amounts it is harmful: heavy exposure may result in inflammatory lung disease, or a Parkinson-type neurological manifestation. (WHO, 1981) Recently it was reported that urinary excretion of manganese was increased in heavy exposure to manganese in manual metal arc welding of mild steel (MMA/MS) (Chandra et al., 1981), although the main excretory route of manganese is through bile to faeces (Piscator, 1979).

The purpose of the present study was to elucidate the relationship between manganese concentrations in the breathing zone and in the blood and urine of manual metal arc welders of mild steel, in order to clarify possibilities of monitoring individual manganese exposure from concentrations in biological fluids.

2. SUBJECTS AND METHODS

Five welders of mild steel in a shipyard, aged 32 to 42 years, formed the group studied. They did not smoke, and were considered healthy based on the occupational health care history. They had been welders of mainly mild steel for 5 to 9 years. The steel welded and the core wire of the welding rod contained less than one per cent of manganese, and the rod coating 2-3 per cent manganese, and 20 per cent calcium fluoride. Each welder used on the average 100 rods daily.

The air samples were collected on membrane fil-

Chemical Toxicology and Clinical Chemistry of Metals
ISBN 0.12.137520.X
Copyright © 1983 by IUPAC
All rights of reproduction in any form reserved

ters within the face masks of the welders. Each day two four-hour samples were collected.

The total dust concentrations were measured gravimetrically. The manganese content of the samples was measured after dissolving in nitric acid and filtering through a polycarbonate filter. X-ray microanalysis of the residues on the filters showed that the manganese in the samples had completely dissolved. The Mn-content of the filtered solution was determined with flame atomic absorption spectrometry (IL 251, Instrumentation Laboratory, U.S.A.)

Spot urine specimens were collected in acid washed polyethylene containers before and after the daily work shifts on Monday, Tuesday, Thursday and Friday. From 7 am on Wednesday till 7 am on Thursday, all the urine excreted was collected in six portions. Special attention was paid to avoiding contamination of the urine specimens.

Venous blood specimens were collected in heparinized prewashed tubes on Monday, Wednesday and Friday morning and afternoon. Stainless steel needles were used in the collection; therefore - in order to avoid contamination - we discarded the first five mL of blood. The concentration of manganese in this discarded portion was 31.8 ±7.3 nmol/L, whereas in the subsequent blood it was 27.7 ± 5.4 nmol/L ($p<0.05$, Wilcoxon test for paired data).

The manganese content of the specimens was measured with electrothermal atomic absorption (Perkin Elmer 4000, HGA 400 graphite furnace (Perkin Elmer, U.S.A.), pyrolytically coated graphite tubes). In the analytical procedure, the samples were diluted in water, acidified (HNO_3), and injected directly to the graphite tube. For urine samples, the quantitation was achieved by the standard addition technique, the blood samples by comparing to standards prepared in blood. The day-to-day variation of the measurements in air, urine, and blood was 5, 10, and 8 per cent, respectively.

3. RESULTS

In spite of similar working conditions, a considerable variation was observed in the airborne concentrations of total dust and manganese in the breathing zone of the different welders. For the worker with the highest level of exposure, the total dust amount varied between 8.9 and 56.3 mg/m^3, and that of manganese between 0.30 and 2.30 mg/m^3, while for the welder with the smallest exposure the respective figures were 2.2 to 5.3 and 0.04 to 0.14 mg/m^3. For the same two welders, the urinary concentration of manganese (corrected to a relative density of 1.024), in the spot afternoon sample varied from 11.7 to 35.6 (the welder with "high" exposure") and from 8.1 to 15.6 nmol/L (the welder with "low" exposure).

In all welders, the highest concentrations of manganese were seen in urine specimens voided during the night (after 7 pm). A similar diurnal variation was evident, although not as prominent, even if the concentrations of manganese were expressed in nmol/gram creatinine.

The concentrations of manganese in the afternoon urine samples of two of the welders showed a good correlation (r=0.89 and 0.63) with the (time weighted average) concentration of manganese in the respiratory zone air samples. For two other welders, no such correlation was seen.

No systematic correlation was detected between the urinary and blood manganese concentrations in individual workers.

The mean concentration of manganese in the spot urine specimens of the welders ranged from 22.8 to 33.7 nmol/L in the morning, and from 13.6 to 24.9 nmol/L in the afternoon. For most of the welders, the mean values differed significantly from the value seen in a small reference sample group without occupational exposure (9.7 ± 11.7 nmol/L, n=24, $p<0.05$, Student's t-test).

The mean concentrations of manganese in blood were between 26 and 32 nmol/L in the morning and between 25 and 29 nmol/L in the afternoon. All these mean values differed significantly from those of the reference sample group without manganese exposure (14 ± 2 nmol/L, n= 64, $p<0.05$, Student's t-test).

4. DISCUSSION AND CONCLUSIONS

The working conditions of the welders studied were similar; they did not smoke, and all were in normal state of health. However, the exposure - the concentration of manganese and total dust in the breathing zone - showed marked differences between the individuals. This probably reflects mainly differences in working habits. Thus the assessment of occupational exposure to manganese in mild steel welding is rather complicated. In addition, marked interindividual differences were seen in the relationship of the concentrations of manganese in the air and in urine samples. Estimation of the individual exposure thus may not be reached from urinary manganese concentrations either. However, the mean Mn concentrations in the urine exceeded those of a reference sample. It is thus evident that in the MMA welding of mild steel some accumulation of manganese in the body takes place.

Only a few blood specimens could be taken during this study, and no firm conclusions on the value of blood specimens in the biological monitoring of manganese exposure in MMA/MS welding may be drawn. The difference in the manganese concentration between the welders and the non exposed reference group was minor, although statistically significant. One confusing factor may be that manganese is ubiquitous in the environment, and thus sources other than occupational exposure may be important.

5. REFERENCES

S.V. Chandra, G.S. Shukla, R.S. Srivastava, H. Singh and V.P. Gupta (1981) Clin.Toxicol., 18, 407-416

M. Piscator (1979) in Handbook of the Toxicology of Metals (Eds, L. Friberg, G.F. Nordberg and V.B. Vouk), Elsevier-North Holland, Amsterdam, pp. 485-501

World Health Organization (1981) Environmental Health Criteria 17: Manganese, WHO, Geneva, pp. 1-110

THE EPIDEMIOLOGY OF IN UTERO AND AIRBORNE LEAD EXPOSURE AND SUDDEN INFANT DEATH SYNDROME

Marilyn M. Erickson and Laura S. Hillman

Department of Pediatrics, Washington University School of Medicine, St. Louis, MO 63110 U.S.A.

1. INTRODUCTION

Sudden Infant Death Syndrome (SIDS) is the sudden death of any infant or young child which is unexpected by history and in which a thorough postmortem examination fails to demonstrate an adequate cause for death. The number of SIDS deaths averages about 2-3 per 1000 live births or about 8000 per year in the U.S. Almost 70% of these deaths occur between 1 and 4 months of age. Most studies show an increased incidence of SIDS in the lower socioeconomic groups, the black race, and among the premature and low birthweight infants. Many of the mothers are under 20 years of age and have at least one other child.

Among the many possibilities explored in searching for a cause of SIDS, investigators have looked for congenital defects, nutritional or hormonal deficiencies or excesses, and drug intoxication. Although certain subtle anatomic differences have been observed in SIDS, none have been severe enough to suggest a mechanism by which death might occur, Beckwith (1973, 1979), Valdes-Dapena (1980).

The epidemiology of toxic metal exposure, especially lead, and of SIDS are similar. The incidence for both is highest in inner cities and among the lower socioeconomic groups. It seemed appropriate, therefore, to investigate the possibility of a relationship between lead exposure and SIDS.

2. METHODS

(a) Tissues

Lung, liver, kidney, and rib specimens were obtained from each infant autopsied at the St. Louis City and County Medical Examiners Offices. The ages ranged from 4 to 26 weeks. There were 66 SIDS cases and 23 non-SIDS cases. The causes of death in the non-SIDS group were pneumonia, asphyxia, menigitis, and congenital heart defects.

Chemical Toxicology and Clinical Chemistry of Metals
ISBN 0.12.137520.X
Copyright © 1983 by IUPAC
All rights of reproduction in any form reserved

Each tissue was packaged separately in plastic bags and stored at -20°C until the analyses were performed. Soft tissue was removed from the ribs; they were split and the marrow washed out with a stream of deionized water. All these tissues were dried and pulverized. NBS 1577 bovine liver was used as a reference tissue. Half-gram portions of dried tissues were digested with nitric acid. Following appropriate dilution lead was measured by electrothermal atomic absorption spectroscopy (AAS). Background correction was used.

(b) Air lead

Air lead levels were obtained from the St. Louis County Air Pollution Control Office. The monthly mean air lead concentrations were compared to the monthly number of SIDS deaths over a 5 year period.

(c) Cord blood lead

Cord blood was obtained from 2284 sequential births in a single hospital over a 1-year period (May 1982 through April 1983). Lead was measured by flame AAS using APDC chelation and MIBK extraction. Hematocrits were measured by centrifugation. Blood lead values were corrected for red cell volume.

3. RESULTS

There were no significant differences between the SIDS and non-SIDS groups in age distribution, sex ratio, or race ratio. The SIDS liver and rib specimens contained significantly more lead than non-SIDS tissues (liver, 1.095 μg/g vs. 0.761 μg/g, $p < 0.05$; rib 1.754 μg/g vs. 1.041 μg/g, $p < 0.01$, respectively). All four SIDS tissues showed significant increases in lead concentrations with age while non-SIDS tissues did not. The intercept of the SIDS liver lead regression line suggests an in utero component of lead exposure. The slope of the SIDS lung lead regression line was significantly steeper ($p < 0.05$) than non-SIDS, suggesting the respiratory route as a major contributor to the SIDS lead burden, Erickson et al. (1983).

When we compared air lead levels to SIDS incidence we found that the highest air lead levels preceded the peak incidence of SIDS deaths by 1-2 months. Similar data have been reported for air pollution in general in Los Angeles, Hoppenbrauwers et al. (1981). However, we found no difference in the tissue lead levels in those infants who died during the peak SIDS incidence period (October-January) and those who died during the low incidence period.

The cord blood lead levels show a late summer/fall peak with a September mean ± S.D. of 15.0 ± 4.5 μg/dl. Alternate months are shown in Table 1. Hospital records for the July births were examined and we found that the cord blood lead was significantly greater than the monthly mean (8.8 μg/dl) for mothers $\leq$ 19 years of age plus $\geq$ 2 parity (10.9 ± 4.4 μg/dl), infants $\leq$ 38 weeks gestation (10.2 ± 4.8 μg/dl) and black infants (9.5 ± 3.6 vs. 6.8 ± 3.0 μg/dl for white infants, $p < 0.001$). The peak September group showed the same pattern but the differences were less pronounced.

TABLE 1

Cord Whole Blood Lead Concentrations By Month, μg/dl
(only alternate months shown)

	Jan	Mar	May	Jul	Sep	Nov
n	175	193	249	221	181	144
$\bar{x}$	10.3	8.3	7.3	8.8	15.0	12.5
S.D.	3.1	2.5	2.9	3.5	4.5	3.2
% $\geq$15 μg/dl	6.3	1.0	3.6	4.5	53.6	17.4
% $\geq$20 μg/dl	1.1	0.5	0.4	1.8	11.6	4.2

4. DISCUSSION

Lead passes freely through the placental barrier exposing the fetus to the lead in the maternal circulation. The very young multigravida woman, known to be at risk for having a SIDS infant, is suspected to have during pregnancy increased mobilization of lead from her bones, the major storage depot for lead. These data are also consistent with theories of increased gastrointestinal lead absorption and release of lead stored in bone during late summer.

Lead deposited in the lungs is said to be completely absorbed. Hence, during periods of high air lead levels, considerably more lead is added to the maternal circulation. Taken together they may explain the increased liver lead, suggestive of increased in utero lead exposure, found in SIDS.

The relationship of air lead peaks in the months immediately preceding SIDS incidence peaks may partly explain the findings of increasing lung lead with age in SIDS infants but not in non-SIDS. Again, the rapid absorption from the lungs postnatally could add to the lead burden acquired in utero and account for the increased bone lead concentrations found in SIDS.

The major portion of the lead in the atmosphere is attributed to the burning of leaded-gasoline in automobiles. Other sources include industrial emissions, the burning of newspapers and magazines in fireplaces, and cigarette smoke. Even with the declining use of leaded-gasoline and a conscious effort to reduce other sources of lead exposure, it probably will require at least another generation to significantly reduce the skeletal lead burden in the prospective mother.

5. CONCLUSIONS

Young multigravidity, prematurity, black race, and late summer/fall birth are risk factors for SIDS and for elevations of cord blood lead. Urban living and lower socioeconomic status also are common both to SIDS and to environmental lead exposure and a temporal relationship between peak air lead concentration and SIDS incidence has been noted. The significance of these epidemiologic relationships between air, cord blood, and tissue lead remains unclear. They may be just expected epidemiologic relationships or they could have physiologic significance. Further study is required.

6. ACKNOWLEDGEMENTS

This work was supported in part by NIH Grant 2R01HD-09998-06 and a grant from the National Sudden Infant Death Syndrome Foundation.

7. REFERENCES

J.B. Beckwith (1973) Curr Probl Pediatr, 3(8), 1-36.
J.B. Beckwith (1979) DHEW Publication No. (HSA) 79-5251.
M.M. Erickson, A. Poklis, G.E. Gantner et al. (1983) Pediatr Res, 17, in press.
T. Hoppenbrouwers, M. Calub, K. Arakawa et al. (1981) Amer J Epidemiol, 113, 623-635.
M.A. Valdes-Dapena (1980) Pediatrics 66, 597-614.

RENAL FUNCTION IN INDIVIDUALS
OCCUPATIONALLY EXPOSED TO MERCURY VAPOUR

Michael D. Stonard, David P. Duffield*,
John J. O'Sullivan* and Gordon T. Steel

*Central Toxicology Laboratory and *Medical Department,
Mond Division, Imperial Chemical Industries PLC, Cheshire, U.K.*

1. INTRODUCTION

The distribution of mercury in the body following inhalation of mercury vapour is generally similar to that of injected salts of ionic mercury compounds, with a marked accumulation of the metal in the kidneys. The ability of ionic mercury compounds to produce overt renal tubular damage in several species including man has been well-established. However, the evidence that exposure to mercury vapour produces renal damage is largely dependent upon a relatively small number of case-histories which describe a nephrotic syndrome (Kazantzis et al. 1962; Mandema et al. 1963).

Several enzymes and proteins have been detected in urine which appear to have diagnostic relevance in the recognition of renal dysfunction in man (Peterson et al. 1969; Price, 1982). The present study, using enzyme and protein measurements, was undertaken to compare renal function in a group of individuals occupationally exposed to mercury vapour in a chlor-alkali plant with a non-exposed group.

2. METHODS AND RESULTS

Urine specimens were collected at monthly intervals from each individual on three separate occasions and a blood sample was collected on the last occasion. The following analyses were carried out on each biological specimen: total protein, albumin, α_1-acid glycoprotein, β_2-microglobulin, N-acetyl-β-glucosaminidase, gamma glutamyl transferase, creatinine and mercury. Since timed urine collections were not made, concentration adjustments have been made on the basis of creatinine excretion.

The basic characteristics (number of individuals, age, duration of exposure and mercury in blood and urine concentrations)

Chemical Toxicology and Clinical Chemistry of Metals
ISBN 0.12.137520.X
Copyright © 1983 by IUPAC
All rights of reproduction in any form reserved

are shown for the two groups in Table 1. The two groups were age-matched and attendance on each occasion was greater than 90%. As expected, mercury in blood and urine concentrations were markedly higher in the mercury-exposed group than in the control group. For the mercury-exposed group at the third visit, there was a highly significant correlation between mercury in blood and urine (Figure 1).

TABLE 1

CHARACTERISTICS OF THE CONTROL AND MERCURY-EXPOSED GROUPS (Values presented are means with ranges in parentheses).

	Control	Mercury-Exposed
Number of individuals	82-87	96-105
Age (yr)	41.4 (20.0-61.3)	38.2 (16.5-61.3)
Duration of exposure (yr)	-	8.0 (1.0-33.0)
Mercury in urine (μg/g creat)	4 (0.4-24)	67 (0.5-275)
Mercury in blood (μg/litre)	< 5	17.5 (< 5-70)

* The number of individuals is given as the minimum/maximum participating at the three sampling periods.

Although the group mean values for urinary enzymes and proteins have been compared at each visit the trends were similar on each occasion. The concentrations of albumin, α_1-acid glycoprotein and β-microglobulin were lower in the mercury-exposed group on each occasion, especially β_2-microglobulin which was statistically significantly decreased ($P < 0.01$) on the second and third occasions. This loss of β_2-microglobulin is probably a consequence of the release of protease enzymes from tubular cells which have been damaged by mercury. No differences of statistical significance were observed when the mean values for the two urinary enzymes and all plasma parameters were compared between the two groups.

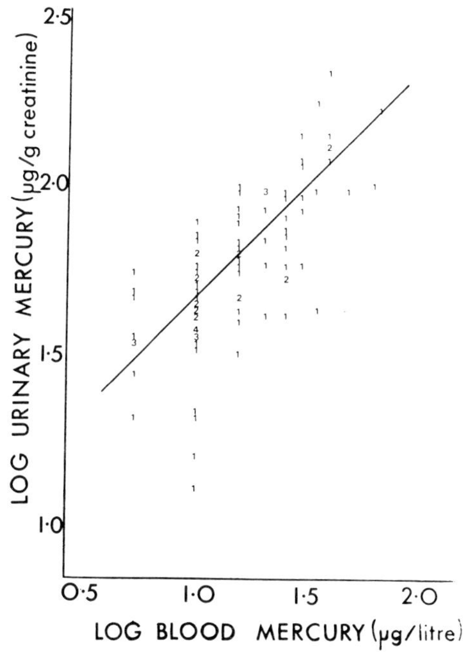

FIG. 1 Relationship between mercury concentrations in blood and in urine for the mercury-exposed group. Log (urinary mercury) = 2.387+0.612 log (blood mercury) N = 91; r = 0.71; P < 0.001. The figures 1,2,3 and 4 refer to the number of observations.

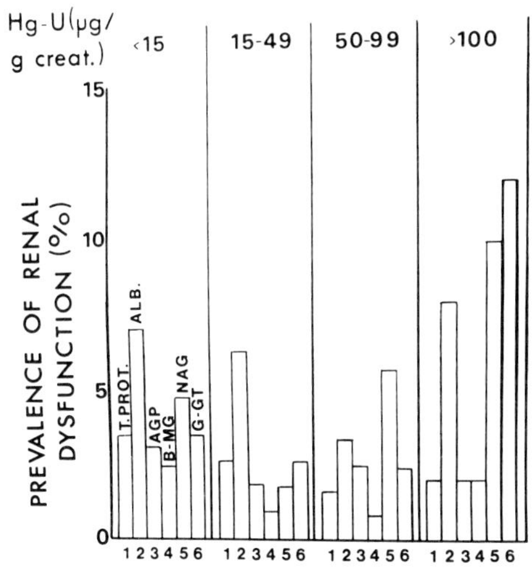

FIG. 2 Relationship between prevalence of renal dysfunction and mercury in urine concentration.

As the prevalence of measurements indicative of renal dysfunction appeared similar from one visit to another, the results were combined. "Upper limit of normal" values were determined by calculating the values corresponding to 2 S.D's above the mean value.

The relationship between the prevalence of renal dysfunction and mercury in urine concentration has been examined (Figure 2) The prevalence of abnormal values indicative of renal dysfunction showed little evidence for a dose (absorption)-response relationship over the full range of mercury in urine concentrations observed; however, a small increase in the prevalence of higher activities of the two urinary enzymes was observed when the mercury in urine concentration exceeded 100μg/g creatinine. These changes in enzyme activities, when considered in the absence of increases in protein concentration, are too small to have health significance.

3. CONCLUSIONS

No evidence has been found for an increased prevalence of renal dysfunction in the mercury-exposed group when compared to a non-exposed group using predominantly measurements of enzymes and proteins in urine as indices of renal dysfunction. Although the present study was unable to confirm previous findings that an increased prevalence of elevated urinary albumin is seen as the mercury in urine concentration increases (Buchet et al. 1980), it did demonstrate that the prevalence of raised urinary enzyme activities increases slightly when the urinary mercury concentration exceeds 100μg/g creatinine.

4. REFERENCES

J.P. Buchet, H. Roels, A. Bernard and R.R. Lauwerys (1980) J. Occup. Med. 22, 741-750.

G. Kazantzis, F.R. Schiller, A.W. Asscher and R.G. Drew (1962) Quart. J. Med. 31, 403-418.

E. Mandema, E. Arends, J. Van Zeijst, G. Vermeer, G.K. Van der Hem and L.B. Van der Slike (1963) Lancet I, 1266.

P.A. Peterson, P.E. Evrin and I. Berggard (1969) J. Clin. Invest. 48, 1189-1197.

R.G. Price (1982) Toxicology 23, 99-134.

CIRCULATING IMMUNE COMPLEXES IN INDIVIDUALS OCCUPATIONALLY EXPOSED TO MERCURY VAPOUR

Michael D. Stonard, Bernard V. Chater, David P. Duffield*
John J. O'Sullivan* and C. Martin Lockwood**

*Central Toxicology Laboratory and *Medical Department, Mond Division, Imperial Chemical Industries PLC, Cheshire*

***Renal Unit, Royal Postgraduate Medical School, Hammersmith Hospital, London, U.K.*

1. INTRODUCTION

Mercury compounds are able to induce not only acute renal tubular necrosis in animals and man but also a nephrotic syndrome in some individuals; mercury in several different forms, including vapour, has been implicated in glomerulonephritis in man (Kazantzis, 1978).

Recent studies have shown that several organic and inorganic forms of mercury can induce a biphasic renal disease in the Brown Norway rat which is characterised initially by the deposition of anti-glomerular basement membrane (GBM) antibodies and secondly by the presence of circulating immune complexes (Druet et al., 1981). Because this experimental animal model may have relevance to autoimmune renal disease produced by mercury in man, renal and immunological investigations have been carried out on individuals exposed to mercury vapour in a chlor-alkali plant. The present paper reports the immunological findings from a main study involving primarily an evaluation of renal function and a smaller follow-up study conducted approximately fourteen months after the main study.

2. METHODS AND RESULTS

The main study was comprised of a group of male workers with occupational exposure to mercury (N=96) and an age-matched group with no exposure to mercury (N=82). The individuals for the follow-up study were drawn from those participating in the main study and included the majority of individuals who at that time were found to have circulating immune complexes (IC's). Also, age-matched individuals from both groups without evidence of IC's were chosen such that the final group sizes were comparable: control (N=28) and mercury-exposed (N=25). Immune

Chemical Toxicology and Clinical Chemistry of Metals
ISBN 0.12.137520.X
Copyright © 1983 by IUPAC
All rights of reproduction in any form reserved

complexes in sera were assayed by a polyethylene glycol precipitation test (PEG) as described by Digeon et al. (1977) and by a platelet aggregation test (PAT) according to Myllyla (1973).

Blood mercury concentrations were determined on all individuals in both studies and were <5μg/litre in the control groups and in the range 5-70μg/litre in the mercury-exposed groups. Both the PEG and PAT assays revealed the presence of IC's in the control and mercury-exposed groups in the main study, with the prevalence being greater in the mercury-exposed group. The control group had 7% and the mercury-exposed group 17% with sera positive for IC's by the PEG assay (Figure 1). The corresponding figures using the PAT assay were 6% and 20% respectively (Figure 2). Both tests showed good agreement.

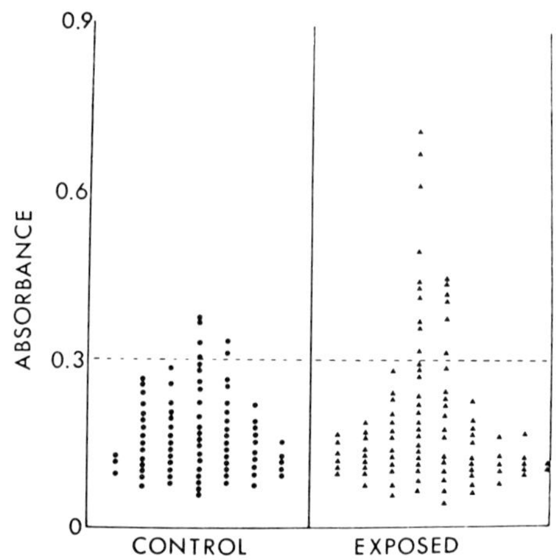

FIG 1. Circulating immune complexes in the control and mercury-exposed groups from the main study as measured by the PEG assay.

A higher prevalence of circulating IC's was again found in the mercury-exposed group when both assays were carried out on the follow-up study. Values obtained by the PEG assay for individuals participating in both studies are shown in Figure 3. Of the original six individuals in the control group with circulating IC's all have been followed up but none now show evidence of IC's. By the same assay, thirteen individuals in the mercury-exposed group with the presence of IC's have been followed up, of whom four individuals still show evidence of IC's. Also, one individual in the mercury-exposed group without IC's in the main study now shows serum positive for IC's by the PEG assay.

FIG 2. Circulating immune complexes in the control and mercury-exposed groups from the main study as measured by the PAT assay □ control; ▨ exposed.

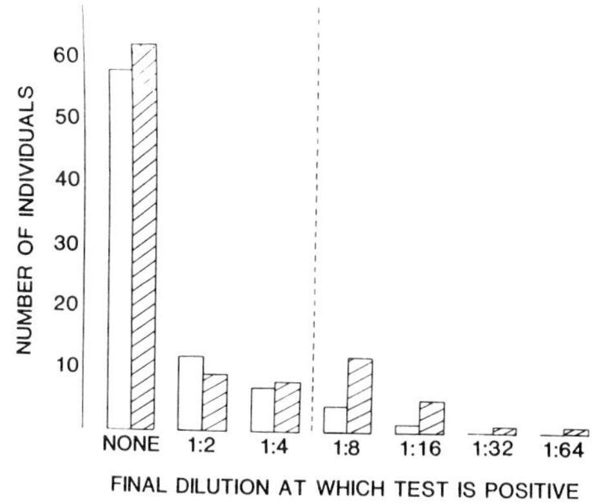

The results of the PEG assay have been confirmed with the PAT assay (Figure 4). All five individuals with evidence of IC's in the control group have been followed up and none of these now show IC's. Fifteen of the original nineteen individuals with the presence of IC's in the mercury-exposed group have been followed up, of whom six individuals still show IC's. Also, one individual in the mercury-exposed group without IC's in the main study now shows serum positive for IC's by the PAT assay.

FIG 3. Comparison of the levels of circulating immune complexes in both studies as measured by the PEG assay. ● control; ▲ exposed.

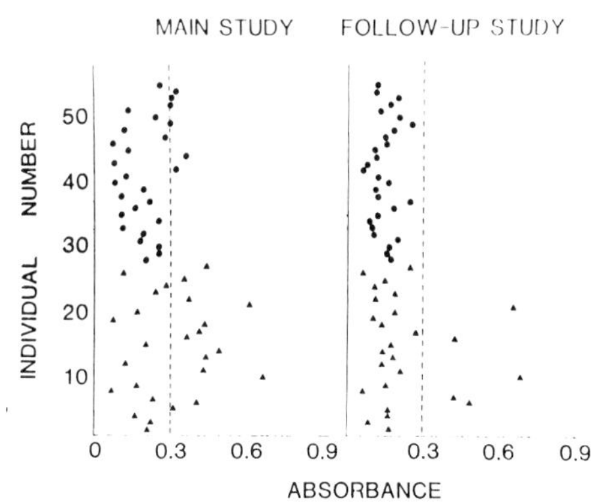

Relationships between the presence of circulating IC's and several parameters have been examined. There is no evidence on an individual basis for a direct relationship between the presence of IC's and i) blood or urine concentrations of mercury and ii) any evidence of renal dysfunction (Stonard et al. 1983).

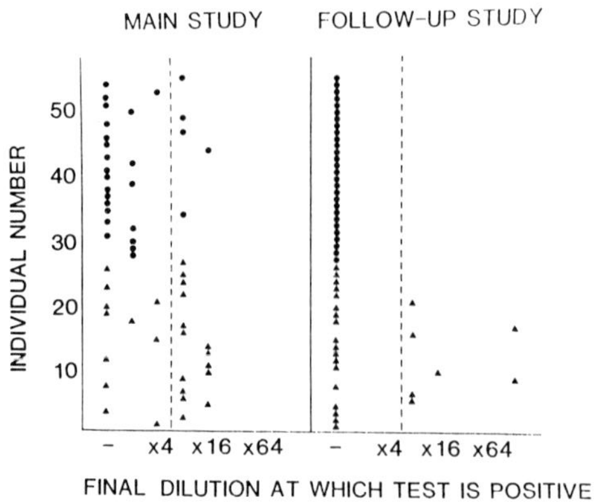

FIG 4. Comparison of the levels of circulating immune complexes in both studies as measured by the PAT assay.
● control; ▲ exposed.

Furthermore, as the presence of an antibody to GBM forms the first phase of immune glomerulonephritis produced by mercuric chloride in the Brown Norway rat, a similar assay was carried out in the present study but failed to detect any positive sera. Because mercury may act as a polyclonal lymphocyte activator, the levels of circulating immunoglobulins E, G, and M were measured. There was no evidence to suggest the occurrence of polyclonal activation in the mercury group as a whole nor in those individuals with the presence of IC's. As might be expected from the absence of polyclonal activation, there were no indications of anti-nuclear autoantibodies in those individuals with the presence of IC's.

3. CONCLUSIONS

An increased incidence of circulating IC's has been found in the mercury-exposed group when compared to the control group in both studies by two different assay methods which show good agreement. The significance of this finding is unclear since no association has been found between the presence of circulating IC's and evidence of renal dysfunction, autoimmune disease or increased mercury in blood or urine concentrations.

4. REFERENCES

M. Digeon, M. Laver, J. Riza and J.F. Bach (1977) J. Immunol. Meths. 16, 165-183.

P. Druet, P. Teychenne, C. Mandet, C. Bascou and E. Druet (1981) Nephron 28, 145-148.

G. Kazantzis (1978) Environ. Hlth. Persp. 25, 111-118.

G. Myllyla (1973) Scand. J. Haematol. Suppl. 19.

M.D. Stonard, B.V. Chater, D.P. Duffield, A.L. Nevitt, J.J. O'Sullivan and G.T. Steel (1983) Int. Arch. Occup. Environ. Hlth. In Press.

RENAL EFFECTS IN WORKERS EXPOSED TO CADMIUM IN A CADMIUM PRODUCING PLANT

N.J. v. Sittert and R. Kummer

Shell Internationale Petroleum Mij, B.V.
Medical and Toxicology Division
The Hague
The Netherlands

1. INTRODUCTION

A study was carried out in a zinc plant in the southern part of the Netherlands producing 150,000 tons of zinc per year. From 1892 until 1973 the so called thermic zinc process was used, in which high temperature zinc ovens were the characteristic feature. With the introduction of the zinc electrolysis process in 1974, representing the most modern technology presently available, the cadmium recovery unit was changed from an original thermic process into an electrolytic one. The objective of the study was to evaluate the workers' cadmium exposure (cadmium in blood and urine) and the status of their renal function.

2. METHODS

(a) *Study Population*

The study population consisted of the following groups: Group A comprised six men who had worked in the old plant for 1-19 years (mean 6 years). Five of those men subsequently worked in the new plant for 3-8 years (mean 7 years). The average age of the workers was 41 years (range: 33-57 years). Group B comprised nine men who had worked in the new plant only for 2-8 years (mean 7 years). Their average age was 37 years and ranged from 27-51 years. Group C comprised 14 men employed in a part of the zinc plant without exposure to cadmium. The average age of Group C workers was 41 years (range: 19-55 years). Most workers in groups A, B and C smoked more than 20 cigarettes per day. The reference Group D comprised 131 men, approximately equally composed of smokers and non-smokers, derived from a petrochemical industry in another part of the country where industrial exposure to cadmium was non-existent.

Chemical Toxicology and Clinical Chemistry of Metals
ISBN 0.12.137520.X
Copyright © 1983 by IUPAC
All rights of reproduction in any form reserved

(b) *Biological Measurements, Sampling- and Analytical Procedures*

Surveys were done on groups A and B in 1980, 1981 and 1982. Group C was measured in 1980 and Group D in 1982 and 1983. The following biological parameters were measured: cadmium in blood (CdB), cadmium in urine (CdU) , β-2 microglobulin in urine (β-2MG) and total protein in urine (TP). Blood was collected by venepuncture in cadmium-free tubes and stored at -20°C. Urine was collected in cadmium-free polyethylene bottles and a portion was taken for pH determination. Urines were stored at - 20°C for the subsequent determination of CdU, β-2MG, TP and creatinine. Approximately 2 hours before urine collection each worker consumed 3 grams of bicarbonate to ensure a urinary pH above 5.7. This procedure was not followed in group D; samples below pH 5.7 were not used for β-2M determination.

The concentrations of CdB and CdU were measured by electrothermal atomic absorption spectrometry (AAS) with maximum power heating supply (Perkin-Elmer model 4000). Blood samples were treated by 1M HNO_3 for deproteinisation and matrix modification, and urines were diluted with water before measurements in the AAS were carried out. The determination of β-2M was carried out by radioimmuno assay and TP was assessed using Tsuchiya's reagent and the biuret reaction.

3. RESULTS

(a) *Reference Intervals*

Reference intervals of CdB, CdU and β-2MG were calculated from group D and factors that influence the reference intervals were determined. Age had a significant influence on CdU, but not on CdB and β-2 MG. Smoking had a significant influence on CdB and CdU but not on β-2 MG.

Reference intervals of CdB and CdU in relation to smoking habits and of β-2 MG are shown in Table 1.

Urinary total protein concentrations, which were not determined in group D, were considered to be abnormal if exceeding 250 μg/g creatinine (Roels et al., 1982).

(b) *Cadmium Exposed Groups*

Two out of six men in Group A showed β-2MG levels of about 2x the upper reference limit (180 μg/g creatinine). All values in Groups B and C were within normal limits. No TP levels above normal limits were found in any group.

TABLE 1

Geometric Means and 95% Confidence Intervals (Group D)

	Total	No smokers	1-20 cigarettes	20 + cigarettes
CdB	-	0.50	1.17	1.71
(μg/l)		0.32-0.95	0.44-3.1	0.72-4.06
N		24	34	15
CdU	-	0.30	0.61	0.78
(μg/g creatinine)		0.01-0.91	0.19-2.0	0.27-2.30
N		24	39	33
β-2 MG	60	-	-	-
(μg/g creatinine)	20-180			
N	61			

The percentage of individuals exceeding the upper reference limits for CdU and CdB was in the order Group A > Group B > Group C. Figures 1-4 show the mean values and the ranges for CdB, CdU, β-2 MG and TP respectively in Groups A,B,C and D and a comparison between the years 1980, 1981 and 1982 (Groups A and B only). It can be seen that CdB and CdU levels in Group A were higher than in the other groups, but the differences between Groups B, C and D ("heavy smokers", >20 cigarettes/day) were small. The concentration of β-2MG was highest in Group A, but there were no differences between Groups B, C and D. There was no difference in TP concentrations between the groups. It can further be seen that there were no major changes in any of the measurements which were made in 1980, 1981 and 1982. The slight increase in β-2MG in Group B (1982) might be due to intra-individual variations. The yearly variation of biological measurements in the two workers in Group A with the highest β-2MG excretion is shown in Table 2. There was no significant change in any of the values in the period 1980 to 1982.

(c) *Correlations*

CdB and CdU values were significantly correlated in Groups A and D, but not in Groups B and C. There were no relationships between CdU or CdB and β-2M or TP concentrations in any group.

4. DISCUSSION

Two out of six men in Group A showed a very slight tubular type of proteinuria with an increased excretion of

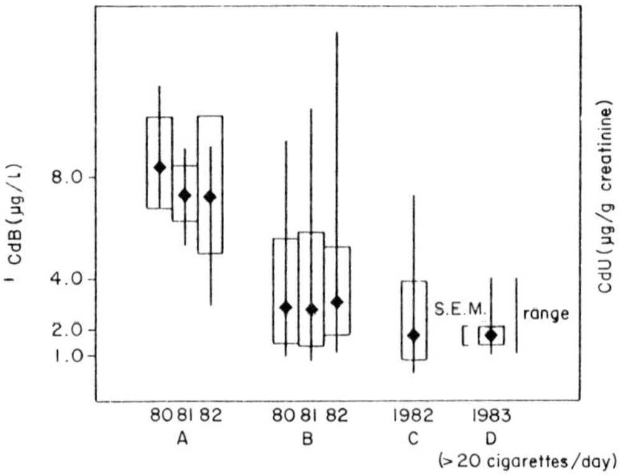

Fig. 1 Ranges, geometric means and standard errors of Cadmium in blood levels of Groups A,B and C (1980, 1981, 1982) and of the heavy smokers in the reference Group D.

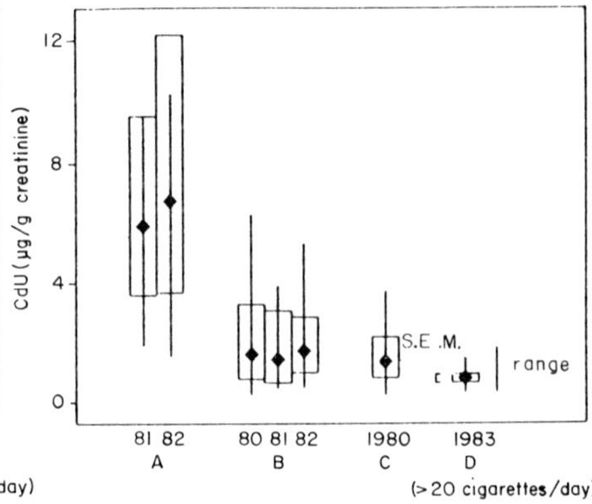

Fig. 2 Ranges, geometric means and standard errors of Cadmium in urine levels of Groups A,B and C (1980, 1981, 1982) and of the heavy smokers in the reference Group D.

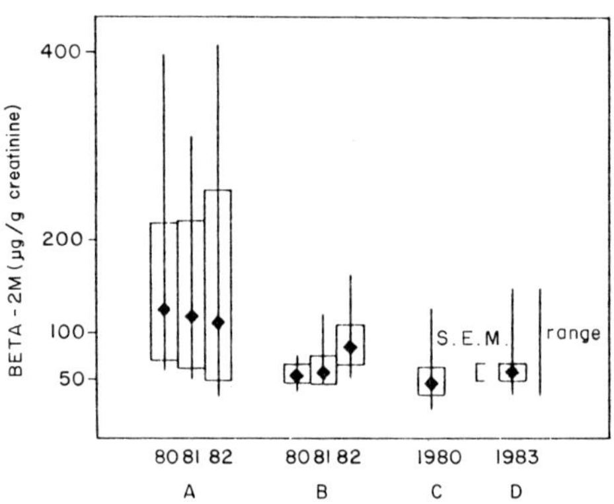

Fig. 3 Ranges, geometric means and standard errors of β-2 microglobulin in urine levels of Groups A, B and C (1980, 1981, 1982) and of the reference Group D.

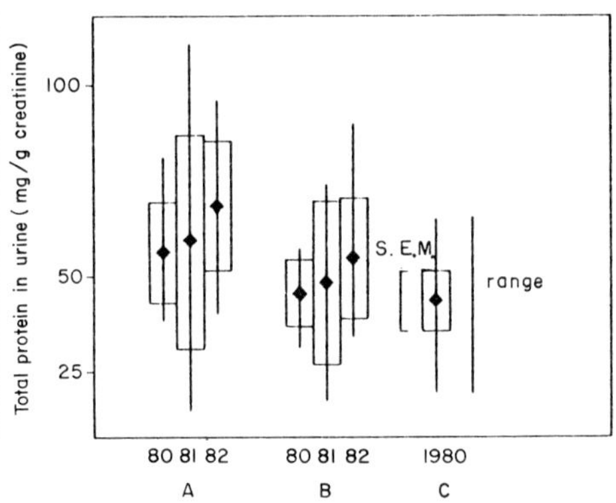

Fig. 4 Ranges, arithmetic means and standard errors of total protein in urine levels of Groups A,B and C (1980, 1981, 1982).

TABLE 2

Values for Biological Measurements in Two Subjects from Group A (1980-1982).

	Subject	1980	1981	1982
CdB	1	7.1	7.1	9.1
(μg/l)	2	-	6.8	7.1
CdU	1	-	7.3	10.3
(μg/g creatinine)	2	-	5.8	8.4
β-2 MG	1	400	243	410
(μg/g creatinine)	2	187	312	335
Total protein	1	62	40	96
(mg/g creatinine)	2	81	70	61

β-2MG, without evidence of a glomerular type of proteinuria. One of these two subjects has not been occupationally exposed to cadmium since 1974, suggesting that the tubular proteinuria might have been caused by excessive exposure in the old plant. There was no evidence of a progression of β-2MG or TP excretion in the two subjects over the years 1980, 1981 and 1982. None of the other workers in groups A, B or C showed signs of renal dysfunction. CdB and CdU levels in Group A did not change significantly over the years 1980, 1981 and 1982, indicating that a steady state might have been reached at a CdB concentration of about 8 μg/l and a CdU concentration of about 8 μg/g creatinine. It was concluded that in subjects with past high exposures to cadmium, such as Group A, CdB and CdU levels reflect body burden and that in these subjects elimination of cadmium from the body occurs at a very low rate. In subjects of Group B, CdB levels reflect recent exposure to cadmium rather than body burden. By comparison of the CdU concentrations of groups B and C with the "heavy smokers" of Group D, it was concluded that occupational and environmental exposure to cadmium may have contributed to half of the body burden in Group B, and to an even smaller amount in Group C individuals. Thus cadmium exposures in the new plant do not seem to contribute much to the body burden.

5. REFERENCE

H. Roels, J. Djubgang, J.P. Buchet, A. Bernard, R. Lauwerijs (1982) Scand. J. Environ. Health, 8, 191-200.

THE USE OF KERATINISED TISSUES TO MONITOR THE DETAILED EXPOSURE OF MAN TO METHYL MERCURY FROM FISH

John S. Hislop, Terence R. Collier, Gaye F. White
Darlando T. Khathing and Eleanor French

Environmental & Medical Sciences Division,
AERE, Harwell, Oxon, OX11 ORA U.K.

1. INTRODUCTION

Of the keratinised tissues, head hair has received the greatest attention as a monitor of human exposure to metals, in particular for mercury and arsenic. Other tissues, such as finger and toe nail, body hair and stubble offer many of the advantages of head hair over use of body fluids, such as ease of sampling and storage and the ability to provide retrospective monitoring, but little work has been carried out in this area. Problems associated with the use of keratinised tissues in, e.g., the monitoring of exposure to methyl mercury (MeHg), include the difficulty of relating the mercury content of the tissue to the blood concentration at the time at which the tissue was formed and also in knowing with any accuracy the level of MeHg to which the subject was exposed.

During the course of a more detailed study to evaluate the dose response relationship of man to MeHg from fish, Sherlock et al. (1983) the opportunity was taken to assess the value of keratinised tissues as monitors of MeHg intake.

2. EXPERIMENTAL

(a) Study Protocol

20 volunteers undertook to consume regularly over a period of 3 months known quantities of halibut which had been analysed previously to establish the naturally elevated MeHg content. Blood samples were collected during the intake period and during the following 3 months when fish consumption had returned to normal levels. Head hair samples were collected for 5 males consuming the highest levels of MeHg, ~1600 μg per week, 2 also provided stubble samples collected daily using a dry razor and 1 volunteer provided finger and toe nail clippings over a period of a year.

Chemical Toxicology and Clinical Chemistry of Metals
ISBN 0.12.137520.X
Copyright © 1983 by IUPAC
All rights of reproduction in any form reserved

(b) Analysis

Samples were analysed for inorganic and total mercury, and hence for methyl mercury by difference, using atomic absorption (Magos 1971). Precautions were taken to minimise contamination in reagents and during sampling of the blood. Blood was analysed as received, hair and nails after dissolution in NaOH without washing. In blood <5% of the mercury was present in the inorganic form, in keratinised tissue a higher percentage was observed resulting from decomposition of MeHg during digestion. The accuracy of the method for total mercury was checked using neutron activation analysis (TNA) non destructively for hair and following radiochemical separation for blood. For accurate data for methyl mercury use had to be made of 45% NaOH solution as recommended by Magos (1971) rather than 30% NaOH as advocated by Clarkson et al. (1977).

3. RESULTS

(a) Head Hair and Stubble

Results for total mercury content of 8 mm segments of head hair from an individual are shown in FIG. 1 along with corresponding data for blood. Samples were cut, close to the scalp, near the end of the study and the data plotted are at the mid points of the growth periods, based on the measured rate of hair growth for 9 volunteers of 11.8 mm/month ($\sigma\%$ = 8).

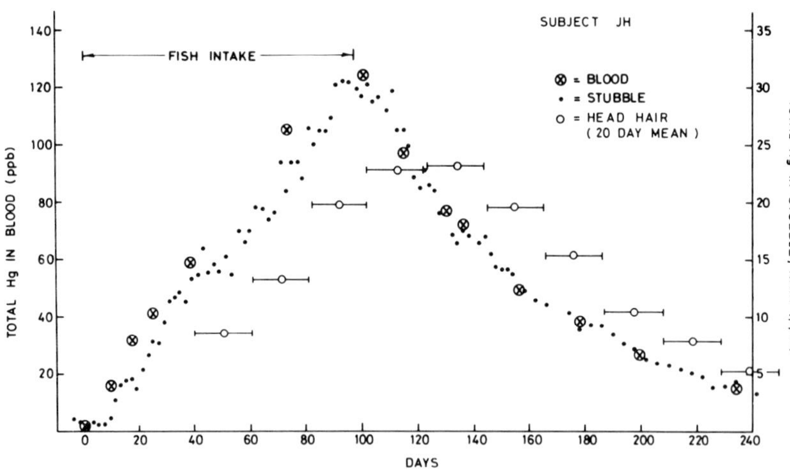

FIG. 1 Hg levels in blood, head hair and stubble.

(b) Finger and Toe Nails

Results for finger and toe nails and head hair for the same individual are shown in FIG. 2. Clippings were collected when approximately 2 mm long and sectioned in the case of toe nails. Results are plotted at the mid point of the growth period.

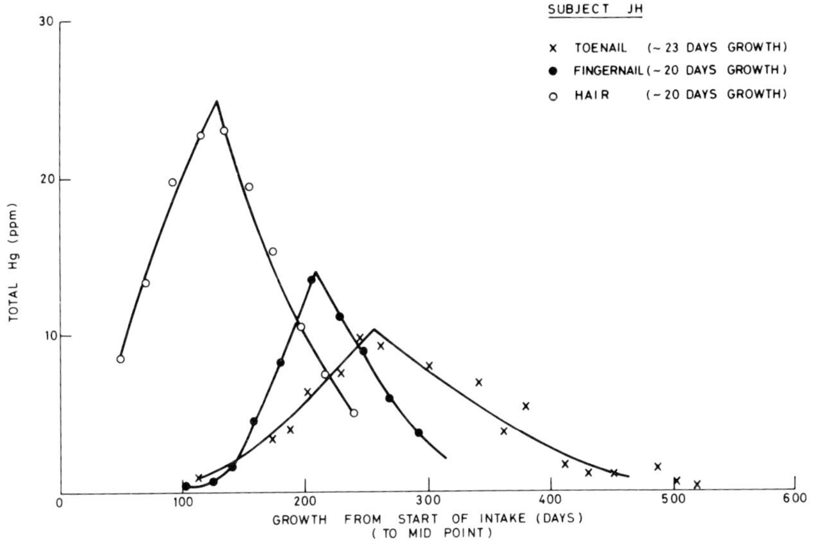

FIG. 2 Hg levels in head hair, toe and finger nails.

4. CONCLUSION AND DISCUSSION

All tissues studied reflect the known pattern of MeHg intake and could potentially be used to monitor this. Depending on the tissue, the time at which monitoring could take place varies from within a few days of exposure to several months.

(a) Head Hair and Stubble

The concentration of mercury in head hair peaked at a time estimated to be approximately 20 days later than that shown by the blood for 3 of the 5 volunteers studied. This implies a delay period before an elevation of MeHg in blood results in an elevation of mercury in hair samples which are cut close to the scalp. Detailed knowledge of hair growth rates and imprecisions in cutting makes accurate estimation of this delay difficult.

As shown in FIG. 1 a much more accurate assessment of the delay in incorporation of methyl mercury was possible in stubble samples. For the two volunteers studied this amounted to 9 days. 4-5 days can be attributed to stubble still below the skin (rate of growth ~1 mm/day), implying a delay of 4-5 days before MeHg in blood is incorporated into the stubble. For volunteer JH (FIGS. 1 and 2) the ratio of mercury in head hair to that in blood was 200:1 and 250:1 for stubble, assuming the peak level in the keratinised tissues was attributable to the peak level in the blood. This suggests a difference in metabolism of MeHg in different hair tissues for an individual. For the 5 volunteers the head hair to blood ratio varied from 200 to 340 confirming that a significant variation also occurs between individuals.

FIG. 1 shows how mercury concentrations in stubble mirror those in blood. Both closely followed calculated values based on 1st order elimination kinetics ($t\frac{1}{2}$ = 45 days) and an accurately known pattern of fish consumption. For male subjects stubble offers a more practical means of sampling for MeHg than blood with a more accurately known and higher time resolution than head hair.

(b) Finger and Toe Nails

FIG. 2 shows that, for an individual, mercury levels in head hair are greater than those in finger nail which in turn are greater than those in toe nail and the delay in incorporation of MeHg is greatest for toe nail although nail growth rates viz 0.8 and 0.4 mm/day, and nail bed lengths, 14 mm and 6 mm respectively for the nails shown have to be considered. The ratio of concentration of mercury in finger nail and toe nail to that observed in blood, again assuming peak levels are equivalent, were 115:1 and 80:1 respectively.

5. ACKNOWLEDGEMENT

The authors gratefully acknowledge support from the Ministry of Agriculture, Fisheries and Food for part of the work described in this paper.

6. REFERENCES

T.W. Clarkson, M.R. Greenwood and L. Magos (1977) in *Clinical Chemistry and Chemical Toxicology of Metals* (Ed, S.S. Brown), Elsevier/North Holland, Amsterdam, New York, London, pp. 201-204.

L. Magos (1971) *Analyst*, 96, 847-853.

J.C. Sherlock, J.S. Hislop, D. Newton, G. Topping, K. Whittle (1983) *submitted for publication*.

URANIUM IN DRINKING WATER - REPORT ON CLINICAL STUDIES IN NOVA SCOTIA

M. A. Moss, R. F. McCurdy, K. C. Dooley, M. L. Givner, L. C. Dymond, J. M. Slayter, M. M. Courneya

Division of Clinical Chemistry, Department of Pathology Victoria General Hospital and Dalhousie University, Halifax, Nova Scotia, Canada.

1. INTRODUCTION

The finding of uranium in drinking water illustrates many of the problems faced by Governmental agencies whose task it is to determine guidelines for drinking water quality.

For some constituents, drinking water guidelines are based upon very limited health data, and they are subject to periodic review as more information becomes available.

Where a metal is typically ubiquitous, such as uranium, it may not be realistic to insist that it is absent from our drinking water. Under these circumstances, a recommendation is based upon a compromise between ideal (in terms of lack of adverse health effects) and what is generally practicable to achieve.

The growing controversy regarding uranium exploration and mining has served to focus public attention on this metal as a nuclear hazard. Our knowledge is much more limited with regard to its health effects as a low level natural environmental contaminant.

Uranium is a non-essential radioactive metal, existing in its natural state as a mixture of three isotopes, of which U^{238} greatly predominates. It constitutes a dual hazard by virtue of both its radioactivity and its chemical toxicity.

In occupational exposure involving particle inhalation, radiological hazards may be of primary concern. In the case of environmental exposure involving ingestion, then chemical toxicity may be determined to be more significant.

The Manhattan Project during the Second World War contributed to our understanding of the effects of large acute exposure, but there is very little published material on the effects of chronic low dose ingestion, and the minimum toxic threshold is not known for man.

Prior to 1978, the Canadian Drinking Water Standards

Chemical Toxicology and Clinical Chemistry of Metals
ISBN 0.12.137520.X
Copyright © 1983 by IUPAC
All rights of reproduction in any form reserved

established the maximum acceptable limit for uranium at 5 mg/L, based only on aesthetic considerations of colour and taste. In 1978, some Russian animal toxicity studies were reviewed, and the guideline for uranium was considerably reduced to a maximum acceptable concentration of 0.02 mg/L. Whilst this recommendation was calculated to provide a safety factor, the actual margin of safety provided is uncertain, owing to the many variables encountered when extrapolating toxicity data from various animal species to man.

It is anticipated that uranium ingestion in drinking water constitutes a chemical hazard primarily, and based upon reported acute and chronic animal studies, it is further anticipated that the kidney is the critical organ, with the earliest effects manifest in the distal third of the proximal tubule.

The discovery of a population in Harrietsfield, near Halifax, Nova Scotia, consuming drinking water containing excessive uranium concentrations provided an opportunity to study this type of exposure. This small community relies on private wells, and, at the time of the study, 42% of 94 wells tested contained excessive uranium, some more than 10 times the maximum acceptable concentration.

The uranium originates from underlying granites, and the deeper drilled wells typically produce a reducing environment, which favours mobility of uranium into the water.

A Task Force was appointed to investigate various aspects of this problem throughout the Province. An extensive well sampling program was carried out, and this helped delineate the population at potential risk.

2. METHODS

In a pilot study, 133 subjects were interviewed by standard questionnaire to ascertain occupational, medical and exposure history. One-third comprised a local control group having acceptable or undetectable levels of uranium in their well water.

The clinical testing carried two objectives. These were, firstly, to establish a satisfactory index of body burden or cumulative exposure; and, secondly, to look for evidence of sub-clinical toxic effects. If such toxic effects could be identified, this might allow an approximation of the minimum toxic threshold, which would be valuable information in the future review of the Drinking Water Guidelines.

Following completion of the questionnaire, participants provided hair and urine samples, which were processed in the following way:

a) Hair uranium concentration - to be assessed as a potential index of body burden.
b) Random urine sample - routine urinalysis and microscopy was

performed on site, designed to screen for gross abnormalities and confounding urinary tract disease. An aliquot was taken for β_2-microglobulin (β_2-m) determination. This is established as a very sensitive indicator of tubular proteinuria, and we therefore considered it a potential index of subclinical toxicity.

c) 24-hour urinary excretion of uranium was determined and assessed as a potential index of body burden.

3. RESULTS OF PILOT STUDY

Data from questionnaires and test results were stored and processed on computer file. The population was stratified into various exposure groups, and an analysis of variance was performed, calculating the mean and 95% confidence intervals for a chosen parameter in each of these groups. The following findings were evident from this pilot study:

a) There was no definitive evidence of toxicity with regard to tubular proteinuria since there was no significant difference between exposed groups and controls. However, there does appear to be a trend of increasing urinary β_2-m excretion with increasing exposure. The group with the highest exposure appear out of keeping with this trend, but many persons in this group, particularly, had heeded advice to stop using their water for drinking purposes, some months prior to the clinic. There is, therefore, the suggestion of a reversible effect.

b) Urinary uranium excretion did not show a significant difference between the various exposure groups. We were interested to note the questionable value of urinary uranium excretion as an index of exposure, since it is used for this purpose to screen employees for accidental exposure in the industry. However, uranium excretion is dependent upon tubular pH, and may, therefore, overlap with the normal range in a relatively low-dose situation such as the drinking water exposure.

c) Hair levels of uranium did appear very promising as an index of exposure. Ideal tissues for body burden studies would be bone or kidney, but these are clearly not accessible for population screening, and hence our interest in developing the hair procedure.

New Ross Study. The well sampling program delineated another community at New Ross, NS, where some wells contained significantly higher levels of uranium than any in Harrietsfield. We were fortunate to be able to continue our studies at that location. A further 191 subjects were recruited, of whom 55% had well levels of uranium in excess of the drinking water guideline. In the series of 324 subjects, approximately two-thirds had been told that their water was unacceptable. Twenty-seven persons had uranium levels at least 10 times higher than the guideline and four of these were 0.7 mg/L, or 35 times higher than the guideline. Our findings were consistent with those of the pilot study, i.e.

a) Urinary uranium excretion showed no clear relationship with exposure.
b) Hair uranium showed a strong relationship with exposure, and would seem to show great promise as an index of exposure. We were interested to note that in each exposure group, females had consistently higher levels of hair uranium, possibly reflecting more time spent at home. These findings are the subject of continuing laboratory studies to evaluate the effects of lifestyle and various other exposure factors.
c) Urinary β_2-m again showed a trend effect of increasing excretion with increasing exposure.

4. SUMMARY

We have described clinical studies of two populations exposed to excessive levels of uranium in drinking water, up to 35 times the current Canadian Guideline. Apart from a much smaller survey in Russia, we believe these are the first studies of their type. We conclude that hair uranium shows promise as an index of exposure, whereas urinary uranium excretion is unreliable.

The significance of the trend effect of increasing urinary β_2-m excretion is that this may represent early tubular proteinuria as a manifestation of subclinical toxicity.

In cadmium poisoning, tubular proteinuria is the first sign of renal injury, and further injury may be irreversible. In view of the similarities between acute toxicity studies of cadmium and uranium in animal models, it seems important to examine this question of tubular proteinuria more closely.

It is unlikely that further clinical studies will provide more useful information, at this stage. We have therefore started further studies using an animal model, to investigate the reversibility of tubular injury once induced by uranyl ion, to establish the threshold of irreversible change, and then relate these findings to the current Drinking Water Guideline. We hope that this will be completed within a year, and we anticipate that these studies will provide valuable health effects information for the next review of the Canadian Drinking Water Guideline for uranium.

REGIONAL REFERENCE RANGES OF ELEMENTAL COMPOSITION LEVELS IN SOME ADULT HUMAN SAMPLES OF CLINICAL INTEREST: A PRELIMINARY EVALUATION

Venkatesh Iyengar

Institute of Medicine
Juelich Nuclear Research Center
D-517 Juelich
Federal Republic of Germany

1. INTRODUCTION

Literature data on "normal levels" for elemental concentrations in biological systems, especially for trace elements, are full of controversial findings (Iyengar et al. 1978). The problems and prospects associated with the use of such data in evaluating environmental quality and human health has been elucidated by Iyengar (1981). In dealing with literature data, gross errors involving several orders of magnitude often stem from the use of wrong analytical methodology. In some cases, it is mainly due to the failure in collecting valid samples for analysis. It should be recognised that obtaining valid biological samples for assay is governed by the influence of presampling factors on one hand (Iyengar 1982), and the demand for quality assurance in sampling, sample preparation, preservation and storage of specimens, on the other. Even when adequate attention is paid to the above mentioned points, one still encounters difficult situations during data handling and interpretation which may be due to sketchy documentation. An additional handicap is that results are seldom accompanied by quality control information by way of analysing standard reference materials (SRM). Although irrelevantly high or unreasonable analytical findings may be dismissed as artifacts, one is still faced with a difficult state of judgement when the reported levels of concentrations are not wholly improbable but there is no guarantee for the analytical accuracy. Unfortunately, most of the literature data fall in this group.

At this stage it is well to ponder on the expression "normal value" itself which is being used somewhat indiscriminately in the world of elemental composition of biological systems. For example, the same expression is applied to Pb and Cd as well as for Cu and Zn. Strictly speaking, true normal values for Pb and Cd e.g. in human blood, should be close to zero. However, envir-

Chemical Toxicology and Clinical Chemistry of Metals
ISBN 0.12.137520.X
Copyright © 1983 by IUPAC
All rights of reproduction in any form reserved

onmental and other factors have contributed to the entry of these two elements into biological systems, and what one means in this case is a certain tolerance limit. On the other hand, elements such as Zn and Cu that are absolutely essential for life need to be present in living systems and are homeostatically regulated. Under ideal conditions, such levels may be expected to fluctuate within narrow limits for a given species, and therefore, one is justified in calling them normal levels. It is highly desirable to establish such normal levels in tissues and body fluids so that alterations during disease conditions can be detected. However, in practice it is a tough task requiring consideration of and compensation for a number of possible concurrent phenomena, and correlations are very complex. Thus baseline values can be deceptive and the question as to what is normal may not be very easy to answer. Therefore, it may be worthwhile to attempt establishing reference ranges of values for distinct global regions, taking the essential regional features such as dietary habits, environmental aspects and geochemical factors into consideration.

2. DATA BASE

To overcome the above mentioned problems to some extent and to salvage good analytical results from the mass of the literature reports, a new approach has been tried. A number of analysts were contacted and requested to identify reliable data from their countries for adult human samples only by screening the literature to the extent possible. Special care was directed to the analytical quality of the results. Some of the collaborators did not provide exclusive data but recommended selected sources and these were followed up by the author. A set of 6 samples of clinical interest (blood and its components, milk, liver and hair) were considered for 15 elements of biological significance: Zn, Cu, Fe, Se, Mn, Pb, Cd, Hg, As, Mo, Cr, Co, I, Ni and F.

About 6o investigators participated by sending selected results. Taking all the input into consideration the results cover over 4o countries and may be grouped as follow: African region (Botswana, Kenya, Nigeria, South Africa and Uganda), Asian(near, middle and far east) region(Bangladesh, China, Egypt, Hongkong, India, Iran, Iraq, Japan, Mangolia, Nepal, Pakistan, Philippines, Russia, Thailand and Vietnam), European region (Austria, Belgium, Bulgaria, Czechoslovakia, Denmark, Finland, France, Germany, Greece, Holland, Hungary, Ireland, Italy, Luxembourg, Norway, Poland, Spain, Sweden, Turkey, United Kingdom and Yugoslavia), American region (USA and Canada, Chile, Cuba, Guatemala, Mexico, Panama, Peru and Venezuela) and Australia and New Zealand.

Concerning the reliability of the data base, the concept that the analyst is the most important part of any analytical system provided the back bone. Some of the participants of this study are outstanding investigators of international reputation. Some have analysed SRMs (e.g. NBS bovine liver, IAEA animal muscle and milk powder) during the course of their analytical work to assess the accuracy and precision of their results. In some of the international collaborative studies (e.g. Cd and Pb in blood and study on milk) extensive quality control steps were initiated using both the available reference samples and developing in-house standards.However, it must be pointed out that some of the crucial SRMs (e.g. serum) are still not available and that in some cases the discretion of the analyst or the author forms the sole basis for inclusion of the results.

Only a few countries were in a position to supply reasonable amounts of data on the samples requested here. This paper covers Zn, Cu, Se, Mn, Pb and Cd in blood, hair and milk.

3. RESULTS AND DISCUSSION

(a) Zinc (Unit μg per mL for fluids, μg per g for hair)

Results from several countries for whole blood, serum and erythrocytes (RBC) are in the range of 6 to 9, o.8 to 1.1 and 9 to 13, respectively. Values tending to be low are seen for whole blood in Bangladesh = 4.4, India = 4.9 and in serum for some groups in Turkey = o.51 to o.58. In milk, high values are seen for Finland = 4, Italy = 3, New Zealand = 3.2, Guatemala = 3.3 and Chile = 7.2. In a number of countries, the concentration ranges from 1.5 to 2. On the other hand, countries such as Turkey = o.8, Sweden = o.8, Hungary = 1.o and Yugoslavia = 1.3 are on the marginal to low side. Hair Zn levels frequently fall in the range of 15o to 25o for many countries. While Poland shows an exception on the high side with 32o, Kenya, Egypt, Turkey, South Africa (Bantu subjects), India, Bangladesh and some parts of USA appear to be in the marginal range of 1oo to 14o.

Based on the picture of Zn in blood, milk or hair, countries such as Bangladesh, Turkey, India, Sweden, Hungary, Yugoslavia, Kenya, Egypt etc, fall in the low to marginal status. However, it may be mentioned that low level in milk may not necessarily reflect the deficiency state of the mother but has more significance to the infant population. The list of Zn deficient groups would have been certainly longer if results were available from more countries. In this context it is interesting to note that among the countries included in the FAO soil survey (1982), Turkey, Hungary, India and Kenya are classified as low Zn areas, while Egypt is termed marginal. On the other hand, soils from Finland, New Zealand and Philippines are grouped under high Zn areas with Italy qualifying as normal to high. This classification of the soil corresponds well with the picture seen for Zn in milk in those countries.

(b) Copper (Unit μg per mL for fluids, μg per g for hair)

Consistent results are seen for all the components of blood. The values range from o.8 to 1.1 for males and 1 to 1.4 for females. Exceptional values are recorded for 2 groups of populations: the aboriginals in some Australian settlements with levels exceeding 1.4 and Tokelau islanders of New Zealand with a concentration of about 1.7. In milk, the values are higher in Australia UK and New Zealand, all having about o.4, Greece = o.46 and Finland = o.6. In several countries such as Hungary = o.21, Sweden = o.2 and Russia = o.2, the results are on the lower side. Most frequent values in hair samples from several countries fall within a narrow range of 15 to 25. Relatively high values are seen for Australia = 39, whereas, Bangladesh = 7, Pakistan = 1o, Nigeria = 1o and Turkey = 11 tend to be on the low side.

Cu is high in blood of aborigines from the Australian settlements. This population also suffers from hypoferraemia. Finland has the highest concentration of Cu in milk, yet when measured in soil it is classified as normal. However, in the low concentration group (Hungary, Zaire, Sweden and Yugoslavia), the soil status of Hungary is known to be Cu deficient. Hair concentrations of Cu in Kenya, Australia and to some extent in New Zealand, are high. The soil status of New Zealand is quoted to be normal to high. Pakistan, Turkey, Nigeria, Thailand, Bangladesh, east Germany and Tokelau islands show low levels in hair. The official soil status in the first 4 countries is given as normal to low in the FAO report.

(c) Manganese (Unit ng per mL for fluids, µg per g for hair)

Results available for blood are too few. In milk, high levels are seen in Philippines = 4o, Nigeria = 16 and Zaire = 11. In a number of countries such as Hungary, Sweden, Yugoslavia and Finland the level is typically between 3 and 6. Higher Mn levels are also seen in hair samples from Kenya = 15. A group of countries such as Nigeria, India, Thailand and Pakistan show about 3 to 4, while most other countries cluster around o.5 to 1.5.

No low concentration level problems are encountered for this element in any of the countries that provided the results. The high values seen in milk from the Philippines and Nigeria the soil Mn profile. Soil status of Zaire is unknown to the author. However, high values seen in hair for a number of countries do not always correlate with the soil Mn status.

(d) Selenium (Unit ng per mL for fluids, µg per g for hair)

Relatively high values are seen in whole blood from Canada (Hamilton area) = 18o, and parts of USA; Wyoming = 234, Cleveland = 157 and California = 162. China (Keshan area) = 21, Finland (some parts) = 58, New Zealand = 59, Egypt = 68 and Italy (Amiata area) = 77 appear on the low side. Other regions of China (non-Keshan) show 95. In many other countries the levels are between 9o and 13o. In milk, Philippines = 33, UK = 31 and Chile = 62 belong to the high level group. A number of countries such as Australia = 12, New Zealand = 1o, USA (Ohio) = 13, Finland = 1o and Hungary = 11, show relatively low levels. Other countries fall in the range of 15 to 25. In hair, some samples from Japan show high values of 3 to 4. China (Keshan area) = o.o7 and parts of Greece = o.o85 belong to the low concentration group. In most other countries the concentration levels are in the order of o.5 to 1.

As seen above, several countries have reported low Se values. The FAO soil study report does not give information on Se. However, it is known from other sources that countries such as New Zealand, Finland and China (some parts) have Se deficient soils. Since Se levels in tissues and body fluids are known to be extremely susceptible to dietary intake and reflect even short term variations in input, changes in Se levels of biological samples should be interpreted with great care.

(e) Lead (Unit ng per mL for fluids, μg per g for hair)

Concentration levels close to 2oo are seen in India (Bangalore), Italy (several cities), Mexico and some parts of USA and Canada. However, in a majority of cases such as the west European countries, most parts of USA, Israel and Peru, the levels are often between 9o and 15o. Parts of Sweden, Japan and China show intermediate values of 5o to 8o. In Nepal and in some remote areas of Venezuela, lowest values of 3o and 8, respectively, are recorded. In milk, high values are found in Philippines = 17, Spain = 17, Sweden = 17, Hungary = 15 and in some parts of USA = 12 to 46. Nigeria, Zaire and Guatemala have reported 3 to 5. High levels in hair are seen in Nigeria = 52, Kenya = 42, some parts of USA = 19 to 45 and Canada = 17. Countries such as India (rural) = 5, Bangladesh = 4 and Iran = 3, represent some of the low exposure areas.

Men have about 2o to 3o per cent more of this element in their blood than women. This is confirmed in several countries in many studies. Since almost all of the Pb in blood is known to be associated with RBC, a substantial part of the sex difference in concentration is explained by the hematocrit. The remaining part of the difference can be related to the variability in the working environment and food consumption (quantity). Therefore, expressing the concentration of Pb on the basis of RBC weight is helpful in identifying other factors that influence the concentration of this element in blood. Smoking elevates blood Pb only to a small extent whereas consumption of wine (e.g. Italy) may elevate the level by a factor of about 1.5.

(f) Cadmium (Unit ng per mL in fluids, μg per g in hair)

Nearly all of the 25 sets of results for blood from 15 countries fall within the range of o.3 to 1.2 for non-smokers. This is quite remarkable in view of the low concentration involved. In milk, Nigeria and the Philippines show about 3. In Zaire, Hungary, Sweden, New Zealand and Guatemala the levels are up to 1.7. Hair Cd in Nigeria, Kenya and in some urban samples of Canada is 2 to 2.4. In most other countries the levels are between o.4 to 1.

Like Pb, Cd also binds almost exclusively to RBC. Yet it is

difficult to demonstrate sex difference because of the low concentration involved. Smoking affects the Cd picture in blood very strongly and up to a factor of about 1o in some cases, presumably depending upon the Cd concentration of tobacco.

4. CONCLUDING REMARKS

Although analytical awareness has been the basis for the results presented here, they must be used with caution. Nevertheless, this survey has been useful in demonstrating certain trends in the trace element picture around the world, at least qualitatively.Both diet and environment have strong influence on the trace element distribution pattern of tissues and fluids as seen for several elements, e.g. As, Cd, Mn, Pb, Se and Zn. These and other factors should be considered while evaluating literature data before rejecting results with substantial variation (Versieck and Cornelis 198o).

This study also revealed the paucity of data in some regions of the world. In view of the nutritional importance of several elements the time is ripe for international organizations to intervene and help produce some reference data for selected global regions which lack data of any kind. Analysis of just whole blood samples to start with can generate a wealth of information to monitor the status of some of the key elements such as Cu, Fe, Se and Zn. Such data can serve as a regional reference bank until each country develops its own data base.

In choosing the reference group of subjects for a data base, a set of healthy persons should be chosen. These results should be used for comparison when random populations are studied.Notable deficiencies or excesses if any, in the reference group, can be identified by comparison with reliable literature data.

Most of the data in this review cover only urban parts which are not ideal study areas in terms of trace element influences on human health. More rural populations should be studied.

5. ACKNOWLEDGEMENT

The author is thankful to all the participants who contributed to this report. A full report containing the names of all the analysts, results and references is in preparation.

6. REFERENCES

G.V.Iyengar, W.E.Kollmer, H.J.M.Bowen (1978) The Elemental Composition of Human Tissues and Body Fluids, Verlag Chemie Publishers, Weinheim, Germany.

G.V. Iyengar (1981) Sci.Total Environ. 19, 1o5-1o9.

G.V. Iyengar (1982) Anal.Chem. 54, 554A-558A.

M.Sillanpää (1982) Micro Nutrients and the Nutrient Status of the soils: A Global Study, FAO Soils Bulletin 48. Rome.

J.Versieck, R.Cornelis (198o) Anal.Chim.Acta 116, 217-254 .

THE "USE" AND "MISUSE" OF HUMAN HAIR IN TRACE METAL ANALYSIS

Howard H. Sky-Peck and Betty J. Joseph

Department of Biochemistry
Rush-Presbyterian-St.Luke's Medical Center
Chicago, Illinois 60612 U.S.A.

1. INTRODUCTION

The analysis of trace elements in hair has become increasingly popular for the evaluation of systemic intoxication, for predictions of exposure to environmental pollutants, for the assessment of nutritional status, and for the diagnosis of disease states. The popularity of hair as a tissue specimen resides, in part, in the ease with which samples can be obtained without trauma to the individual, which is important in sampling large populations, and in the relative stability of specimens which require no special equipment or storage conditions except to prevent post-sampling contamination (Katz, 1979).

These properties coupled with sensitivity of newer analytical techniques including atomic absorption, neutron activation and x-ray fluorescence have been partly responsible for this current surge of applied research on trace elements in hair. In spite of these advantages, the ranges of values reported for many of the trace elements found in hair have varied widely. The reasons for this include sampling techniques, sample preparation, external contamination arising from hair treatment such as shampoos, cosmetic preparations, and atmospheric fallout.

The present study was designed to show some major reasons for the wide range of values found in a healthy population.

2. METHODS AND MATERIALS

Scalp hair was generously donated by 458 healthy individuals of both sexes ranging in age from 3 to 70 years. Six strands of hair were pulled out with the root and washed in a non-ionic detergent, distilled water, isopropanol, distilled water and air-dried. The dried strands of hair were suspended across a 2 x 2 in. plastic holder and held in place by a 1.0 in. Teflon ring. The samples were placed in a sample chamber of the x-ray system which was then air-evacuated prior to analysis.

Chemical Toxicology and Clinical Chemistry of Metals
ISBN 0.12.137520.X
Copyright © 1983 by IUPAC
All rights of reproduction in any form reserved

A quantitative energy dispersive x-ray fluorescence (EDXRF) system has been developed for rapid simultaneous multi-element analysis of small samples. The principles of the EDXRF system have previously been described by Sky-Peck and Joseph (1981).

2. RESULTS

There were significant differences in the distribution of trace elements in hair with respect to sex, as seen in Table 1.

TABLE I

DISTRIBUTION OF TRACE ELEMENTS IN HAIR ACCORDING TO SEX*

	MALES 155	FEMALES 215	P***
SULFUR**	38.2±6.4	36.1±6.4	0.014
CALCIUM	501.4±394.0	1280.8±1130.5	0.0001
IRON	14.1±10.3	11.5±9.9	0.043
NICKEL	0.76±0.76	1.09±1.10	0.005
COPPER	19.90±9.8	23.9±16.2	0.025
ZINC	189.1±39.3	194.4±43.2	0.023
SELENIUM	0.72±0.32	0.58±0.23	0.0004
BROMINE	4.40±3.60	2.14±1.81	0.0001
STRONTIUM	6.40±12.40	6.50±6.10	0.47
CHROMIUM	1.16±0.66	1.01±0.62	0.56
MANGANESE	1.14±0.65	1.17±1.27	0.43
ARSENIC	0.50±0.61	0.38±0.80	0.13
MERCURY	1.03±2.40	0.89±0.58	0.28
LEAD	12.44±15.01	7.25±11.80	0.005
RUBIDIUM	0.74±0.33	0.74±0.40	1.00

*MICROGRAMS PER GRAM DRY WEIGHT.

**MILLIGRAMS PER GRAM DRY WEIGHT.

***STUDENT'S *t* DISTRIBUTION AND CHI-SQUARE PROBABILITY.

The S, Fe, and Se contents were lower in females while Ca, Ni, Cu, and Zn were significantly elevated. However, some of these differences such as S, Ca, Ni, and Zn may in part actually reflect hair treatment by women. Some of these effects of hair treatment are presented in Table II. Both peroxide bleach and permanent wave markedly altered the contents of S, Ca, Fe, and Ni in hair. In addition, peroxide affected Zn levels and permanents affected Cu and As levels. These changes due to hair treatment complicated the interpretation of sexual differences.

Natural hair color also played a significant role in the interpretation of trace element values. Blonds had less Fe and Mn than brunettes, while redheads had more Fe, Cu, and Zn in their hair. In addition, separating individuals according to age indicated significant decreases in the S, Ca, and Sr contents with the older adults.

A comparison between caucasian, black and oriental females indicated some statistically significant differences in trace

TABLE II

DISTRIBUTION OF TRACE ELEMENTS ACCORDING TO HAIR COLOR (FEMALES)*

	BRUNETTES 32	BLONDS 30	P	REDHEADS 30	P**
SULFUR	36.8±5.3	35.8±3.7	0.29	38.5±4.5	0.16
CALCIUM	953.6±778.6	768.0±462.3	0.23	937.4±686.7	0.48
IRON	9.6±2.4	7.6±1.4	0.007	17.8±6.8	0.0001
NICKEL	0.53±0.26	0.44±0.20	0.15	0.75±0.35	0.02
COPPER	17.3±4.2	17.2±3.6	0.47	24.2±8.2	0.0014
ZINC	187.9±39.0	181.4±38.3	0.33	211.8±32.1	0.03
SELENIUM	0.58±0.21	0.53±0.15	0.30	0.55±0.19	0.33
BROMINE	1.64±0.88	1.53±0.61	0.41	1.25±0.69	0.08
STRONTIUM	5.25±3.30	3.80±1.4	0.08	5.35±3.15	0.45
CHROMIUM	1.12±0.43	0.90±0.44	0.09	0.95±0.41	0.12
MANGANESE	1.03±0.54	0.72±0.31	0.04	1.18±0.54	0.21
ARSENIC	0.30±0.61	0.30±0.25	1.00	0.39±0.48	0.32
MERCURY	1.00±0.56	0.99±0.58	0.31	0.63±0.31	0.01
LEAD	3.44±2.90	6.60±5.70	0.03	4.80±3.56	0.11
RUBIDIUM	0.79±0.39	0.72±0.35	0.44	0.70±0.48	0.27

MEAN AND STANDARD DEVIATION, MICROGRAMS PER GRAM DRY WEIGHT; SULFUR MILLIGRAMS PER GRAM DRY WEIGHT.

**STUDENT'S *t* DISTRIBUTION AND CHI-SQUARE PROBABILITY.

element concentration between the races. These are presented in Table III. Striking differences between caucasians and blacks

TABLE III

DISTRIBUTION OF TRACE ELEMENTS IN HAIR ACCORDING TO RACE (FEMALES)*

	CAUCASIAN 50	BLACK 24	P	ORIENTAL 17	P**
SULFUR	36.8±4.9	32.5±8.7	0.005	36.5±4.3	0.41
CALCIUM	986.3±642.8	1550.3±1132.2	0.005	468.2±267	0.002
IRON	11.6±5.4	20.9±9.9	0.0001	7.5±2.0	0.002
NICKEL	0.70±0.48	2.16±1.50	0.0001	0.45±0.19	0.03
COPPER	21.4±7.1	21.8±11.2	0.43	15.4±5.4	0.002
ZINC	194.6±39.6	181.8±68.9	0.16	197.6±33.0	0.39
SELENIUM	0.55±0.18	0.62±0.29	0.37	0.65±0.21	0.04
BROMINE	2.15±1.80	2.49±1.37	0.21	1.87±1.0	0.28
STRONTIUM	5.69±4.00	6.73±5.10	0.17	3.3±2.3	0.03
CHROMIUM	0.99±0.42	1.48±0.71	0.0003	0.88±0.29	0.17
MANGANESE	1.13±0.56	1.68±1.05	0.003	0.68±0.36	0.002
ARSENIC	0.33±0.44	1.10±1.58	0.002	0.15±0.18	0.07
MERCURY	0.87±0.48	0.40±0.44	0.0001	0.81±0.45	0.33
LEAD	5.51±5.97	28.40±32.40	0.0001	1.56±1.47	0.005
RUBIDIUM	0.74±0.41	0.51±0.24	0.008	0.66±0.38	0.24

*MEAN AND STANDARD DEVIATION, MICROGRAMS PER GRAM DRY WEIGHT; SULFUR MILLIGRAMS PER GRAM DRY WEIGHT.

**STUDENT'S *t* DISTRIBUTION AND CHI-SQUARE PROBABILITY.

were found with increased levels of Ca, Fe, Ni, Cr, Mn, As, and Pb in blacks. Here again, some of these apparent racial differences may in part be a reflection of hair treatment and environmental exposure.

Long strands of hair, obtained from 12 healthy young ladies whose only treatment was shampoo and brushing, were serially analyzed at 4 cm intervals starting at the root. The results, similar to those reported by Valkovic *et al.* (1975) are shown in Figure 1. The elements Zn, Cu, Fe, Se, and S were relatively

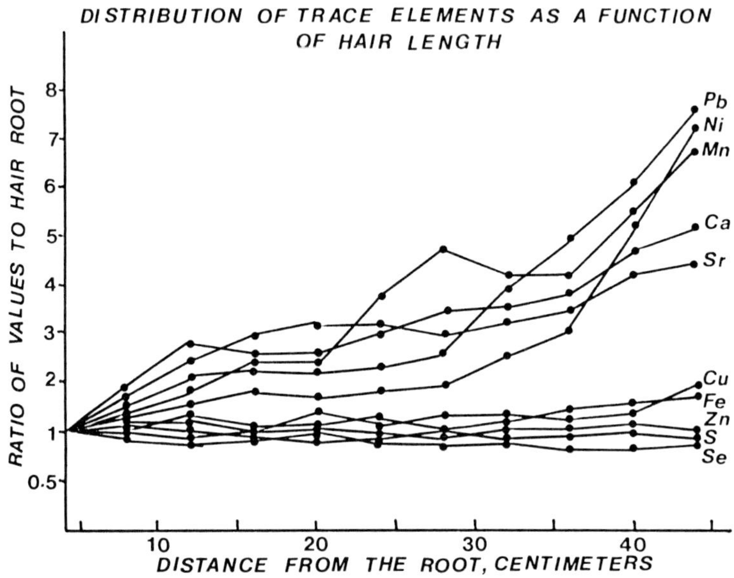

FIG. 1. The relative concentrations of trace elements in hair as a function of distance from the root. Absolute values (μg/gm) were normalised to the value at 4 cm., from the root.

constant throughout the hair length, while Pb, Ni, Mn, Ca, and Sr significantly increased in concentration as a function of distance from the root. These changes with length were interpreted to indicate environmental exposure as a function of time.

4. DISCUSSION

From the results presented above it is obvious that there are wide variations in the distribution of trace metals in hair. This is due to many factors, among which are: age, sex, race, and especially environmental exposure, e.g., cosmetics, aerosols, detergents and shampoos. Even the essential trace elements S, Fe, Cu, Zn, and Se which appear to be relatively constant throughout the hair length (Fig.1) show wide variation when analyzed as absolute values (Table 1). Thus, the composition of hair is not only determined by incorporation of trace elements from the blood, but by external environmental factors (Ryabukhin, 1980; McKenzie, 1978).

Despite the many drawbacks hair analysis has been successfully used in forensic medicine, as a screening tool for the detection of heavy metal poisoning, and for monitoring levels of environmental pollutants. According to the U.S. Environmental Protection Agency (1980) hair analysis is an effective screening test system for toxic heavy metal contamination especially for lead, arsenic, cadmium, and mercury.

Unfortunately in recent years a number of self-styled nutritional authorities and several commercial laboratories have touted hair analysis for the diagnosis of nutritional status of individuals and have openly solicited mail-order analysis of hair samples. On the basis of their results, often on inappropriate samples and inadequate histories, these practitioners have prescribed elaborate vitamin and mineral dietary supplements, or chelation therapy in the case of excess trace metals, Fletcher (1982). Such analyses provide numerical results of very little value, which are frequently used to support questionable diagnoses based on the assumption that hair analysis alone is a diagnostic rather than a screening test.

5. CONCLUSION

If used in conjection with other clinical information and other laboratory tests, hair analysis may become a useful screening tool for the determination of certain nutritional disorders and chronic diseases. Its usefulness at present is questionable, due to a lack of knowledge about the many factors which influence the distribution of trace elements in hair. Until the roles of essential trace elements and the effects of environmental factors on human health and disease are better elucidated, the analysis of hair for evaluation of an individual's metabolic state will remain experimental as well as controversial.

6. REFERENCES

D.J. Fletcher (1982) *Post Grad.Med.*, 72, 79-88.

S.A. Katz (1979) *Amer.Lab.*, 11, 44-52.

J.M. McKenzie (1978) *Amer.J.Clin.Nutr.*, 31, 470-476.

Y.S. Ryabukhin (1980) in *Hair, Trace Elements and Human Disease* (Eds, A.C. Brown and R.C. Crounce) Praeger Publ., N.Y., pp. 34-36.

H.H. Sky-Peck and B.J. Joseph (1981) *Clin.Biochem.*, 14, 126-131.

U.S. Environmental Protection Agency (1980) Publication No. 600/3-80-089, Washington, D.C.

V. Valkovic, D. Rendic, and G.C. Phillips (1975) *Environ. Sci.Tech.*, 9, 1150-1152.

SPECIATION AND INTERACTIONS

COMPUTER MODELS OF METAL BIOCHEMISTRY AND METABOLISM

David R. Williams

Department of Applied Chemistry, University of Wales Institute of Science and Technology, Cardiff CF1 3NU U.K.

1. INTRODUCTION

The subject of computer models of metal biochemistry and metabolism is just one aspect of the far larger field of chemical information technology that has been expanding rapidly during the latter part of this century. In bio-inorganic chemistry, computer simulation models are able to operate at concentrations orders of magnitude below those normally available from analytical chemistry data, thus revealing species compositions and concentrations which are much better indicators of biological or industrial activity and of bioavailability. Such "speciation"* information can now be computed from information obtained from a wide range of analytical approaches such as titrimetric procedures, atomic spectroscopy, ion-selective electrodes, and enzymology. The data from very many individual experiments, usually expressed in terms of derived equilibrium constants, is accumulated in data banks which can then be used by large programs to simulate whole systems so that the "knock-on" effects of changing the concentration of a chemical can be computed for steady state conditions [Berthon et al.(1978), May et al.(1976,1977)].

Computer simulation is used as a tool to assist our understanding of total chemical systems in a wide range of industrial and environmental scenarios. In industry it is assuming an increasingly dominant role, both for optimizing production processes to give the maximum yield, and for choosing "lead" directions for investigation by research and development teams.

Our own experience arises from dealing with large equilibrium solutions at, or near to, steady state conditions, and we have,

---

*Speciation may be defined as a knowledge of the individual physical chemical forms in which a metal exists and the concentrations of each of these forms. Commonly, the forms are composed of a number of metal ions and ligands producing an overall charge on each of the species.

Chemical Toxicology and Clinical Chemistry of Metals
ISBN 0.12.137520.X
Copyright © 1983 by IUPAC
All rights of reproduction in any form reserved

to date, examined reactions occurring in blood plasma (involving more than 10 000 species), in intestinal fluids, in milk, in soil-water relationships, and in generic models of radioactive waste disposal silos. The necessary data bank for each of these models has to be carefully constructed and research programs to improve each of these models, in some respects, is based upon improving the data that is not well supported in the literature. Earlier attempts at computer simulation modelling in various spheres have tended to be over-confident and thus predictions based upon such models have been in error or subject to claims of unwarranted precision. Nowadays, the modern generation of chemical information technologists carefully integrates models into the practical, chemical, environment by spot calibration experiments specifically designed to reveal weaknesses in the model.

Models of trace element biochemistry are based upon the following concepts. The transition metals exist in biological fluids in a steady state condition close to equilibrium whereby they are in a labile equilibrium between aquated ions (at an extremely low concentration), low molecular weight complexes and labile exchangeable protein [Perrin and Agarwal (1973)]. There are also inert metalloproteins present in which the metal is not available for exchange. The biological processes involve a lively competition for complexing between different ligands and different metal ions <u>in vivo</u>. Nevertheless, such competing biochemistry is reflected in a steady state existing which, to a thermodynamicist, suggests that solutions are pretty close to thermodynamic equilibrium. Unfortunately, it is not possible to measure the individual species concentrations without upsetting the system and, further, many are present at concentrations orders of magnitude below the limits of laboratory analysis.

The advent of large and powerful computers and their accompanying programs has meant that we can take thousands of constants and set up equilibrium models of very many low molecular weight species such as amino acids, carbohydrates, carboxylic acids, etc., in complexing competition with proteins for the biometals. In order to allow for non-equilibrium factors, the output is plotted as a percentage of the total metal or ligand present, rather than as absolute concentrations. The complex appearing at the top of the percentage league will, in all probability, still be the major species concerned when the prevailing conditions are steady state. Naturally, one tries to take account of such possibilities by the judicious choice of spot-checks and calibration experiments using the more prevalent species that just come within range of ion-selective electrode measurement.

The influences of a range of polluting metals or of pharmaceutical ligands upon such equilibria are then quantifiable in terms of new complexes formed, of existing complexes depleted, and of the resultant charges on species present. Low-charge-density species tend to diffuse through cell membranes and, for example, can thus pass from the intestine into blood plasma, whereas charged complexes either remain in the aqueous fluids in vivo, or are excreted from blood through the renal system. The aims of such computer simulation modelling include that of imitating reality in order to make predictions concerning the likely side effects of new drugs; these suggestions are then screened through animal tests. As time progresses, predictions are being made with increasing conviction. Confidence has built up markedly over the last few years arising, in the main, from the use of more precise data and more "user friendly" computer programs such as ECCLES [Linder et al.(1977)].

Speciation knowledge is particularly useful since it is usually the speciation which determines whether a complex will be toxic or harmless. For example, cyanide is invariably toxic unless complexed in a species such as that of Vitamin B12 [Williams (1976)]. Similarly, chromium(III) is now regarded as an essential metal ion, whereas chromium(VI) is regarded as highly toxic. Florence (1982) has claimed that it is inevitable that in the near future water quality legislation for heavy metals will include statements relating to their speciation. Total metal analyses are insufficient knowledge to reach a decision.

Our emphasis upon low molecular weight complexes arises because it is well established that intestinal absorption, reactions in blood plasma, deposition into tissues, passage through the blood-brain barrier, renal and biliary excretion, and even the synthesis of metalloprotein, all involve low molecular weight complexes of the transition metal ions, usually of metal amino-acid anion ternary complexes [Williams (1976), Fiabane and Williams (1977), Halstead and Williams (1983)].

It is convenient to depict the distribution of a metal between the various states in vivo as a pie diagram as shown in Figure 1. Such figures indicate, by magnifying the low molecular weight complexes portion for clarity, the predominant low-molecular weight complexes normally present, and then those resulting from administration of ligands such as pharmaceuticals. Figures 1a-h illustrate recent computations for blood plasma and for milk and also depict the calculated disturbances caused by a range of pharmaceuticals.

2. MODELS OF ESSENTIAL AND TOXIC METALS in vivo

(a) Copper

The copper found in human biological fluids is mainly in four fractions: most of the metal is present as metalloproteins from which the copper cannot be reversibly removed; this is termed 'thermodynamically inert' metal [Furnival et al.(1982)]. Some five or ten percent of the metal is present as a labile protein complex of serum albumin. This, in turn, is in equilibrium with

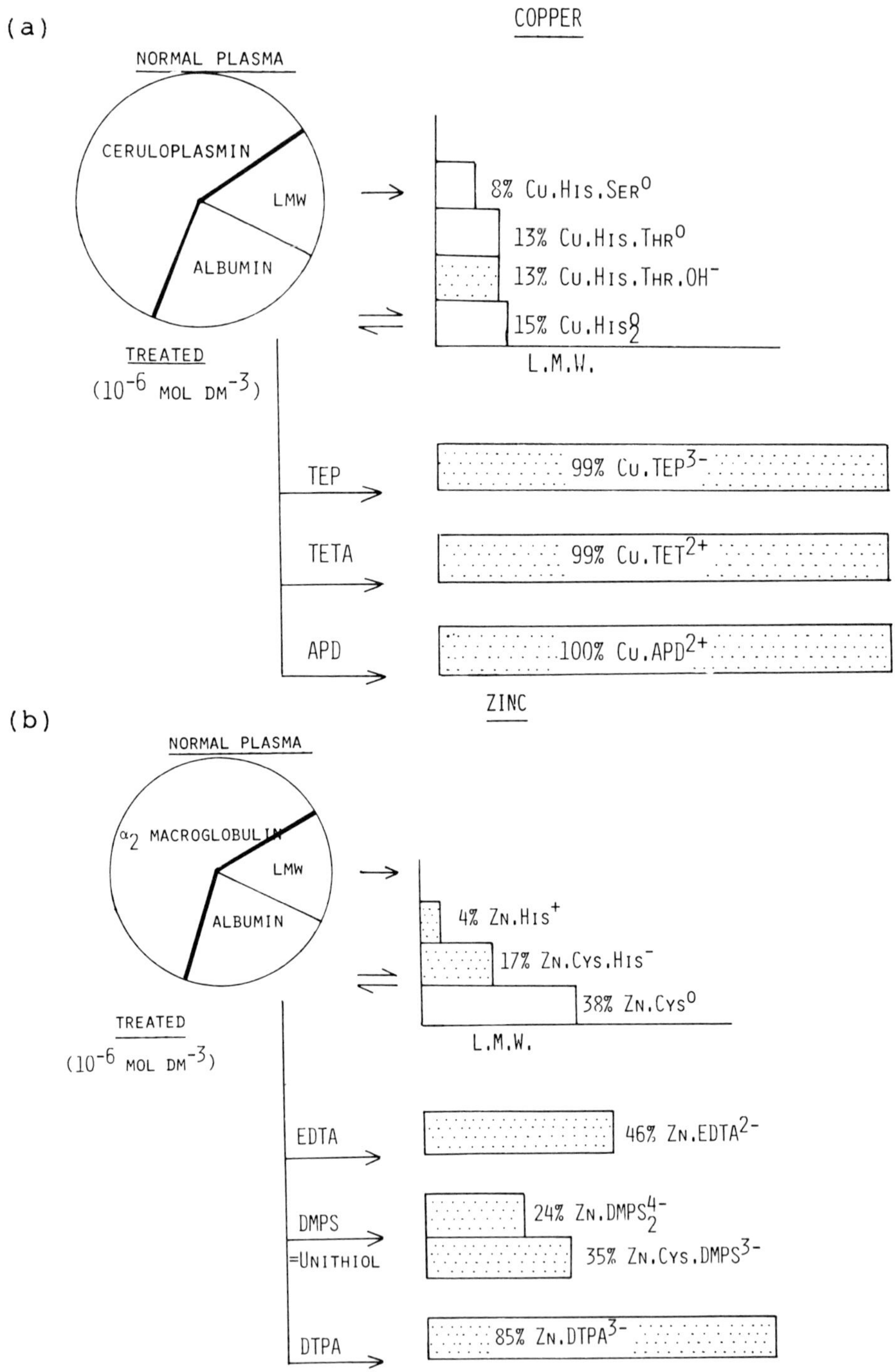

Figure 1. Low molecular weight complex distribution in plasma (a) for copper, and (b) for zinc

low molecular weight copper complexes and with aquated metal ions. Under normal conditions, the major fraction of copper in blood plasma is bound inertly to ceruloplasmin and most of the remainder is bound to serum albumin in labile equilibrium with the most useful form of copper, which although present in a minority grouping, is that of its low-molecular-weight complexes.

Figure 1a indicates this distribution of copper in normal blood plasma. The plasma albumin copper complexing serves to act as a metal ion buffer and since blood plasma is at a fixed pH of 7.4, it is possible to calculate the distribution of low molecular weight species present under normal, or pharmaceutically-treated, conditions. Several hundred conceivable low-molecular-weight copper complexes are considered by the ECCLES program but, for clarity, only the top few, i.e. the most prevalent, complexes are depicted in the "blow-up" of the low molecular weight slice of the pie diagram in the right hand side of Figure 1a. Those complexes which are neutral are depicted as clear bars, whereas those which are charged are stippled in the diagram. Such charged complexes are capable of being excreted through the kidneys, whereas the neutral complexes will pass through cell membranes into tissue or be excreted through the bile duct. Towards the bottom of Figure 1a one sees a typical case of the disturbance caused in low molecular weight copper complexes by the addition of three drugs, TEP, TETA, and APD (tetraethylenepentamine, triethylenetetramine, N,N'-bis(2-aminoethyl)-1,3-propanediamine). At micromolar drug concentrations, these effectively take over most of the low-molecular-weight copper complexing and form charged complexes appropriate to excretion through the renal route.

D-penicillamine has been used in the treatment of copper overload occurring in Wilson's disease for more than two decades. The toxicity of the condition arises due to excessive amounts of copper in the labile equilibrium system in blood plasma, and D-penicillamine is able to promote the copious cupruresis although treatment takes a considerable period to initiate copper removal. Unfortunately, the source of the copper liberated by D-penicillamine therapy eludes us. We have recently reviewed therapy by this drug and have demonstrated that the copper is not removed from the circulating plasma albumin, nor, indeed, from the ceruloplasmin [Jackson et al. (1978)]. It has been suggested that D-penicillamine may remove the copper from liver metallothionein but this has yet to be proved. Paradoxically, D-penicillamine is able to remove zinc and lead from circulating plasma albumin and can be involved in the treatment of overload by these metals. A corollary of this is that any situation treated with penicillamine will possibly lead to zinc depletion of the patient. Our review of the *modus operandi* of D-penicillamine *in vivo* sets aside the possibility that a simple reductive chelation process may be involved and critically assesses the possibility of the mixed oxidation state red-violet complex being crucial to this mechanism. Although there is much that can be written concerning copper and penicillamine, the major contribution of computer models to this discussion has been in negating the suggestion that it is the plasma albumin bound copper that is excreted. Readers are referred to our reviews for a more extensive appraisal [May et al.(1978), Furnival et al.(1982)].

The use of penicillamine for the treatment of both Wilson's disease and for rheumatoid arthritis seems paradoxical since treatment of the former involves copper removal, whereas the latter involves mobilizing copper such that it passes through into inflamed tissues. In fact, if penicillamine is viewed as a metal mobilizing agent, this paradox is readily understandable. By increasing the low molecular weight copper complex fraction, the pressure to pass out charged copper complexes through the kidneys and the pressure to pass neutral copper complexes through cell membranes into tissues are both increased by the enhanced copper low molecular weight fraction. However, in Wilson's disease, where the tissues are already heavily loaded with copper, the concentration gradient mitigates against the movement of even more copper into the tissues.

(b) Zinc in Blood Plasma

The distribution of zinc in blood plasma is similar to, but simpler than, that for copper [Figure 1b]. The inert fraction, which is the major proportion of zinc, is present as an α_2-macroglobulin, and the three labile equilibrium fractions are, once again, an albumin-zinc complex (not necessarily involving the same donor sites as for the copper albumin complex), low molecular weight zinc amino-acid complexes, and an extremely low concentration of aquated zinc ions. The low molecular weight zinc is present as both charged zinc-cysteine-histidine species and as a neutral zinc-cysteine complex. This permits the zinc to be excreted through the kidneys or to pass into tissues.

Although instances of zinc overload are exceedingly rare in medicine, the use of drugs which excrete zinc through the renal route is far more common. These drugs are administered in order to remove other undesirable metal ions and produce the co-excretion of zinc as a side effect. At a therapeutic concentration of 1 micromolar EDTA, Unithiol, and DTPA (used respectively in the treatment of calcium, cadmium, and plutonium overload) all produce a sizeable increase in the proportion of low molecular weight charged zinc complexes present as indicated in Figure 1b; these pass out through the kidneys. When protracted therapy with such mobilizing ligands is used, either topping-up therapy with zinc salts or the administration of the EDTA or DTPA as a zinc complex is desirable in order to keep the patient in zinc balance.

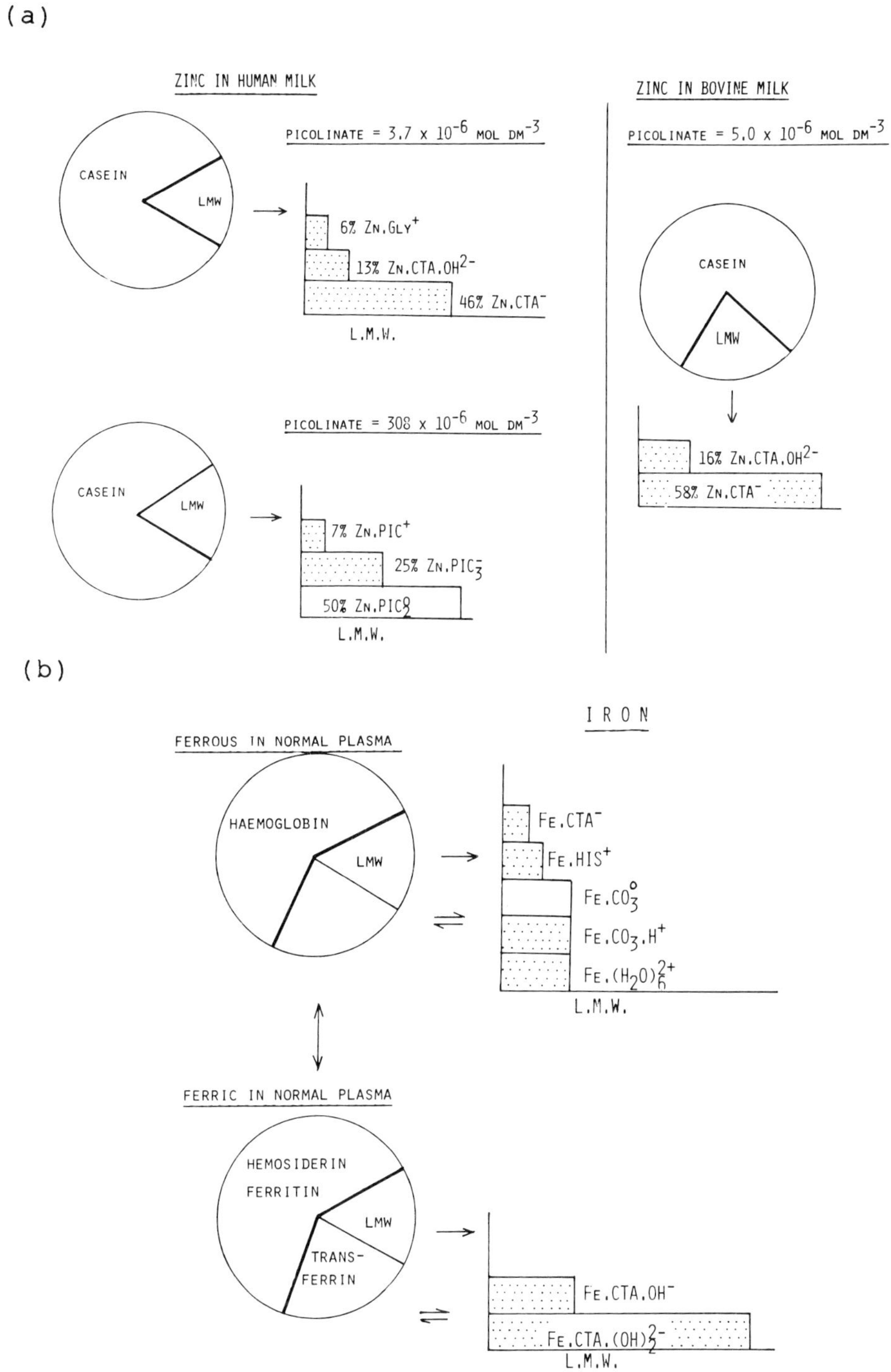

Figure 2. Low molecular weight complex distribution of (a) zinc in milk, and (b) iron in plasma

(c) Zinc in Milk

Recently, there has been a great deal of interest in the low-molecular-weight zinc complexes present in milk [May et al.(1982)]. Whereas Evans and Johnson (1980) have suggested that picolinic acid was the major binding ligand, Lönnerdal et al.(1980) have claimed that it was citrate. Such a zinc complex is important to the absorption of milk from the diet and it has further been suggested that the low-molecular-weight zinc complexes present in human milk are better equipped to encourage zinc absorption than those from cows milk.

Computer simulation models of speciation occurring in human and in bovine milk have been published. There is a choice of two different human milk models depending upon whether Evans' total picolinate acid concentration or Lönnerdal's figure is used (Table 1). Table 2 lists the predominant complexes formed in both types of human milk, differing only by their total picolinic acid contents, and in bovine milk. Figure 2a depicts these conditions.

TABLE 1 Constituent Concentrations in Human and Bovine Milk [May et al.(1982)]

	Total concentration (mol dm^{-3})	Free concentration (mol dm^{-3})
HUMAN MILK:		
Citric acid	1.0×10^{-3}	
Glutamic acid	1.0×10^{-3}	
Picolinic acid	3.7×10^{-6} or 3.0×10^{-4}	
Zinc		1×10^{-13} - 1×10^{-5}
Calcium		2.5×10^{-3}
Magnesium		6.0×10^{-4}
BOVINE MILK:		
Citric acid	9.5×10^{-3}	
Glutamic acid	4.3×10^{-5}	
Picolinic acid	2.0×10^{-5}	
Zinc		1×10^{-13} - 1×10^{-5}
Calcium		2.0×10^{-3}
Magnesium		8.1×10^{-4}

It is exciting to find that the low-molecular-weight distribution is very dependent upon the picolinic acid concentration. Hurley's far lower concentration than that of Evans dictates that the zinc is mainly bound to citrate, whereas the higher picolinic acid concentration, as might be expected, has the zinc bound up mainly to picolinate. Further, the latter is a neutral complex whereas the citrate complexes are predominantly ionic.

TABLE 2

Major Low-Molecular-Weight Zinc Complexes

Human milk with 3.7×10^{-6} mol dm^{-3} picolinic acid		Human milk with 3.0×10^{-4} mol dm^{-3} picolinic acid		Bovine milk	
Complex formed	% LMW zinc	Complex formed	% LMW zinc	Complex formed	% LMW Zinc
$Zn(CTA)^-$	58	$Zn(PIC)_2^0$	63	$Zn(CTA)^-$	44
$Zn(CTA)(OH)^{2-}$	22	$Zn(PIC)^+$	13	$Zn(CTA)_2^{4-}$	28
$Zn(PIC)^+$	4	$Zn(PIC)_3^-$	12	$Zn(CTA)(OH)^{2-}$	17
$Zn(CTA)_2^{4-}$	4	$Zn(CTA)(PIC)^{2-}$	7	$Zn(CTA)(PIC)^{2-}$	8
$Zn(CTA)(PIC)^{2-}$	2	$Zn(CTA)^-$	2.5	$Zn(PIC)^+$	1.5
$Zn(GLU)^0$	1.3	$Zn(CTA)(OH)^{2-}$	1	$Zn(PIC)_2^0$	0.5
$Zn(CTA)(GLU)^{3-}$	0.6				
$Zn(PIC)_2^0$	0.2				

Clearly, ionic complexes are at a disadvantage in terms of absorption from the gastrointestinal tract, whereas the neutral complexes are more readily absorbed. The bovine milk containing picolinic acid at 5 micromolar is distinctly similar to that of the Hurley version of human milk containing picolinic acid at 3.7 micromolar.

This all suggests that if the free zinc concentrations are the same for both human and bovine milk, then the zinc complexes ought to be approximately equal in terms of bioavailability. In practice, this is not reported to be the case and so one must investigate the high-molecular-weight binding of casein to zinc in both types of milk. This sort of approach highlights the facts that the models are only as reliable as the input data and that an analytical knowledge of ligands such as picolinate present in different samples of milk is critical to the calculations.

(d) Iron in Blood Plasma

The fact that primitive life appears to have evolved under a reducing atmosphere for a time and then changed to an atmosphere containing oxygen means that early biomolecules were essentially dependent upon the lower oxidation state of iron, ferrous, whereas nowadays, the ferric state is far more readily attainable *in vivo*. Unfortunately, such iron(III) complexes very readily form polymeric oxides and hydroxides which are soon precipitated and so the chemistry of iron(III) in *homo sapiens* is one of maintaining the metal tightly complexed at all times.

This close watch which has to be kept over ferric ion speciation is also augmented by the readiness with which the ferrous and ferric states redox between each other [May and Williams (1980)]. In blood plasma, the classical division of transition metal complexes into four categories regard iron as being kinetically inert in both haemoglobin and myoglobin and, to a certain extent, in ferritin, whereas the three phase labile equilibrium involves transferrin, low-molecular-weight ferric complexes and aquated ferric ions [Figure 2b]. The kinetics of redoxing andthe kinetics of complexing of iron to transferrin both lead to a rather complicated picture. Transferrin binds iron reversibly and rapidly but its dissociation to yield this iron may be far slower unless there is a suitable intermediate low-molecular-weight chelating agent available to mediate in the transfer operation. Although there is a long history of scientists believing that an active transport mechanism exists to move iron across all biological membranes, there is a reasonable groundswell of opinion in favour of passive transfer occurring when low charged complexes can be encouraged.

(e) Cadmium in Blood Plasma

The computer simulation models of Smith (1983) indicate that

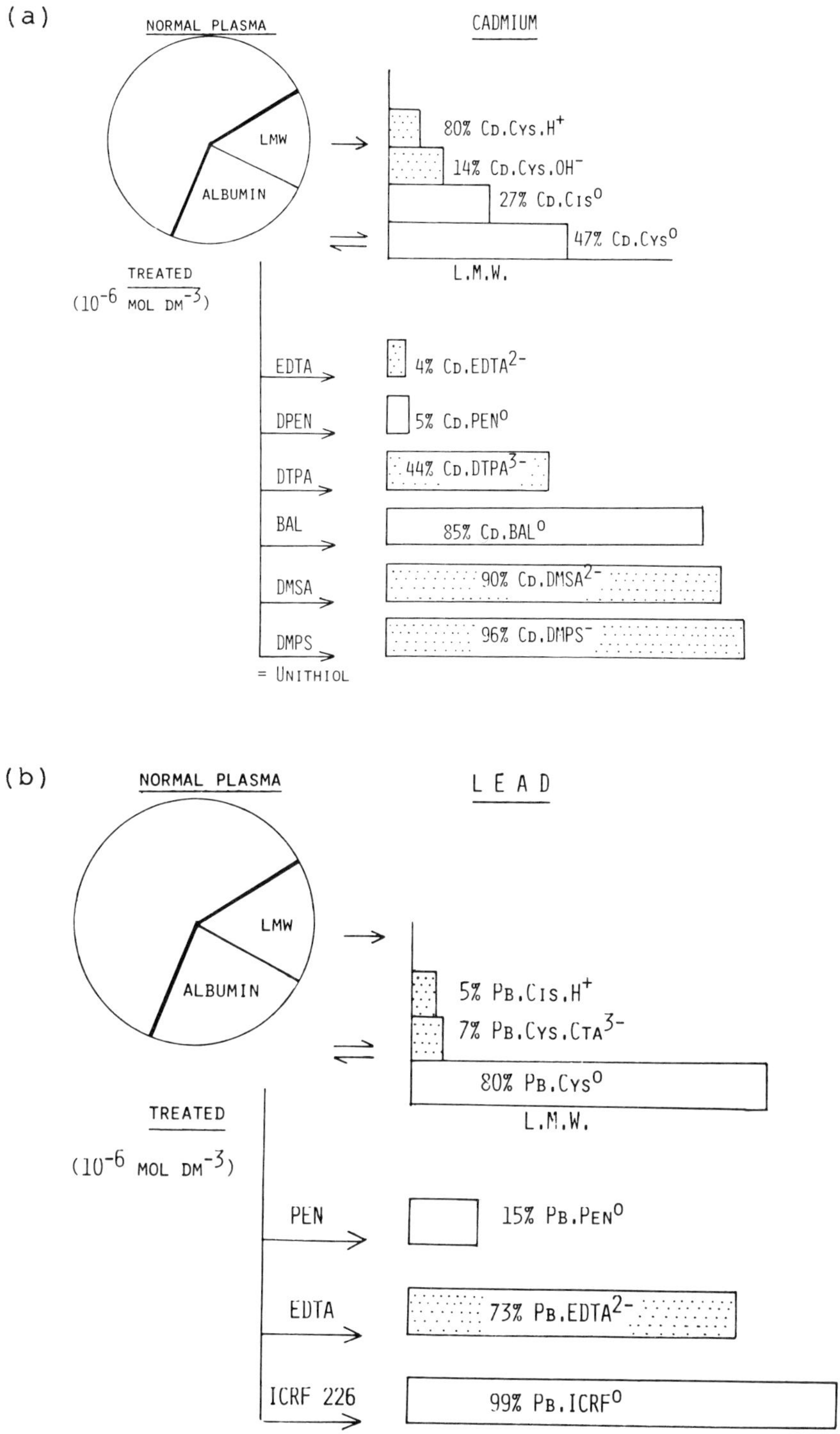

Figure 3. Low molecular weight complex distribution in plasma of (a) cadmium, and (b) lead

the low molecular weight cadmium in blood plasma is distributed between cysteine and cystine complexes, some being charged and some being neutral [Figure 3a]. The presence of the latter suggests that any cadmium which happens to get into blood plasma adventiously will eventually pass through membranes into tissues.

The treatment of cadmium-containing subjects requires synergistic chelation therapy whereby one drug is used in order to remove the cadmium from tissues as a neutral complex, and then a second, more powerful, binding drug is administered which takes over the metal and forms a charged complex which can then be excreted through the kidneys. Over the last decade, a wide range of ligands has been screened and the relative efficacies of six of them are shown in Figure 3a. This diagram shows the ability of agents to increase the low molecular weight fraction and to take over the cadmium from albumin and from cystine/cysteine. The most useful agent appears to be Unithiol (2,3-dimercaptopropane-1-sulphonate) or dimercaptosuccinic acid. Both of these form charged complexes which may be passed out through the kidneys.

Unfortunately, as yet there is no agent available for removing the cadmium from tissues into blood plasma. Thus, therapy, to date, must rely upon the slow degradation of cadmium-containing tissues and then this protracted leakage into plasma can be seized upon by chelating agents which move the metal out through the kidneys. Some animal and clinical trials have been arranged concerning Unithiol but the results have not yet been published.

(f) Lead in Blood Plasma

Lead becomes complexed as metallothionein-lead, or as bone lead after protracted exposure. However, in blood plasma the labile fraction is distributed between albumin-bound lead and low-molecular-weight cystine and cysteine complexes similar to those of cadmium. The ability of three agents to mobilize lead from albumin is indicated in Figure 3b. The last agent mentioned, ICRF 226, which is related to the anti-cancer drug ICRF 159, hydrolyzes <u>in vivo</u> to produce ICRF 198, which our models predict is <u>an avid</u> lead-complexer. Unfortunately, since this agent has anti-cancer activity, it is cytotoxic and so cannot be used in lead-removal experiments. However, its ethyl homologue, ICRF 226, is non-cytotoxic and, according to our computer simulation models, is equally capable of forming powerful complexes with lead [Willes and Williams, unpublished data].

The removal of lead from patients who had accummulated it over a considerable period of time was the original source of a widely quoted example by May and Williams (1977) of synergistic chelation therapy. Essentially, D-penicillamine is administered which, being neutral, is able to penetrate through from blood plasma into tissues whence it forms the neutral lead penicillamine complex indicated in Figure 3b. This then passes the lead back out into blood plasma. The second drug administered, for example, EDTA, can then form a more powerful lead complex which is charged and so lead removal through the kidneys is possible. Alternatively, if the second drug used is ICRF 226, one forms a neutral complex of lead which can be passed out through the bile duct. Clearly, animal and human screens will be necessary before this new therapy using ICRF 226 can be safely used in the clinic.

(g) Nickel in Blood Plasma

As might be expected from its Hard and Soft Acids and Base character, nickel - being a somewhat harder acid - complexes in blood plasma with histidine as much as with cysteine (Figure 4a). The majority of these complexes are neutral and so the metal may pass quite readily into tissues. Plasma nickel is distributed between low-molecular-weight, albumin-bound, and α_2-macroglobulin-bound nickel. Jones et al.(1980) have reported the first two of these states as being in labile equilibrium with free aquated nickel ions and when treated with TET, APD, or TEP, an increase in the low molecular weight fraction can be achieved at a micromolar dose. All three complexes formed are charged and would seem to form ideal candidates for the second drug which promotes renal excretion during synergistic chelation therapy.

As was the case for cadmium, the problem is finding the first agent in the pair. For many years, diethyldithiocarbamate has been used to treat workers contaminated with nickel carbonyl (a completely uncharged complex which rapidly penetrates to the interior of cells). Unfortunately, the setting up of computer simulation models of diethyldithiocarbamate has been harassed because the complex formed by the agent is so remarkably insoluble. Very recently it has been suggested that the copper complex of DDC may assist its passage into the cell. If the modus operandi through which this agent removes nickel does involve aqueous nickel complexes, then it must surely be a metabolic product of DDC rather than the parent compound. Other alternative means of removing the metal have recently been suggested.

(h) Plutonium in Blood Plasma

Plutonium, in addition to being a toxic heavy metal, is also radioactive and Duffield et al.(1983) have urged caution concerning leaving it in bones adjacent to bone marrow: also one ponders whether passing it through the kidneys may well render those organs more likely to undergo radiation damage.

In blood plasma, plutonium mimics the higher oxidation state of iron and is found as inert metalloprotein forms such as haemosiderin and ferritin, and as a labile equilibrium between a

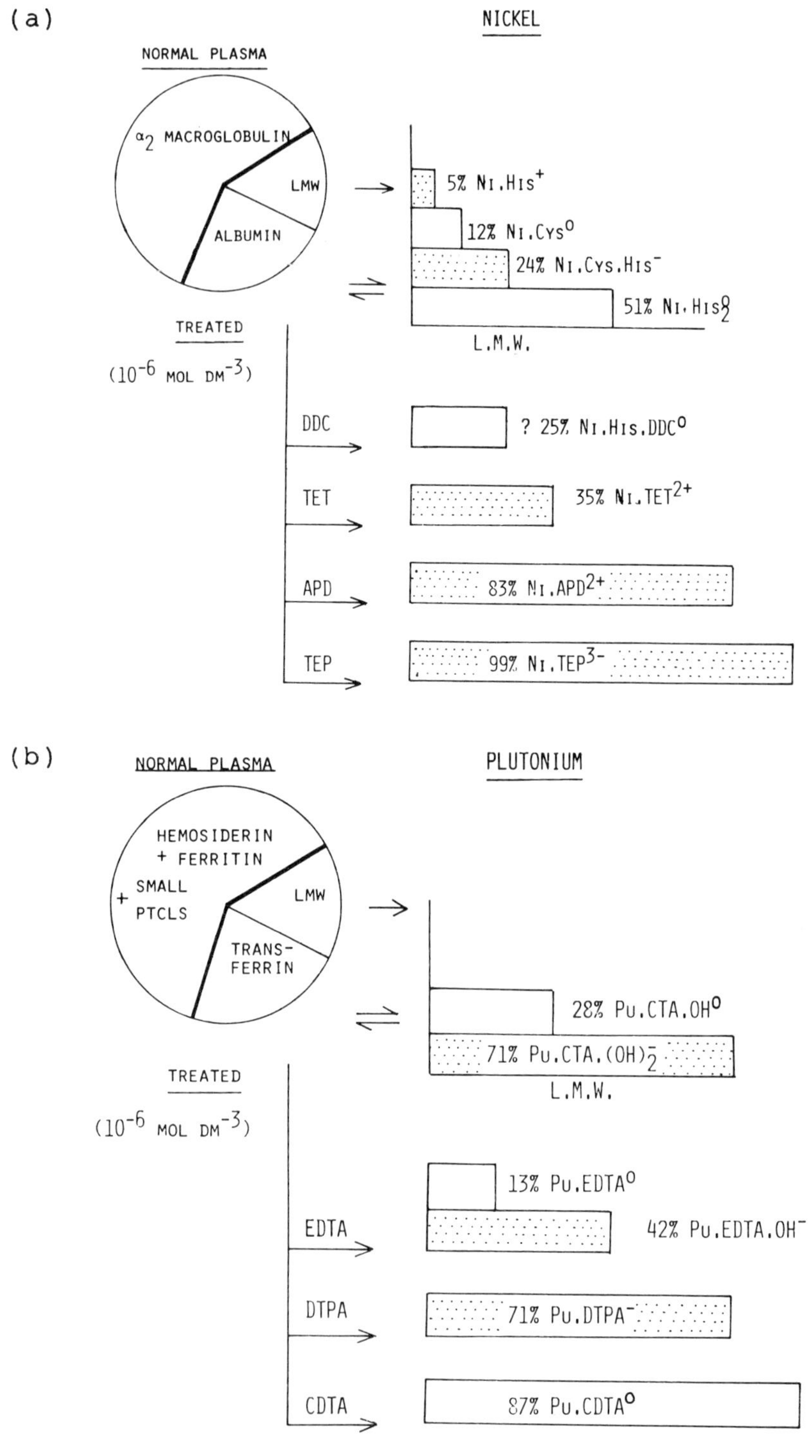

Figure 4. Low molecular weight complex distribution in plasma of (a) nickel, and (b) plutonium

transferrin-bound form and low-molecular-weight species (Figure 4b). The latter are complexed to the citrate as was found for iron. Additionally, plutonium tends to form small hydrous oxide particles which circulate in biofluids. These may be considered to be an inert form of the metal in that they are not involved in the labile equilibrium. They are difficult to dissolve with drug therapy and they do not suffer the normal end-of-lifetime degradation process that metalloproteins, such as haemosiderin and ferritin, undergo, thus releasing the metal in due course. The removal of these particles is a major problem being tackled by the health authorities serving the nuclear industries.

Aside from this circulating particle problem, the low-molecular-weight form of plutonium may be enhanced by administering any of the polyaminocarboxylic acid drugs, such as EDTA, DTPA, or CDTA. The first two tend to encourage charged complexes which will pass out through the renal route, whereas CDTA produces a neutral complex which may pass out through the bile duct. A lot of speculation has concerned the new agent called Puchel which has pK values similar to those of CDTA and a structure similar to that of DTPA with a large lipophilic side chain which serves, allegedly, to facilitate membrane penetration but our simulation models have shown that Puchel, in fact, binds less than DTPA. Another new agent is called 3,4,3-LICAMS and although it complexes well, serious toxicity problems have been revealed. Also, it is rather too large to penetrate successfully through cell membranes and so has to "stalk" the plutonium circulating in blood plasma.

3. CONCLUDING REMARKS

Trace elements are far more important *in vivo* than their total concentrations suggest. Unfortunately, straightforward analysis of the speciation present is not possible in most instances because concentrations are way below those approachable by even the most sophisticated analytical laboratory techniques and also because preconcentration approaches to analysis tend to disturb the delicate equilibria present. However, a balanced knowledge of a wide range of trace element dependent side effects and of toxic reactions from polluting elements is very essential and computer simulation models are one means of approaching this goal.

Most of us might react rather strongly against the use of such models for interpreting physiological reactions mainly because most of human biology is so tremendously sophisticated and interdependent upon other biochemical cycles. Nevertheless, it is under such circumstances that a good simplification is a distinct step forward since a model, if well designed, can present the most important features without clouding the issue with secondary details.

REFERENCES

G.Berthon, P.M.May and D.R.Williams (1978) J.Chemical Society Dalton, 1433-1438.

J.R.Duffield, P.M.May and D.R.Williams (1983) J.Inorganic Biochemistry, at press.

G.W.Evans and P.E.Johnson (1980) Pediatric Res., 14, 876-880.

A.M.Fiabane and D.R.Williams (1977) The Principles of Bio-inorganic Chemistry, Royal Society of Chemistry, London, Monograph No.31.

T.M.Florence (1982) Talanta, 29, 345-364.

C.Furnival, P.M.May and D.R.Williams (1982) in Inflammatory Diseases and Copper, Ed. J.R.J.Sorenson, Humana Press, New Jersey, pp.45-59.

B.W.Halstead and D.R.Williams (1983) Clinical Toxicology 19(7) at press.

G.E.Jackson, P.M.May and D.R.Williams (1978) FEBS Letters, 90, 173-177.

D.C.Jones, P.M.May and D.R.Williams (1980) in Nickel Toxicology, Eds. S.S.Brown & F.W.Sunderman Jr., Academic Press, London, pp.73-76.

P.W.Linder, P.M.May and D.R.Williams (1977) J.Chemical Society, Dalton, pp.588-595.

B.Lönnerdal, A.G.Stanislowski and L.S.Hurley (1980) J.Inorgic Biochemistry, 12, 71-78.

P.M.May, P.W.Linder and D.R.Williams (1976) Experientia, 32, 1492.

P.M.May and D.R.Williams (1977) FEBS Letters, 78, 134-138.

P.M.May, D.R.Williams et al.(1978) J.Inorg.Nucl.Chem., 40, 1189, 1195, 1201, 1209, 1221, 1227; (1981) 43, 825.

P.M.May and D.R.Williams (1980) The Inorganic Chemistry of Iron Metabolism, in Iron in Biochemistry and Medicine II, Eds. A.Jacobs and M.Worwood, Academic Press, London, pp.1-28.

P.M.May, G.L.Smith and D.R.Williams (1982) J.Nutr., 112, 1990-1993.

D.D.Perrin and R.P.Agarwal (1973) in Metal Ions in Biological Systems, Ed. H.Sigel, Marcel Dekker, New York, Vol.2, pp.168-206.

G.L.Smith (1983) Speciation of Trace Metals in Biofluids, Ph.D Thesis, University of Wales, p.101.

D.R.Williams (1976) (Ed.) An Introduction to Bio-inorganic Chemistry, C.C.Thomas, Springfield, Illinois.

ACKNOWLEDGEMENTS

We gratefully acknowledge the work of the bio-inorganic research group in UWIST and especially Dr.P.M.May.

SPECIATION AND FRACTIONATION OF TOXIC METALS IN BIOLOGICAL MATERIALS

John Savory, Roger L. Bertholf and Michael R. Wills

Departments of Pathology, Biochemistry and Internal Medicine
University of Virginia Medical Center
Charlottesville, Virginia 22908, U.S.A.

1. INTRODUCTION

The biochemistry of a metal ion depends chiefly on its bioavailability, which in turn depends on the chemical form the metal ion adopts in biological systems. Hence, an understanding of the biochemical effects of a metal which result in toxicity is firmly predicated on a description of the chemical form(s) adopted by the metal in plasma and tissues. The extent to which a metal ion in plasma is protein bound or free, and the nature of the protein(s) which bind the metal ion determine the bioavailability to other tissues such as brain, bone, kidney where toxic effects can be observed.

The purpose of this present review is to survey methods which have been used to separate and identify various metal compounds which exist in plasma and tissues. Some specific examples of such separations will be reviewed.

Among the earliest work to be carried out on the physiochemical states of a metal ion in plasma was in 1911 when Rona and Takahashi (1911) measured diffusible calcium using dialysis of serum. Most of the pioneering work on the state of calcium in plasma was carried out by McLean and Hastings (1935) who defined the relationship of the ionic calcium and the protein-bound calcium as being an equilibrium which can be expressed by equations for association or dissociation. McLean and Hastings (1935) proposed that ionic calcium was the biologically active

Chemical Toxicology and Clinical Chemistry of Metals
ISBN 0.12.137520.X
Copyright © 1983 by IUPAC
All rights of reproduction in any form reserved

species and that the protein-bound form was inactive but functioned as an immediately available, first reserve for the ionic form. The techniques used in some of the earlier and even more recent studies on calcium have been applied to other metal ions.

2. ULTRAFILTRATION AND DIALYSIS

Ultrafiltration and dialysis are similar techniques used to obtain a protein-free filtrate of plasma. Ultrafiltration uses a high pressure process while dialysis refers to a diffusion process at low pressures. In principle, all ultrafiltration procedures utilize a membrane of small pore size and supply filtration pressure to obtain a protein free filtrate. Either centrifugation or pumping is used to supply positive pressure to the sample and the process requires several hours for completion. Obviously, careful control of pH, temperature, buffers etc. are needed in these procedures. As stated earlier, Rona and Takahashi (1911) separated diffusible and non-diffusible calcium fractions by dialysis experiments on serum.

(a) Aluminium

Ultrafiltration techniques, of course, have been applied to studies of many toxic metals and one of current interest is aluminium. Aluminium toxicity has mainly been associated with patients on long-term hemodialysis and has serious consequences such as apparently being a major cause of dialysis encephalopathy (Alfrey et al., 1976) and dialysis osteodystrophy (Parsons et al., 1971). It is also associated with a reduction in bone formation rate in patients receiving chronic parenteral nutrition (Ott et al., 1983). Also, it has been proposed that aluminium may play a toxic role in the pathogenesis of Alzheimer's disease of the senile and presenile type (Crapper et al., 1973). It is highly likely that the ultrafiltrable portion of aluminium contributes more to the toxicity of this element than the protein bound form. However, very few studies have been carried out on the fractionation of aluminium species in plasma.

Lundin et al. (1978) studied the partition of aluminium in the plasma of 10 normal individuals using ultrafiltration with membranes with a cutoff of 6000 - 8000 daltons and obtained an average of 59% protein binding. Elliott et al. (1978) in a brief communication reported that more than 70% of the aluminium in blood was present in the plasma compartment. In normal subjects, they reported that a very small proportion of the plasma aluminium was ultrafiltrable which was not entirely consistent with the findings of Lundin et al. (1978). In the studies of Elliott et al. (1978) the tendency was for the ultrafilterable fraction to decrease as the total plasma aluminium concentration fell below 200 μg/L. At higher plasma aluminium levels, studies using polyethylene glycol and direct ultrafiltration indicated that 60-70% of the aluminium was bound to high-molecular weight proteins, 10-20% was bound to albumin and 10-30% was ultrafiltrable. Graf et al. (1981,1982) performed <u>in vivo</u> ultrafiltration studies on patients during hemodialysis where the ultrafiltrate was obtained by disconnecting the dialysate supply from the dialyzer and applying a reduced pressure of 200 mm Hg by a suction pump. Their results revealed an ultrafiltration fraction of about 20% of total plasma aluminium, suggesting that 80% of the aluminium was protein bound. Additional studies on the fractionation of plasma aluminium species by gel filtration chromatography have been carried out in the author's laboratory and will be discussed later.

(b) Gold

The applications of ultrafiltration and dialysis procedures have been used in studies of many other trace and toxic metals in plasma. Biggs et al. (1979) studied the plasma protein binding of three ^{198}Au labelled gold compounds (sodium aurothiomalate, sodium aurothiosulfate and aurothioglucose) using ultrafiltration. The studies were carried out with a plasma pool diluted 1:5 with saline and then diluted 1:1 with a saline solution of the gold compound. Ultrafiltration was carried out by centrifuging the final solution at 2000 rpm for 10 minutes through Amicon Centriflow Membrane Cones. In this study no attention was paid to control of the ultrafiltration process with respect to pH, ionic strength, and temperature which probably affect the true binding of the gold compounds to plasma proteins in vivo.

(c) Radiopharmaceuticals

Similar approaches have been used to study various radioisotopes used for in vivo radiological procedures. Protein binding of three rare earth isotopes, yttrium, lanthanum and scandium, were studied with ultrafiltration and equilibrium dialysis with purified protein and DNA solutions, but again no effort was made to simulate true physiological conditions (Rosoff and Spencer, 1979). An even more extreme approach has been used by Schumichen et al. (1980) to study plasma protein binding of a radiopharmaceutical containing technetium-99m. Here protein precipitation with reagents such as trichloracetic acid or perchloric acid was used to study protein binding. The effect of protein precipitation on any protein binding must be considerable and makes these studies extremely difficult to interpret.

(d) Nickel

Ultrafiltration using a Diaflo pressure dialysis apparatus has been employed to study nickel binding in serum from human adults, rats, lobsters, dogs and rabbits (Hendel and Sunderman, 1972). Potential problems with this study are that during the active ultrafiltration process variations from true physiological conditions are likely. These studies were extended by the same workers to in vitro nickel binding of purified serum albumins using equilibrium dialysis (Callan and Sunderman, 1973). Here the conditions could be controlled to represent physiological conditions with respect to buffer composition and temperature. Also the binding could be monitored accurately since a $^{63}Ni(II)$ salt was used, thus eliminating the imprecision of the earlier studies which necessitated atomic spectroscopic determinations of nickel in the separated fractions. Additional studies were carried out using equilibration of rabbit serum with $^{63}Ni(II)$, followed by pressure ultrafiltration and thin-layer chromatography of the ultrafiltrates to further elucidate the nature of nickel binding (Asato et al., 1975). Detection of the various binding species was carried out by ninhydrin staining and also autoradiography.

Equilibrium dialysis has been employed in some elegant studies by Glennon and Sarkar (1982) to investigate the Ni(II)-binding site of human serum albumin. The binding studies were carried out using purified human serum albumin using a dialysis buffer of 0.1 M N-ethylmorpholine/HCl, pH 7.53 and I 0.16. The binding was followed using $^{63}Ni(II)$ and some studies were carried out in the presence of various concentrations of L-[^{14}C] histidine. These studies demonstrated a specific Ni(II)-binding site on the protein. L-Histidine was shown to have a greater affinity of Ni(II) than human serum albumin. Also, a small amount of ternary complex albumin-Ni(II)-L-histidine was detected. These relatively simple studies provided information allowing the authors to propose a mechanism for the transportation of Ni(II) in the serum based on the ability of histidine to remove nickel from albumin via the ternary complex, and on the ability of the low molecular weight L-histidine complex to traverse the biological membrane.

These various studies point out that it is possible to obtain preliminary information of metal binding using ultrafiltration and equilibrium dialysis techniques. However, it is difficult to adjust the conditions to accurately represent true *in vivo* conditions. Thus, the data obtained by these approaches should be interpreted with caution as representing the true *in vivo* situation.

3. ELECTROPHORESIS - AUTORADIOGRAPHY

Electrophoretic methods have been applied to the study of protein binding characteristics of a variety of metals. Generally, these methods involve electrophoretic separation of proteins on an agarose or polyacrylamide gel, followed by determination of metal localization which correlates with areas of protein concentration. A relatively straightforward application of electrophoretic methods was reported by Conradi et al. (1978). These investigators studied the serum protein binding characteristics of lead by incubation of the serum with ^{210}Pb, separation of the proteins by isoelectric focusing, and assay of small sections of the gel for the presence of lead. Using this technique to study serum protein binding of lead in amyotrophic lateral sclerosis (ALS) patients, Conradi et al. (1978) found that lead is concentrated primarily in the orosomucoid fraction in both ALS patients and controls. Lead was also observed to associate weakly with α_1-antichymotrypsin and IgG. The lead binding capacity of orosomucoid was confirmed by these investigators in experiments using purified orosomucoid. Bruns et al. (1977) used sodium dodecyl sulfate-polyacrylamide gel electrophoresis to isolate a calcium-binding protein from rat intestine. This method involves the use of ^{45}Ca to localize the protein, and includes several electrophoretic steps in the protein purification scheme. A unique feature of this method is inclusion of a step in which the protein concentrate from the first electrophoretic separation is electrophoresed on a gel containing EDTA. The EDTA chelates the calcium bound to the protein, and the calcium-free protein migrates at a different rate than the calcium-saturated protein, allowing it to be separated from other proteins that have similar electrophoretic properties under calcium saturation conditions. One of the more specific and elegant techniques used in the study of metal-protein binding was described by Bradwell and Burnett (1975). Their technique is termed two-dimensional immunoelectrophoresis and involves electrophoresis in one direction on an agarose gel, and immunoelectrophoresis in a perpendicular direction. In the latter step, the gel is impregnated with antisera and the presence of protein can be established by the appearance of precipitin lines as antibody-antigen complexes form during immunoelectrophoresis. This method has been combined with autoradiography (Scott and Bradwell, 1983) using ^{59}Fe, ^{65}Zn, ^{109}Cd, ^{63}Ni and ^{45}Ca to study the protein-binding characteristics of these metals. This procedure, while quite sensitive and quantitative to the extent of demonstrating the number of proteins to which a metal will bind, requires further procedures to identify specifically what proteins are binding the metal ion. Scott and Bradwell (1983) used radial immunodiffusion to identify the metal binding proteins. In this procedure, serum which has been incubated with a radioisotope of the metal of interest is allowed to radially diffuse into a gel impregnated with antisera specific to a particular protein. The appearance of precipitin, due to antibody-antigen complexes, combined with autoradiographic localization of the metal radioisotope allows identification of the protein which is binding the metal ion. This technique applied to Fe, Zn, Cd, Ni and Ca has demonstrated binding to some 16 plasma proteins.

4. GEL FILTRATION

One of the most versatile and useful techniques for studying metal binding in biological samples is gel filtration chromatography. Although the analysis of the column fractions can be extremely tedious, much valuable information can be obtained concerning the nature of the various protein and low molecular weight metal binding species.

(a) Calcium

Two early studies (Toffaletti et al., 1976; 1977) were carried out in the author's laboratory using gel filtration to study the various moieties in human serum that bind calcium. These studies will be described in some detail since the principles of the separation which were carried out under truly physiological conditions should be applied to toxic metal fractionation. The key to the success of the fractionation was that conditions affecting the binding of calcium were maintained constant throughout the separation. These factors were ionic calcium concentrations, buffer pH, temperature, ionic strength and magnesium ion concentration. The column used for the first separations was a packed with Biogel P-2 (100-200 mesh) and maintained at 37°C. The column eluent contained per liter, 140 mmol of sodium, 1.10 mmol of $CaCl_2$, 0.50 mmol of $MgCl_2$ and 10 mmol of tris buffer adjusted to pH 7.40 at 37°. The fractions collected from the column were analyzed for bicarbonate, citrate, phosphate, total protein, lactate and sulfate which are all species known to bind calcium. Potassium was also measured as a marker for the elution of ionic calcium. The results of this separation are shown in Figure 1.

The first calcium peak was protein-bound calcium followed by calcium complexes of phosphate, citrate, sulfate and lactate. To examine whether the observed concentrations of citrate, bicarbonate, phosphate, lactate, and sulfate accounted for the bound calcium measured in the individual tubes, we determined the association constants for the calcium complexes of these compounds by chromatography of appropriate aqueous solutions. These associated constants agreed with those obtained by other methods, which supports the validity of the technique in its capability of providing information which represents the true physiological state of the various calcium species in plasma.

These studies were extended to provide a fractionation of high molecular weight plasma calcium species using a Sephacryl S-200 column and the separation is shown in Figure 2. The earlier fractions (tubes 3-24) contain protein, which appear to bind a considerable amount of calcium, as seen by comparing the peaks for calcium in these same fractions. The nonprotein-bound calcium complexes elute as a single peak (tubes 32-40), which contains citrate, phosphate, and bicarbonate. Closely following is a peak of uncertain origin (tubes 40-43). Finally, the ionic calcium from the serum ultrafiltrate (tubes 44-48) elutes immediately after the unidentified calcium peak. Specific protein analysis of the protein-containing fractions (tubes 3-24 in Figure 2) shows that the initial protein peak is composed of IgM, α_2-macroglobulin, haptoglobins, and orosomucoid. The second protein peak contains IgA, C_3, C_4 and IgG. The third peak is composed largely of albumin but also contains transferrin and α_1-antitrypsin. The validity of the technique is reinforced by our determination of a mean intrinsic association constant for calcium/albumin for healthy individuals, of 101 ± 5 liter/mol which agrees with the 101 ± 6 liter/mol reported by Pederson (1971).

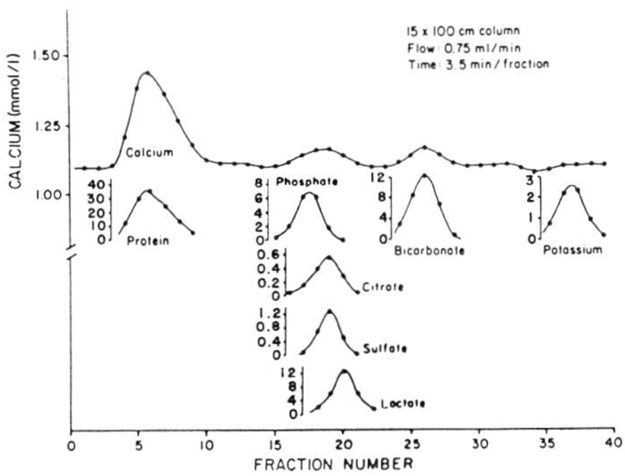

FIG. 1. Representative elution pattern of serum calcium-containing constituents chromatographed on Biogel P-2 (All concentrations in mmol/liter, except protein in g/liter). (Reproduced by permission, Clin. Chem. 22, 1969, 1976).

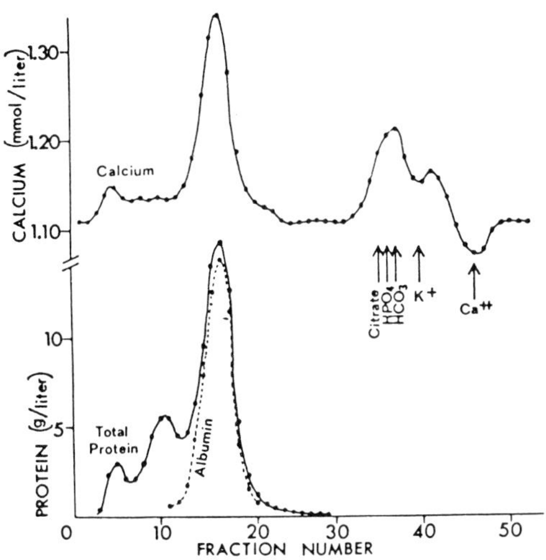

FIG. 2. Elution pattern of serum calcium-containing constituents chromatographed on Sephacryl S-200 (The arrows indicate position of the tube containing the maximal concentration of that constituent). (Reproduced by permission, Clin. Chem. 23, 2307, 1977).

(b) Aluminium

These techniques have been applied in the author's laboratory to the fractionation of aluminium species in human plasma. As stated earlier, aluminium toxicity has serious consequences for patients on long-term hemodialysis in being an apparent major cause of dialysis encephalopathy and dialysis osteodystrophy. The distribution of aluminium among various plasma proteins and the existence of aluminium complexed in some way to low molecular weight species could be of considerable importance in elucidating the nature of the toxicity of aluminium in these pathological conditions.

The fractionation of plasma aluminium species has been carried out in the author's laboratory using the same conditions as those for calcium. The aluminium concentration in the eluting buffer was kept as low as possible and a concentration of 11 μg/L was achieved. In the initial study (King et al., 1979) a separation was carried out on blood collected from a normal volunteer. The separation was carried out on a Sephacryl S-200 column and the fractions analyzed for total protein, aluminium, and albumin content. The results are shown in Figure 3. It can be seen that a small portion of aluminium elutes very early (Peak A) and is probably in the void volume of the column. This peak has been shown (Toffaletti et al., 1977) to contain macroglobulins, IgM, haptoglobin and orosomucoid, but whether aluminium in this fraction is associated with any of these constituents or is perhaps in colloidal form is unknown. A second aluminium peak (B) is apparently bound to albumin. When an excess of calcium was added to the eluting buffer, aluminium was no longer found in the albumin-containing fractions. This observation suggested that the aluminium-binding sites on albumin are easily saturable and calcium may compete for the same sites. The majority of the aluminium in the sample eluted in Peaks C and D. This result is inconsistent with the data previously described which indicated that only 20-40% of plasma aluminium was ultrafiltrable (Elliott et al., 1978). However, apparently one of these peaks is retarded because of the type of gel used, since SDS gel electrophoresis

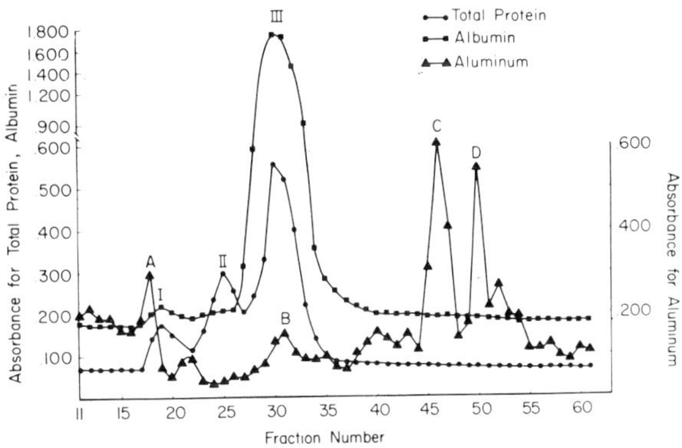

FIG. 3. Elution profile for normal serum fractionated by gel chromatography showing appearance of proteins, albumin and aluminium. (Reproduced by permission, Res. Comm. Chem. Pathol. Pharmacol., 26, 166, 1979).

indicated the presence of a protein having a molecular weight of about 66,000 daltons. This protein is probably a glycoprotein having considerable affinity for the Sephacryl S-200 gel. When aluminium was added to the buffer, little increase was seen in the albumin-associated aluminium peak (Peak B), a small increase was seen in Peak A, and an increase was seen in Peaks C and D. This observation is consistent with the data presented by Elliott et al. (1978) which indicated that the fraction of ultrafiltrable aluminium increases with increasing total aluminium. It seems apparent that the aluminium binding sites on plasma proteins are saturated at relatively low total plasma aluminium concentrations. At higher concentrations aluminium is distributed primarily in the ultrafiltrable fraction of plasma.

Fractionation of a plasma sample obtained from a dialysis patient has been accomplished in the author's laboratory (King et al., 1982). The fractionation pattern was similar to normal plasma except that there appeared an additional aluminium peak in the void volume, and a relatively greater amount of aluminium was associated with albumin. This may account for some of the conflicting data on ultrafiltrable aluminium in normal and uremic plasma.

The preceeding discussion makes clear the fact that the understanding of aluminium speciation in plasma is far from complete. Information on the ultrafiltrable fractions is of considerable interest because it is this component(s) which is probably bioavailable to brain and bone tissue producing the neurotoxic and osteomalacic effects. Complete identification of this aluminium fraction should be of importance in our basic understanding of aluminium toxicity.

(c) *Nickel*

Gel filtration chromatography has been employed in a few studies directed towards the fractionation of nickel species. Van Soestbergen and Sunderman (1972) studied nickel binding in rabbit serum obtained following injection of $^{63}NiCl_2$. A Sephadex G-200 column was used with a pH 7.4 Tris-HCl buffer to obtain protein bound and protein-free fractions. These fractions corresponded approximately to the proportions obtained by ultrafiltration. Serum ultrafiltrates from the same rabbits were separated on a Sephadex G-25 column and gave five nickel-containing fractions. The same column was used for rabbit urine separations demonstrating three distinct nickel complexes. A more recent report from the same laboratory (Sunderman et al., 1981) employed chromatographic and electrophoretic fractionation of ^{63}Ni-constituents of renal cytosol samples from rats following administration of $^{63}NiCl_2$. Here the separations were carried out on Sephadex G-200 or Sephacryl S-300 columns at 8°C using a tris-HCl buffer pH 7.4 containing NaCl. This technique was able to resolve five nickel-containing macromolecular constituents. A large percentage (68%) of the nickel eluted in association with a low molecular weight component.

5. HIGH PERFORMANCE CHROMATOGRAPHY

A major difficulty with the gel filtration technique is

the amount of work and the length of time required for each separation. An obvious development is the use of high performance chromatography to shorten the procedure and provide better resolution. The conventional soft gels such as dextran, agarose or polyacrylamide gels used in gel permeation chromatography cannot be used under high pressure because of their low mechanical strength. Early developments made two types of supports available. One type comprised the hydrophilic organic polymer gels such as poly(ethylene glycol dimethacrylate) gel (Merckogel PGM), poly(2-hydroxyethyl methacrylate) gel (Spheron P) and sulphonated polystyrene gel (TSK-GEL LS210, Aquapak). The other type consisted of inorganic based supports such as controlled porous glass-based support (Glyceryl PG) and silica-based support (μ Bondagel). However, few gels exhibited the desired properties of low adsorption and high resolution. Moreover, inorganic supports generally have high mechanical strength but show strong adsorption. Many attempts to diminish the strong adsorption have been made.

Recently, a new chemically modified silica gel-based aqueous gel permeation chromatography support has been marketed by Toyo Soda (Yamaguchi, Japan) as the TSK-GEL SW type column. The characteristics and applications of this new support have been investigated (Fukano et al., 1978; Pfannkoch et al., 1980). Here the separation is based upon steric-exclusion particles of macromolecules between a silica gel bonded phase (TSK-GEL SW-2000 or SW-3000) and a aqueous mobile phase. This method has been used to separate metallothionein in liver cytosol (Nomiyama and Nomiyama, 1982). Recently the technique has been applied to the separation of ^{63}Ni-constituents in renal cytosol and microsomes from $^{63}NiCl_2$-treated rats (Sunderman et al., 1983). The separations were monitored using a spectrophotometric monitor (280 nm) and the fractions (0.2 ml) were collected directly into liquid scintillation vials. The resolution of the ^{63}Ni-containing fractions was much improved over a similar study from the same laboratory using conventional gel filtration chromatography on Sephadex G-200 (Sunderman et al., 1983). The use of these high performance chromatographic techniques will contribute considerably to our understanding of metal metabolism since the resolution is excellent and separations can be carried out much faster than with the older conventional methods. Continuous monitoring of the metal elution is still a problem, especially for metals such as aluminium where no radioisotope is readily available. Continuous electrochemical detection has been applied to nickel and copper in quality control or effluent monitoring programs associated with industrial plants (Bond and Wallace, 1983). Detection limits, however, are well above those needed for studies in biological materials.

6. SUMMARY

Fractionation of toxic metal compounds in plasma and tissues is an important technique in our understanding of metal toxicology since the bioavailability and cellular localization of the metal can be determined. Monitoring of any separation is made much simpler if a radioisotope of the toxic metal is available. Studies on toxic metals such as aluminium are made extremely complicated since a convenient aluminium radioisotope is not available. Simple techniques such as ultrafiltration or dialysis can provide some preliminary useful information. Better resolution can be obtained using either electrophoresis-autoradiography and column chromatography. With these techniques attention must be paid to the conditions under which the separation is made in order that these closely resemble physiological conditions. Particular attention must be paid to pH, temperature, and ionic composition. Newer techniques using high performance chromatography with molecular exclusion columns offer many advantages over the older gel filtration systems.

7. REFERENCES

A.C. Alfrey, G.R. LeGendre and W.D. Kaehny (1976) N. Engl. J.Med., 294, 184-188.

N. Asato, M. Van Soestbergen and F.W. Sunderman, Jr. (1975) Clin. Chem., 21, 521-527.

D.F. Biggs, D.M. Boland, P. Davis and J. Wakaruk (1979) J. Rheumatol., Suppl. 5, 68-73.

A.M. Bond and G.G. Wallace (1983) Anal. Chem., 55, 718-723.

A.R. Bradwell and D. Burnett (1975) Clin. Chim. Acta, 58, 283-290.

M.E.H. Bruns, E.B. Fliesher and L.V. Avioli (1977) J. Biol. Chem., 252, 4145-4150.

W.M. Callan and F.W. Sunderman, Jr. (1973) Res. Comm. Chem. Pathol. Pharmacol., 5, 459-472.

S. Conradi, L. Ronnevi and H. Stibler (1978) J. Neurol. Sci., 37, 95-105.

D.R. Crapper, S.S. Krishnan and A.J. Dalton (1973) Science, 180, 511-513.

H.L. Elliott, A.I. MacDougall, G.S. Fell and P.H.E. Gardiner (1978) Lancet, II, 1255.

K. Fukano, H. Komiya, H. Sasaki and T. Hashimoto (1978) J. Chromatogr., 166, 47-54.

J.D. Glennon and B. Sarkar (1982) Biochem. J., 203, 15-23.

H. Graf, H.K. Stummvoll and V. Meisinger (1981) Kidney Int., 19, 587-592.

H. Graf, H.K. Stummvoll and V. Meisinger (1982) Lancet, I, 46-47.

R.C. Hendel and F.W. Sunderman, Jr. (1972) Res. Comm. Chem. Pathol. Pharmacol., 4, 141-146.

S.W. King, M.R. Wills and J. Savory (1979) Res. Comm. Chem. Pathol. Pharmacol., 26, 161-169.

S.W. King, J. Savory and M.R. Wills (1982) Ann. Clin. Lab. Sci., 12, 143-149.

A.P. Lundin, C. Caruso, M. Sass et al. (1978) Clin. Res., 26,636A.

F.C. McLean and A.B. Hastings (1935) J. Biol. Chem., 108, 285-322.

K. Nomiyama and H. Nomiyama (1982) J. Chromatogr., 228, 285-291.

S.M. Ott, N.A. Maloney, D.L. Klein, A.C. Alfrey, M.E. Ament, J. Coburn and D.J. Sherrard (1983) Ann. Int. Med., 98, 910-914.

V. Parsons, C. Davies, C. Goode, C. Ogg and J. Siddiqui (1971) Br. Med. J., 4, 273-275.

K.O. Pederson (1971) Scand. J. Clin. Lab. Invest., 28, 459-469.

E. Pfannkoch, K.C. Lu and F.E. Reqnier (1980) J. Chromatogr. Sci., 18, 430-441.

P. Rona and D. Takahashi (1911) Biochem. Zeit., 31, 336-344.

B. Rosoff and H. Spencer (1979) Clin. Chim. Acta, 93, 311-319.

C. Schumichen, K. Koch, A. Kraus, G. Kuhliche, K. Weiler, A. Wenn and G. Hoffman (1980) J. Nuclear Med., 21, 1080-1085.

B.J. Scott and A.R. Bradwell (1983) Clin. Chem., 29, 629-633.

F.W. Sunderman, Jr., E.R. Costa, C. Fraser, G. Hui, J.J. Levine and T.P.H Tse (1981) Ann. Clin. Lab. Sci., 11, 488-496.

F.W. Sunderman, Jr., B.L.K. Mangold, S.H.Y. Wong, S.K. Shen, M.C. Reid and I. Jansson (1983) Res. Comm. Chem. Pathol. Pharmacol., 39, 477-492.

J. Toffaletti, H.J. Gitelman and J. Savory (1976) Clin. Chem., 22, 1968-1972.

J. Toffaletti, J. Savory and H.J. Gitelman (1977) Clin. Chem., 23, 2306-2310.

M. Van Soestbergen and F.W. Sunderman, Jr. (1972) Clin. Chem., 18, 1478-1484.

METAL-PROTEIN COMPLEXES IN DETOXIFICATION PROCESSES

Robert A. Goyer

National Institute of Environmental Health Sciences
Research Triangle Park, NC 27709, U.S.A.

1. INTRODUCTION

The concept that formation of protein-metal complexes may have a role in detoxification or prevention of toxicity is relatively new. Pathological effects of toxic metals have traditionally been regarded as a function of dose and potency for toxicity of the metal itself. This notion has been based largely on empirical or clinical observations, that is, case reports and epidemologic studies. And there is validity to this approach since it has provided guidelines for safe levels in food, for occupational exposures and for permissible levels in the environment. Nevertheless, wide margins are required for safety and there are occasional exceptions to expectations.

Over the past few years, basic knowledge of the metabolism and mechanisms of injury, particularly at the cellular or molecular level, has been greatly expanded. It has been learned that individual metals do indeed have some specificity to their metabolism and that they are involved in molecular processes within cells, particularly as cofactors for enzymes.

This paper will summarize studies concerning proteins involved in detoxification or protection from toxicity of a few metals. Morphologically discernible cellular inclusion bodies are present with exposures to lead, bismuth and a mercury-selenate mixture. Metallothioneins form complexes with cadmium, zinc, copper and other metals and ferritin and hemosiderin are intracellular iron-protein complexes. Before discussing the details of these examples, some general comments can be made. For one, the protein complexes formed by lead, bismuth, mercury-selenate and iron, at least for hemosiderin, have attracted interest because these complexes are insoluble in tissues and can be observed histologically. However, it is this lack of solubility that has made detailed biochemical study very difficult. On the other hand, for those metal-protein

Chemical Toxicology and Clinical Chemistry of Metals
ISBN 0.12.137520.X
Copyright © 1983 by IUPAC
All rights of reproduction in any form reserved

complexes that are stable and soluble in aqueous media, such as metallothionein and ferritin, there is considerable biochemical information. The absence of a morphologically visible form of metallothionein has left questions about intracellular localization but the recent availability of an antibody suitable for labelling has been very helpful. We know perhaps as much about ferritin as any of these complexes but that is probably to be expected since it is soluble and at the same time it has a unique ultrastructural appearance so that it can be readily identified in cells and organelles. A final general comment is that none of these proteins or metal protein complexes have any enzymatic activity and in a sense they might be regarded as storage proteins.

2. INCLUSION BODIES

Nuclear inclusion bodies in renal tubular lining cells are a distinguishing histologic feature of lead toxicity. They occur most frequently in the proximal portion of the tubule (pars recta) and by light microscopy appear as dense spherical eosinophilic bodies (Figure 1). They can be differed from viral inclusion by being acid-fast when stained with the

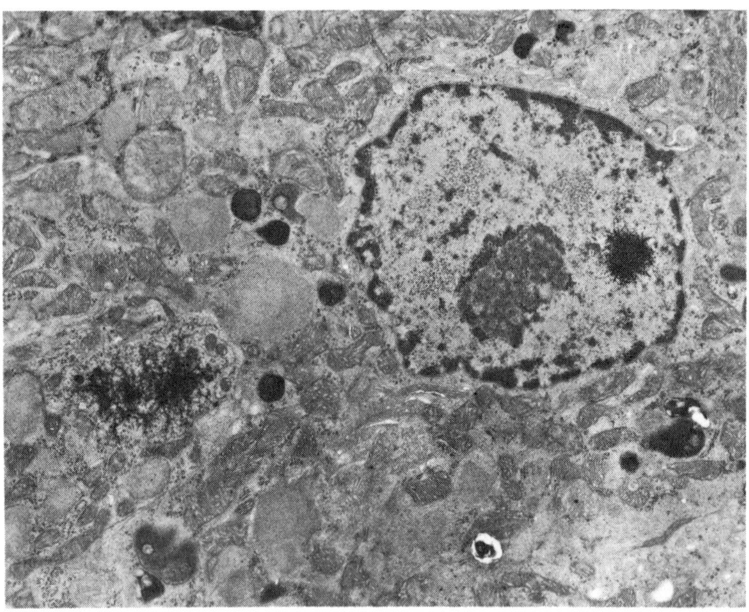

FIG. 1 This renal tubular lining cell is from a rat injected intraperitoneally with lead acetate. Lead-containing inclusion bodies are present in the nucleus (separate from the nucleolus) and in the cytoplasm and consist of an outer fibrillary margin and dense central core.

Ziehl-Neelsen's technique. By electron microscopy the inclusions consist of a dense central core and outer fibrillary zone. They vary considerably in size and the fibrils do not have periodicity.

Study of the lead-induced inclusion bodies in experimental animals has provided some insight into their composition and role in lead poisoning. A concentration of the lead-protein complexes can be obtained from kidneys of lead toxic rats by differential centrifugation on sucrose gradient as one would isolate nucleoli. The nucleoli are disrupted by sonication and the remaining inclusion bodies are cleaned by rinsing in saline (Moore et al., 1973).

Lead analysis of the isolated inclusion bodies has shown that, even in kidneys of control animals, most of the lead is contained in nuclei. The increment in lead content of mitochondria in lead-poisoned rats is small. This is consistent with other studies showing that mitochondria from lead toxic animals are functioning but have reduced oxidative and phosphorylative abilities (Goyer, 1971). The lead in the nucleus is concentrated in the inclusion bodies (Goyer et al., 1970a; Horn, 1970).

Efforts have been made to characterize the protein by amino acid analysis and by electrophoresis. The protein has been found to be a non-histone acidic protein rich in glutamic and aspartic acids and glycine. Electrophoresis of protein dissolved in 6 M urea or sodium dodecyl sulfate on polyacrylamide gels showed multiple bands but only one band retained lead and this band was determined to have a molecular weight of about 25,500 (Moore and Goyer, 1974). More recently others have continued efforts to characterize the inclusion body protein. Shelton and Egle (1982) of the Medical College of Virginia, using similar methods for isolation, report a molecular weight of about 32,000, close enough to the earlier measurement to suggest the numbers may be credible. The formation of the lead-protein complexes or inclusion bodies contribute to the metabolism of lead in at least two ways. They are protective or, in a sense, detoxify lead and they may have a role in the excretion of lead. In other experiments it has been shown that inclusion bodies are present in kidneys before other signs of lead toxicity occur (Goyer et al., 1970b).

Lead is excreted through the kidney in the glomerular filtrate and from peritubular capillaries by transtubular transport. This later mechanism is thought to assume greater importance as blood lead levels are increased.

Excretion of lead, particularly in the presence of elevated blood levels, may be transtubular from peritubular capillaries to tubular lumen. In the course of the transtubular flow of lead, a portion of the lead enters the cell where it forms a lead-protein complex and is no longer diffusible. The inclusion bodies may actually be formed within the cytoplasm initially and then transported into the nucleus by invagination of nuclear membranes.

Reversal of this process, that is, extrusion of inclusion bodies by evagination of nuclear membranes occurs following chelation therapy (Goyer and Wilson, 1975).

Support for the cytoplasmic formation of inclusion bodies is found in studies showing that the administration of a simple dose of lead to primary cultures of rat kidney epithelial cells results in formation of typical inclusion bodies in the cytoplasm within four hours and that the inclusion bodies become almost all localized in the nucleus by twenty-four hours (McLaughlin et al., 1980). De novo synthesis of inclusion body protein with exposure to lead is suggested in that the formation of inclusion bodies in the cell cultures is blocked by the addition of metabolic inhibitors such as cycloheximide and actinomycin D to the culture medium.

Participation of the lead-protein complexes in the excretion of lead is suggested from a couple of observations. For one, inclusion bodies have been found in urine of children with acute lead toxicity (Landing and Nakai, 1959), and may also be found in urine of workmen with elevated blood lead levels (Schumann et al., 1980). An interesting set of data was collected by Barry (1972) in the course of routine monitoring of blood and urine lead levels in workmen that suggest inclusion bodies enhance lead excretion. Barry compared blood and urine lead levels of workers prior to employment and after working for months or years. At a particular blood lead level, say 20 or 25 μg/dl, workers have a considerably greater excretion of lead. This suggests greater activity of the transtubular transport mechanisms and perhaps lead-protein complexes formation accounting for the increased excretion.

As has been stated previously, lead-protein complexes or inclusion bodies have been observed in a number of different species of animals, in other tissues including liver and osteoclasts, and in some plants that show resistance to lead (Goyer, 1977).

Other metals also form intranuclear inclusion bodies but less is known about their biologic significance than is known about lead. Bismuth forms a dense appearing body in renal tubular lining cells of experimental animals and in persons with excess exposure to bismuth. This is uncommon today but was noted some years ago when bismuth was a popular choice for treatment of syphilis. The bodies have been isolated by the method used to isolate lead-induced inclusion bodies. The protein is acidic and has an amino acid composition similar to that found in the lead-induced bodies. The bismuth content of the bodies has been demonstrated by direct measurement and by x-ray microanalysis (Fowler and Goyer, 1975).

Nuclear inclusion bodies similar to those formed by lead or bismuth are also seen in experimental animals following exposure to mercuric chloride and sodium selenate (Carmichael and Fowler, 1980). Mercuric chloride is a severe renal toxin and a low dose given in drinking water to rats over a 22 week period produces fatty degeneration and cellular necrosis in the liver and tubular cell necrosis in the kidney, as well as, severe weight loss. Sodium selenate does not produce histological changes but does cause severe weight loss. However, when given together there are no histologic changes and less severe weight loss. This suggests that there is a synergistic, protective effect. And when animals are given both metals, inclusion bodies, as seen in the nucleus of tubular cells (Figure 2), have been shown by x-ray microanalysis to contain both mercury and selenium. The matrix has not been studied so it is not known whether or not this is another example of an acidic protein-metal complex, but the metals are complexed in some way in these bodies and toxicity is reduced.

3. METALLOTHIONEIN

Metallothionein is the prime example of a metal-protein complex with a major detoxification or protective role. Metallothioneins are actually a group of small isoproteins with

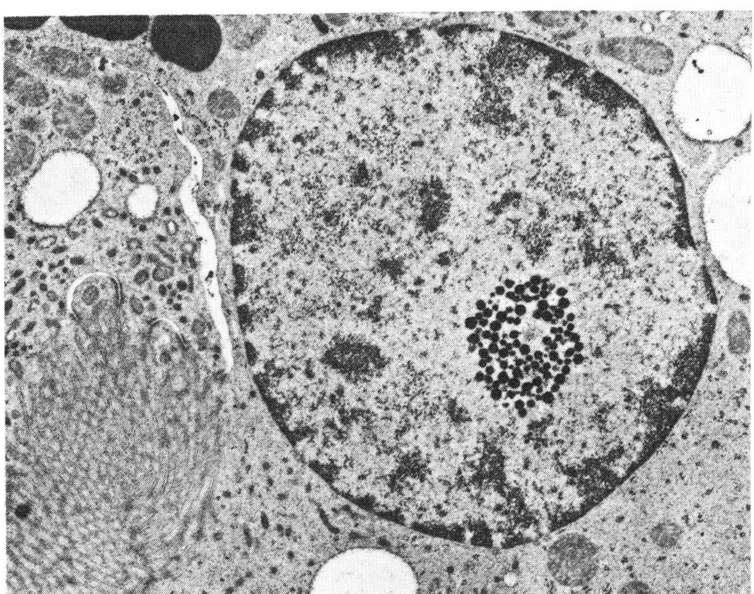

FIG. 2 Renal tubular lining cell of rat given sodium selenate and mercuric chloride. The cluster of small, dense nuclear inclusions has been shown by x-ray microanalysis to contain selenium and mercury (photomicrograph from Dr. Bruce Fowler, NIEHS).

a number of common physical characteristics. Molecular weights are about 6500 and they usually have a high content of either zinc and/or cadmium or copper when isolated. The amino acid composition is characterized by approximately 30 percent cysteine and the absence of aromatic amino acids. Specific optical absorption is due to location of metal thiolate complexes in the protein. Zinc, cadmium, copper and mercury thioneins are identified at 225, 250, 275 and 300 nm respectively. The metallothioneins contain 61 amino acids and 20 are cysteine. Structural studies using nuclear magnetic resonance spectroscopy and electron spin resonance spectroscopy have identified two distinct metal clusters in mammalian metallothionein. The clusters seem to have significant differences in their affinity for different metal ions; one of the clusters has a high level of specificity for zinc. Metal binding is by trimercaptide bridges (Boulanger, 1983).

Metallothionein is primarily a tissue protein and is ubiquitous in most organs but is in highest concentration in liver, particularly following recent exposure, and in kidney where it accumulates with age in proportion to cadmium concentration. Piscator (1964) found that metallothionein was increased in livers of rabbits exposed to cadmium. He suggested that the biosynthesis of the protein was inducible by cadmium and that the metal bound to the protein offered protection from cadmium toxicity. Since that time there have been a number of studies from experimental animals, as well as, tissue culture models that confirm the protective role of metallothioneins. It has been found that synthesis of metallothionein in tissues is directly related to exposure to metal. Toxicity of cadmium to kidney only occurs when exposure exceeds the ability of that organ to either synthesize or store additional cadmium. Toxic cell injury is thought to be caused by unbound cadmium or free cadmium ion (Suzuki, 1983).

Pretreatment of experimental animals with small doses of cadmium has been shown to prevent acute toxic effects of a large dose of cadmium. This property is not restricted to protection from cadmium toxicity alone. Pretreatment of experimental animals with small doses of cadmium or mercury salts can prevent the nephrotoxic effects of high doses of mercury chloride (Magos and Webb, 1974). There is also some experimental evidence that suggests that the teratogenic effects of cadmium in Golden hamsters is prevented by pretreatment with zinc salts or small amounts of metallothionein (Ferm and Carpenter, 1967). Other studies have shown that certain sulfhydryl requiring enzymes are inhibited *in vitro* by small amounts of cadmium but not affected *in vivo* where intracellular cadmium is bound to metallothionein (Merali and Singhal, 1976). Human cells in tissue culture, in which metallothionein has been induced by pretreatment with cadmium, become resistant to previously lethal exposure to cadmium (Cherian, 1980).

The isometallothioneins or isoproteins have only slight differences between species and even within the organs (Karin and Richards, 1982; Klauser et al. 1983) Generally, the protein exists in two major isoprotein forms, metallothioneins I and II, which differ only very slightly in amino acid composition and in charge. The two isoprotein forms are usually present in comparable amounts but their ratio may be affected by metal induction (Suzuki and Yamamura, 1980).

Regulation of synthesis occurs at the transcription level and involves multiple genes that are tightly clustered (Cherian et al., 1981; Durnam and Palmiter, 1981). Synthesis of metallothionein is induced not only by heavy metal ions but also by glucocorticoid hormones (Karin and Richards, 1983). However, the response differs. Induction by metals is rapid and mediated by removal of a repressor which is thought to be the apothionein. When presented with available metal, it forms metallothionein with loss of the repressor action of the apothionein with induction of new synthesis. This is followed by repression of synthesis again. The reaction has a high degree of specificity since only metal ions that bind to the apoprotein can lead to induction of new synthesis. Induction of metallothionein has been shown to occur following treatment with a number of divalent cations including Zn^{++}, Cu^{++}, Cd^{++}, Hg^{++}, Bi^{++} and Ag^{++} (Cherian and Nordberg, 1983). Although nickel may bind with metallothionein _in vivo_, _de novo_ synthesis is only weakly stimulated by nickel exposure (Sunderman, 1983). There is no evidence to date that metallothioneins have any role in lead metabolism or toxicity.

The mechanism of induction by glucocorticoids is different in that it by-passes apothionein and interacts with the hormone receptor complex bound to DNA resulting in persistent activation of the genes and is not dependent on concomitant protein synthesis (Karin and Herschman, 1979; Mayo and Palmiter, 1981). Therefore, induction of metallothionein is not necessarily dependent on presence of excess metal in the cell, suggesting that this protein may have a role in regulation of trace metal or essential metal metabolism apart from the protective role it conveys for cadmium exposure.

In fetal and neonatal liver, zinc and copper are the major constituents of metallothionein. Cytohistochemical studies using a peroxidase-antiperoxidase method with an antibody to polymerized metallothionein have shown that in fetal mouse liver metallothionein is localized in the nucleus (Figure 3) (Panemangalore et al., 1983).

This suggests that in fetal and perhaps in neonatal life the major function of metallothionein is storage of copper and zinc, and the protective role becomes more important in adult life. It should be pointed out, however, that protection from potential toxicity is important during gestation and that metallothionein is inducible in the placenta by exposure to cadmium (Hanlow et al., 1982).

In liver of the adult mouse exposed to cadmium, metallothionein is confined to the cytoplasm (Figure 4).

4. FERRITIN, HEMOSIDERIN

The protective/regulatory functions of metallothionein have many biologic analogies to the role of ferritin in the

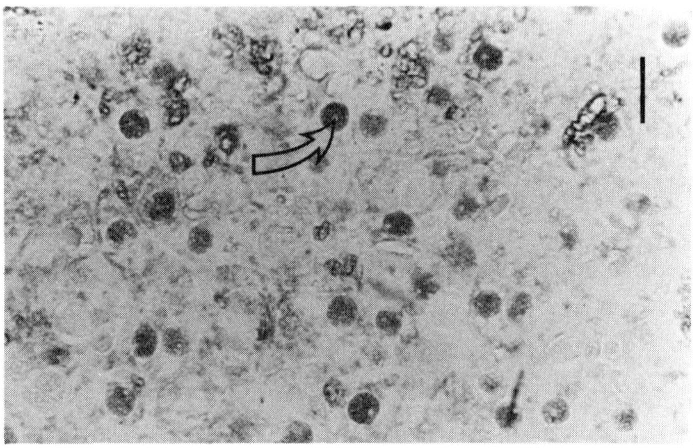

FIG. 3 Histochemical localization of metallothionein in nuclei of fetal mouse liver (arrow from Panemangalore, et al., 1982).

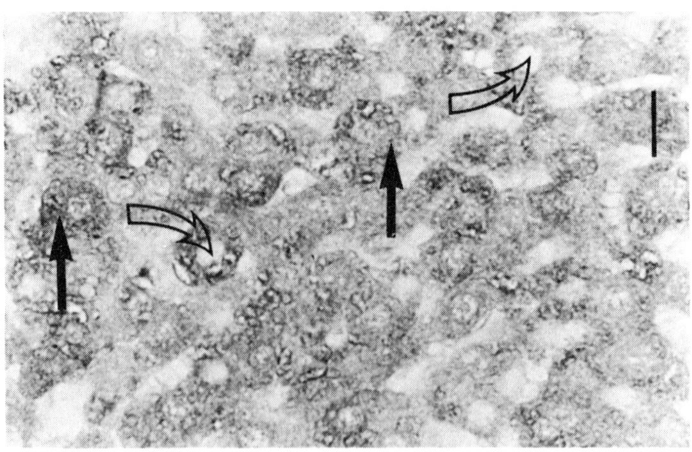

FIG. 4 Histochemical localization of metallothionein in cytoplasm of adult mouse liver (arrow from Panemangalore, et al., 1982).

metabolism of iron. Ferritin, like metallothionein, is an intracellular protein. It regulates iron metabolism in that it provides storage for iron when needed for heme synthesis or for non-heme requirements. On the other hand, while binding to ferritin, the iron is in a non-toxic form and its presence is tolerated in the cytoplasm and by the organelles of the cell. Completely assembled ferritin is a large molecule but it is composed of subunits. When large amounts of iron are made available to the cell, existing apoferritin molecules are immediately filled with iron, which in turn results in the assembly of subunits to form apoferritin and ferritin (Bezkorovainy, 1980).

More is known about the process for induction of new synthesis of apoferritin than is known about induction of metallothionein. According to a scheme proposed by Zähringer and coworkers (1976), some of the unassembled ferritin subunits are attached to free messenger RNA. As ferritin is assembled, messenger RNA becomes derepressed and becomes available for combination with ribosomes and apoferritin biosynthesis.

Ferritin metabolism is largely in parenchymal cells; the liver is the organ with the greatest content of iron. Ferritin is water soluble but has a distinct ultrastructure which has a tetrad appearance by electron microscopy.

Excess ferritin is converted by lysosomes to hemosiderin, probably by a process of denaturation and polymerization. The product has an irregular globular structure and is insoluble, so it has been difficult to study in greater detail. Nevertheless, hemosiderin is also a site for storage of iron and, like ferritin, serves a protective role in that it maintains intracellular iron in a bound form. Hemosiderin is the final intracellular depot for excess iron and in conditions of iron overload, cells of the reticuloendothelial system become filled with hemosiderin.

5. SUMMARY

Intracellular metal-protein complex formation has been observed for a number of potentially toxic metals. Toxicity of a metal or potency is determined by the intracellular dose of the toxic form of the metal at the target site. The toxic form is usually the free cation as determined for cadmium and possibly mercury and other metals. Metal-protein complex formation may serve as a protective or detoxification process by reducing the availability of free metal ion.

For at least two of the apoproteins that form protective complexes, induction of synthesis involves specific genetic mechanisms. It is interesting to speculate whether nature has provided genetic information specifically for the purpose of detoxification. This seems to be true for metals that are essential, ferritin for iron, metallothionein for zinc and copper. Without evidence that cadmium or mercury or any of the other toxic metals that form complexes with metallothionein are indeed essential, it must be postulated that the occurrence of these complexes is both fortuitous and fortunate. In terms of the acidic proteins that form complexes with lead and bismuth, there is not sufficient information available about these proteins to form any genetic hypothesis.

6. REFERENCES

P.S.I. Barry (1972) in Lead in the Environment (Ed, P. Hepple), Appl. Sci. Publ., Essex, England, 72.

A. Bezkorovaing (1980) in Biochemistry of Nonheme Iron, Plenum, New York, 207.

Y. Boulanger, C.M. Goodman, C.P. Forte, S.W. Fesik and I.M. Armitage (1983) Proc. Nat'l. Acad. Sci., 80, 1501.

N.G. Carmichael and B.A. Fowler (1980) J. Environ. Pathol. Toxicol., 3, 399.

M.G. Cherian (1980) Toxicol., 17, 225.

M.G. Cherian and M. Nordberg (1983) Toxicology, 27 .

M.G. Cherian, S. Yu and C.M. Redman (1981) Can. J. Biochem., 59, 301.

D.M. Durnam and R.D. Palmiter (1981) J. Biol. Chem., 256, 5712.

V.H. Ferm and S.H. Carpenter (1967) Nature, 216, 1123.

B.A. Fowler and R.A. Goyer (1975) J. Histochem. Cytochem., 23, 722.

R.A. Goyer (1971) Am. J. Pathol. 64, 167.

R.A. Goyer (1977) in Clin. Chem. Toxicol. of Metals (Ed, S.S. Brown), Elsevier/North-Holland, Amsterdam, 93.

R.A. Goyer, P. May, M. Cates and M.R. Krigman (1970) Lab. Invest., 22, 245.

R.A. Goyer, J.F. Moore, B. Rhyne and M.R. Krigman (1970) Arch. Environ. Hlth., 20, 705.

R.A. Goyer and M.H. Wilson (1975) Lab. Invest., 32, 149.

D.B. Hanlow, C. Specht and V.H. Ferm (1982) Environ. Res., 27, 89.

J. Horn (1970) Virchows Arch. Ab. B. Zellpath., 6, 313.

M. Karin and H.R. Herschman (1979) Science, 204, 176.

M. Karin and R.I. Richards (1982) Nature, 299, 797.

M. Karin and R.I. Richards (1983) Environ. Health Perspect., 54,

S. Klauser, J.H.R. Kägi and K.J. Wilson (1983) Biochem. J., 209, 71.

B.H. Landing and H. Nakai (1959) Am. J. Clin. Pathol., 31, 499.

L. Magos, M. Webb and W.H. Butler (1974) Brit. J. Exp. Path., 55, 589.

K.E. Mayo and R.D. Palmiter (1981) J. Biol. Chem., 256, 2621.

J.R. McLachlin, R.A. Goyer and M.G. Cherian (1980) Toxicol. Appl. Pharmacol., 56, 418.

Z. Merali and R.L. Singhal (1976) Brit. J. Pharmacol., 57, 573.
J.F. Moore and R.A. Goyer (1974) Environ. Health Perspect., 7, 121.
J.F. Moore, R.A. Goyer and M. Wilson (1973) Lab. Invest. 29, 488.
M. Panemangalore, D. Banerjee, S. Onosaka and M.G. Cherian (1983) Dev. Biol., 97, 95.
M. Piscator (1964) Nord. Hyg. Tidskr., 45, 76.
G. Schumann, S.I. Lerner, M.A. Weiss, L. Gawronski and G.K. Lohiya (1980), Am. J. Clin. Pathol., 74, 192.
K.R. Shelton and P.M. Egle (1982) J. Biol. Chem., 257, 11802.
F.W. Sunderman Jr. (1983) Environ. Health Perspect., 54,
K.T. Suzuki (1983) Environ. Health Perspect., 54, .
K.T. Suzuki and M. Yamamura (1980) Biochem. Pharmacol., 29, 2407.
J. Zähringer, B.S. Baliga and H.N. Munro (1976) Proc. Nat'l. Acad. Sci. 73, 857.

MOLECULAR TARGETS OF METAL TOXICITY

Thomas W. Clarkson

Division of Toxicology, University of Rochester School of Medicine, Rochester NY 15642, U.S.A.

1. INTRODUCTION

The idea of a molecular target comes primarily from the area of pharmacology where drugs are tailor-made to produce a specific response. When the drug attaches to its target molecule, a series of events is initiated which ultimately leads to the observed effect of the drug in the whole animal. The target molecule is frequently referred to as the receptor. Indeed, volumes have been written on drug receptor reactions.

In order for the drug to interact with its target, it must be absorbed into the blood stream and transported to the vicinity of the target. This transport process may involve the reaction of a drug to specific carrier molecules. The molecules, however, are not normally regarded as the target molecules. Likewise, enzymes that may transform the drug into a reactive species are also not normally regarded as the target molecules.

The question arises as to whether the concept of a molecular target is useful in understanding metal toxicity. In general, metals differ greatly from man-made drugs in both their chemical and biological properties. Toxic effects of metals are often not specific but cover a wide variety of signs and symptoms as opposed to the specific actions of most drugs. Indeed, in the whole field of toxicology itself, most poisons produce a wide range of effects. Nevertheless, there is at least one example where the effects of the poison are probably understood in terms of the reaction with a specific protein receptor in the cytoplasm of the target cells. The example is dioxin. In fact, the specificity of the reaction of dioxin with its receptor are truly remarkable. Over one hundred isomers of this compound have been examined. It was found that only a very few of these different stereochemical configurations react with this protein receptor and produce the toxic effects characteristic of this molecule. Do some metals match the dioxin-like selectivity for a particular target molecule?

Chemical Toxicology and Clinical Chemistry of Metals
ISBN 0.12.137520.X
Copyright © 1983 by IUPAC
All rights of reproduction in any form reserved

The possibilities for the metal receptor interaction need not be limited to the high specificity of the dioxin type. Indeed, the metal may react with more than one class of molecular targets to produce its spectrum of toxic effects. Alternatively, it may behave in a hit-and-run fashion affecting one target at one time and moving on to another target later on. However, before we examine particular metals to see if the molecular target concept is useful, by way of introduction, I believe it would be useful to examine some of the chemical and biological properties of metals as a background to this discussion.

Cations

$$M^{+} + L^{-} \rightleftharpoons ML$$

$$M_1L + M_2^{+} \rightleftharpoons M_2L + M_1^{+}$$

Oxyanions

$CrO_4^{=}$

Oxidation States

As^{3-} As^{+} As^{3+} As^{5+}

Organometallics

CH_3Hg^{+}

Catalysts

Pb^{++}

FIG. 1 Toxicologically important properties of metals.

Figure 1 lists five properties of metals that are important to our understanding of possible molecular targets and tend to set aside metals as a distinct class from organic poisons. Most metal salts form cations once they dissolve in solution. The cations, being positively charged, have a high affinity for a wide variety of organic ligands, i.e., organic molecules or molecular substructures capable of donating electrons. Reversible complexes are formed having very high affinity constants; for example, affinity constants of mercury cations (Hg^{++} or CH_3Hg^{+}) with ionized sulfhydryl groups (S^{-}) are of the order of 10^{15} or 10^{20}. Other organic ligands such as chloride, hydroxyl, carboxyl, phosphoryl, etc. all have substantial affinity constants with mercury, of the order of 10^{15} or more. Other metals show varying patterns of affinities with different ligands. The point is that metal cations have many possibilities for interacting with all kinds of molecules in the cell.

One metal may substitute for another metal in any given complex and we will see examples where the "molecular target" is occupied by an "essential" metal which is replaced by a "non essential" metal resulting in toxic effects.

Certain metals form what are termed oxyanions depicted also in Figure 1. These are coordinate complexes of the metals with oxygen having a specific stereochemical configuration that lead to interaction with the molecular targets and transport proteins in a very special way.

A number of metals have a variety of oxidations states. These oxidations states, in general, differ in both chemical and toxicological properties. Arsenic (Fig. 1) is a good example of this. It is likely, therefore, that oxidation will influence the type of molecular target associated with the metal.

Certain metals form organometallic compounds in which the metal atom forms at least one covalent bond with the carbon atom. This bond tends to be stable, to be ruptured only by special enzymic action, and confers on the metal distinct toxicological properties different from those of the inorganic species of the metal. Again, we can expect a different molecular target for the organometallic forms of the metal.

Certain metals and certain oxyanions of metals can act as catalysts. This has long been known, of course, in the field of chemistry. Some of the biological effects of metals might be due to their acting as catalysts, not just as part of the active group of an enzyme, but catalysts in their own right.

In short, metals have considerable possibilities for reacting with a wide variety of molecules in the cell. As we shall discover, this is one of the key difficulties in trying to identify specific molecular targets for metals. The second difficulty is in the area of analytical chemistry. Toxic concentrations of most metals are usually in the micromolar or sub-micromolar range. The concentrations of most common organic ligands found in the cell or in biological media are in the millimolar range. This presents the analyst with a tremendous challenge to try to identify and measure specific metal ligand complexes in the micromolar concentration range when that same complex is surrounded by similar ligands a thousand-fold in excess. The attachment of the metal to these ligands is often rapidly reversible as in the case of mercury and sulfur groupings so that attempts to isolate it are fraught with difficulties of the redistribution of metal to other ligands. Thus our knowledge of the bio-inorganic chemistry of metals is useful background to us. In general, it cannot predict for any particular metal what the biological target will be. Frequently we have to work backwards to look for the biological effect and then see what is the most likely target to explain that effect. In the case of lead, for example, the discovery of porphyrin metabolites in blood and urine indicated that the molecular targets for lead are enzymes in the sequence of synthesis of hemoglobin. The initial clue was biological effect; it was not a special knowledge of the chemistry of lead and how it interacts with organic ligands.

Following this brief introduction, I should like to explore some of the general mechanisms of toxicity of metals and to see how the concept of molecular targets fits into these mechanisms. As in any other field of toxicology or pharmacology, there are only a very few examples where we have some understanding of the mechanism of toxicity of metals. Having considered these examples, I should then like to discuss some current researches into the mechanism of toxicity of one metal, namely mercury, and to indicate some of the major challenges we face when trying to understand the toxicity of metals and, particularly, trying to identify their molecular target.

2. GENERAL MECHANISMS

(a) *Selective Effects Due to Selective Accumulation*

The action of uranium on the proximal tubular cells of the kidney nephron is an example of how selectivity comes about because of selective accumulation of a metal at its site of action. Uranium is present, in part, as a bicarbonate complex in plasma which is filtered at the glomerulus and thereby enters the proximal tubular fluid. The reabsorption of bicarbonate liberates the highly reactive uranyl cation which subsequently is concentrated by the reabsorption of water in the proximal tubular part of the nephron. This highly reactive cation is then free to react with the proximal tubular cells resulting in the destruction of these cells. With this type of mechanism, it is not useful to assume a specific interaction with a molecular target. It suffices to say that the highly reactive metal cations are directed in sufficiently high concentrations to a group of target cells (Fig. 2).

It is possible that the reno-toxic action of other metals such as chromate and, possibly, inorganic mercury occurs by selective accumulation in the proximal tubule. However, uranium remains about the only really well-documented example of this type of phenomenon.

The organic mercury compound, mersalyl, is believed to act as a diuretic by a mechanism indicated in Fig. 3. The mercurial is actively secreted into the proximal tubular fluid and then carried to its site of action in the ascending limb of the loop of Henle. At this point, it interacts with a specific receptor, diagrammatically indicated in Fig. 3, resulting in the blockade of sodium chloride absorption. Its reaction with this receptor is so selective that no other effects are produced on kidney function. Furthermore, its effects may be reversed by parachloromercuribenzoate, which will react with the receptor but

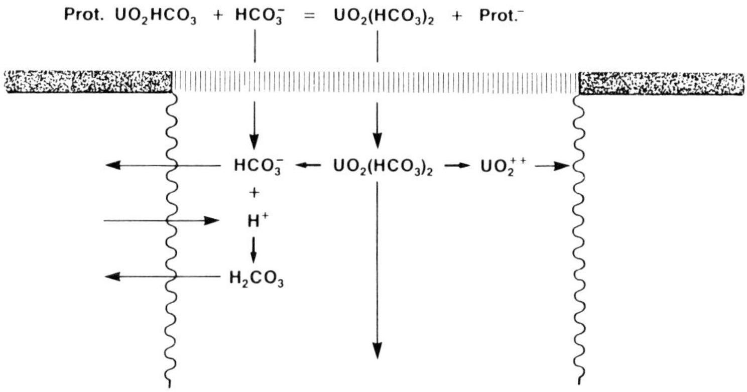

FIG. 2 A schematic representation of the selective accumulation of uranyl cations in the proximal tubule of the kidney (from fig. 4 in Clarkson and Shaikh 1982).

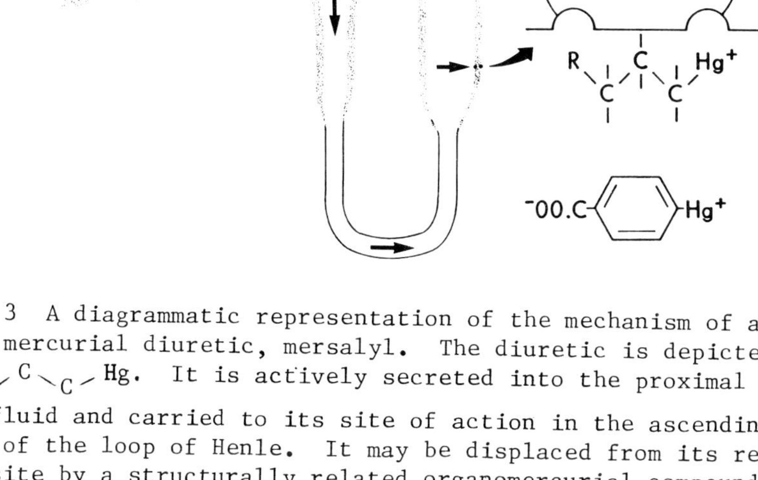

FIG. 3 A diagrammatic representation of the mechanism of action of a mercurial diuretic, mersalyl. The diuretic is depicted as R⟍C⟋C⟍C⟋Hg. It is actively secreted into the proximal tubular fluid and carried to its site of action in the ascending limb of the loop of Henle. It may be displaced from its receptor site by a structurally related organomercurial compound - p-mercuribenzoate (from Fig. 1 in Clarkson and Shaikh, 1982).

not activate it. In this example, the concept of a specific target is useful in understanding why the mercurial produces such a specific effect, that is, the selective inhibition of sodium chloride, and why other organomercurials can antagonize and reverse its action.

Specific interactions with target molecules. Perhaps the best documented and, historically, the best known example of a specific interaction, is that of trivalent arsenic with the molecule α lipoic acid as depicted in Figure 4. Trivalent arsenic exists as an oxyanion with a pyramidal structure. This allows the two oxygen atoms to come into close proximity with the two sulfhydryl groups present at the end of the carbon chain of lipoic acid. The bond lengths and bond angles are highly favorable to the reaction depicted in this figure, resulting in the formation of a stable six member ring compound containing arsenic, resulting in inactivation of α lipoic acid.

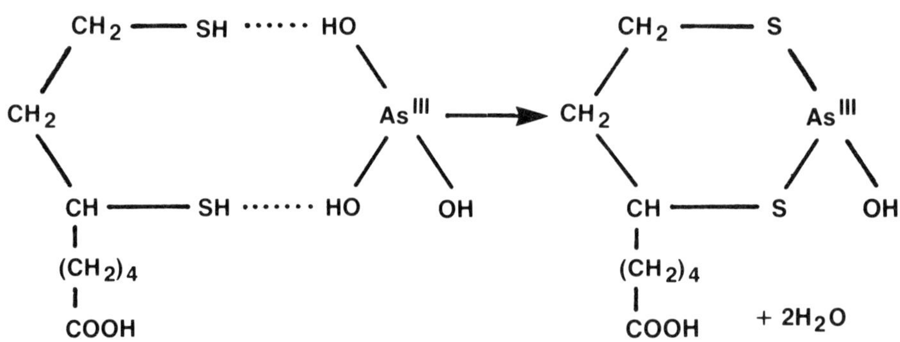

FIG. 4 The reaction of the arsenite anion with the sulfhydryl group of α lipoic acid (adapted from Jennette 1981).

The ability of trivalent arsenic, as an oxyanion, to react with neighboring SH groups was known since the 1940s, a reaction discovered by Peters' group (for review see Oehme 1972). It was known at that time that the action of arsenic in the whole animal resulted in elevated pyruvate levels. The compound, α lipoic acid, was indeed not discovered until ten years later (about the mid-50s). Indeed, the study of the toxicity of arsenic prepared the way for the discovery of this important co-factor in dehydrogenase reactions. Examples of specific interactions between metal and ligands in the target molecule are given in contributions to these proceedings.

(b) *Specificity; Interaction with Essential Elements*

The biochemical and transport mechanisms in the body which require certain metals might be fooled into accepting another metal which will partly fill the role of the essential metal but not completely so, finally resulting in toxic effects. Probably, the mechanism of action of many metals might involve this kind of exchange between the essential and non-essential metal. We know many examples of antagonisms where the toxicity of one metal is modified by another (for review, see Jennette, 1981). Zinc, for example, counteracts the toxicity of cadmium. Iron and calcium affect the reabsorption of lead, possibly also its toxicity. Molybdenum antagonizes copper; tungstate is believed to replace molybdate in certain enzymes such as xanthine oxidase.

Lithium, for example, is known to effectively mimic sodium in transport into mammalian cells (Tosteson 1981). Lithium is carried into the red blood cell by the sodium carrier as depicted in Fig. 5. The actual receptor for lithium is not known

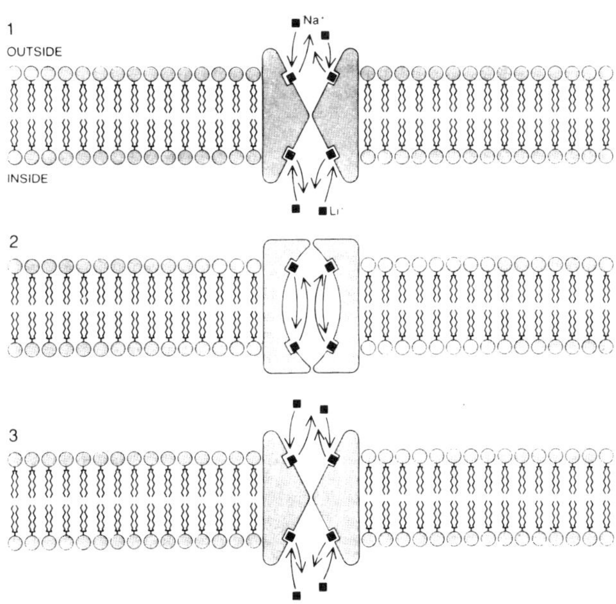

FIG. 5 A diagrammatic representation of the transport of lithium and sodium into the red blood cell (adapted from Tosteson 1981).

but both the pharmacological effects on the central nervous system and the toxic effects, such as diuresis, are probably mediated by the ability of lithium to participate in transport and enzymic processes normally utilizing sodium. Another example is the toxicity of thallium. Thallium is accumulated by cells by processes that normally accumulate potassium (Gehring and Hammond 1967). Again, as in the case of lithium, we do not know the final target or receptor molecule for thallium but its toxic effects on the nervous system are probably due to its mimicking of potassium.

Mimicking essential substrates. The interactions of metals with target molecules might also be brought about by mimicking not just essential metals but also by mimicking essential substrates in the cells which are non-metallic. Examples of this type of reaction are those metals that are present as oxyanions as depicted in Figure 6. The oxyanions of chromium, molybdenum and pentavalent arsenic are compared with the anions sulfate and phosphate. The stereochemical configuration of these anions are very similar. As a result, the oxyanions of metals can fool the cells transport machinery and cellular enzymes into thinking that they are either sulfate or phosphate anions.

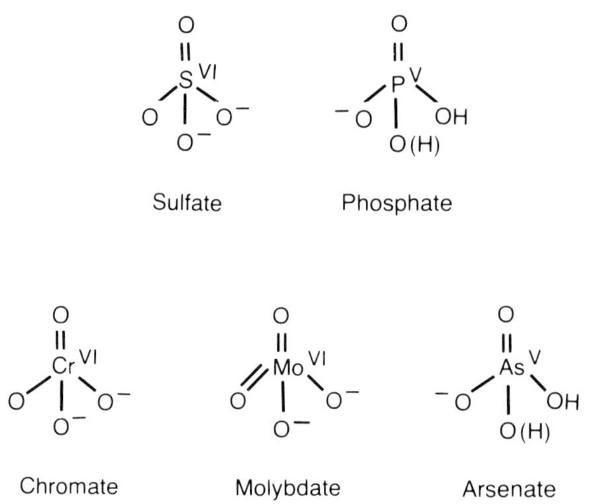

FIG. 6 Oxyanions and oxo-complexes of certain toxic metals are compared with sulfate and phosphate anions (adapted from Jennette 1981).

Arsenate can substitute for phosphate in the cell's transport and enzymic machinery. The result is the formation of arsenate compounds in place of high energy phosphate compounds. These arsenate compounds are unstable and break down in a reaction referred to as arsenolysis, which is illustrated in Fig. 7. In this example, glyceraldehyde is normally phosphorylated in a two step reaction with inorganic phosphate and with ADP forming adenosine triphosphate. However, if arsenate anions happen to be present, the enzyme in step 1 accepts arsenic in place of phosphate, an unstable intermediate is formed and rapidly hydrolyzes with the release of inorganic arsenic which is recycled, and ATP is not formed. Arsenolysis reactions of this type explain why inorganic arsenate acts as an uncoupling agent.

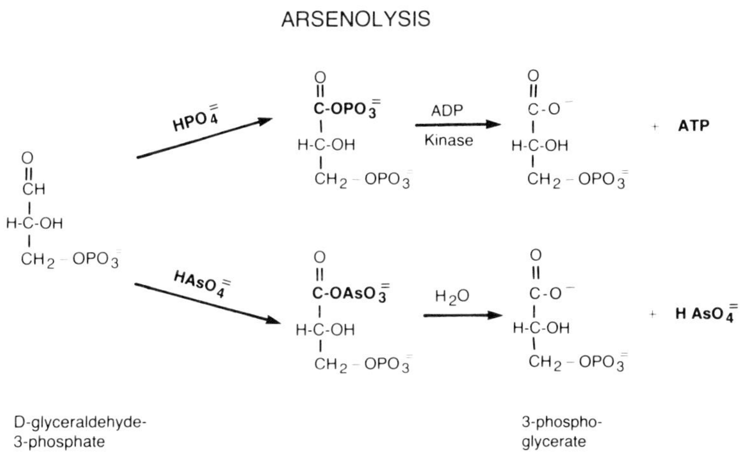

FIG. 7 Arsenolysis reaction: the arsenate anion substitutes for phosphate to form an unstable compound (adapted from Jennette 1981).

Extensive studies by Jennette (1981) have shown that chromate enters the cell on the sulfate transport system. Chromate is reduced to trivalent chromium by P-450. Trivalent chromium is known to react with DNA to produce mutagenic effects. Indeed, this scheme may explain the carcinogenic action of hexavalent chromium. Although DNA may be the ultimate target molecule for chromium, a key step in this interaction is the mimicking of sulfate by the hexavalent chromate anion which allows it to get into the cell in the first place. Trivalent chromium itself is unable to cross the cell membrane.

In summary, the formation of specific complexes, such as we have seen between α lipoic acid and trivalent arsenic, the mimicking reactions of metals whereby they may displace an essential metal of similar properties, or the oxyanions of metals mimicking essential substrates such as sulfate and phosphate, appear to be the best documented mechanisms whereby metals will react with target molecules. The remainder of the discussion will deal with much more speculative examples, which will be concerned with the toxicity of mercury and its compounds.

(c) *The Putative Mechanisms of Toxicity of Mercury*

Mercury exists in a number of distinct physical and chemical forms each having characteristic toxic effects. Mercury vapor (i.e., the vapor from metallic mercury) and methyl mercury compounds are of special interest in view of their effects on the human central nervous system. Studies on the mechanism of action of these two forms of mercury present a specially difficult challenge as the signs and symptoms of poisoning are not accompanied by measureable biochemical disturbances.

The signs and symptoms of mercury vapor poisoning have been well described for over 2000 years. One of its most interesting effects are the psychological disturbances. These take a rather bizarre form best illustrated by the Mad Hatter in *Alice in Wonderland.* The person may become highly anxious, irritable, excessively shy, emotionally unstable. This fact has been known for a long time but we are almost as far away from understanding the molecular targets involved as we were when these effects were first described (Ramazzini, 1713).

How inhaled mercury vapor produces these bizarre psychological effects on the central nervous system remains a mystery. A special characteristic of inhaled mercury vapor from the biochemical standpoint is that it can undergo cyclic oxidation and reduction as illustrated in Fig. 8. The catalase-hydrogen peroxide pathway oxidizes the elemental vapor to the ionic form. An enzyme known to be present in the cytoplasm of liver cells is able to reduce the divalent ionic mercury back to the elemental form. This cycling of mercury between the two oxidation states could be imagined to interrupt normal electron flow, for example, in the mitochondria and result in an uncoupling. A number of metals besides mercury vapor may undergo cyclic oxidation but the role of this reaction in the toxicity of the metal is still poorly understood.

Methyl mercury also produces a disturbance of the central nervous system but of a different type than does mercury vapor. Methyl mercury produces severe pathological changes in specific areas of the brain, particularly in the cerebellum and in the sensory areas of the brain such as in the visual cortex. A special characteristic is the very long latent period before the

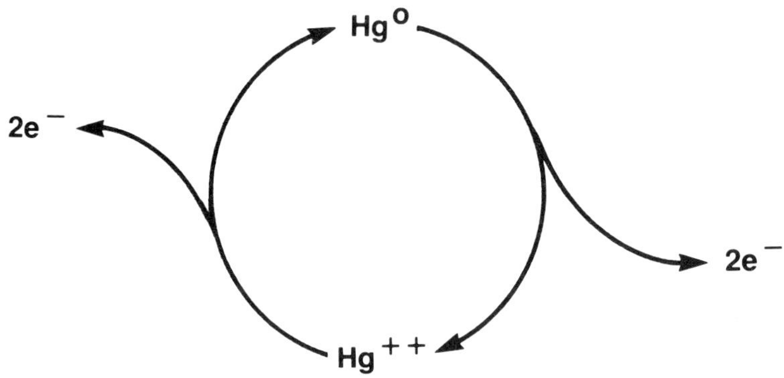

FIG. 8 The cyclic oxidation of inorganic mercury in mammalian cells (for further details see Clarkson et al. 1980)

signs and symptoms of poisoning appear. For example, in cases in Japan the first signs and symptoms of poisoning have been delayed up to 1 year after the cessation of exposure. In animal experiments by Evans and colleagues (1977), monkeys who were maintained at the constant blood level did not exhibit the first symptoms for periods of up to two years (Fig. 9).

This long latent period presents a formidable challenge to those interested in unravelling the mechanisms of toxicity and in identifying the molecular targets.

A number of mechanisms have been proposed (Ganther 1978) but only two will be mentioned. One mechanism involves the homolytic cleavage of the methyl mercury cation (Fig. 10). The methyl radical may then go on to alkylate macromolecules or it may react with lipids to initiate a chain of peroxidation reactions such as has been described in carbon tetrachloride poisoning.

Another mechanism whereby methyl mercury can affect brain lipids has been proposed by Segall and Wood (1974). In vitro experiments demonstrated that methyl mercury can catalyze the hydrolysis of important structural phospholipids essential to membrane structure and function (Fig. 11). This loss of plasmogens might be particularly important in the nervous system which depends very much on membrane integrity for its normal function. Indeed, some recent data (Grundt et al. 1980) has confirmed that

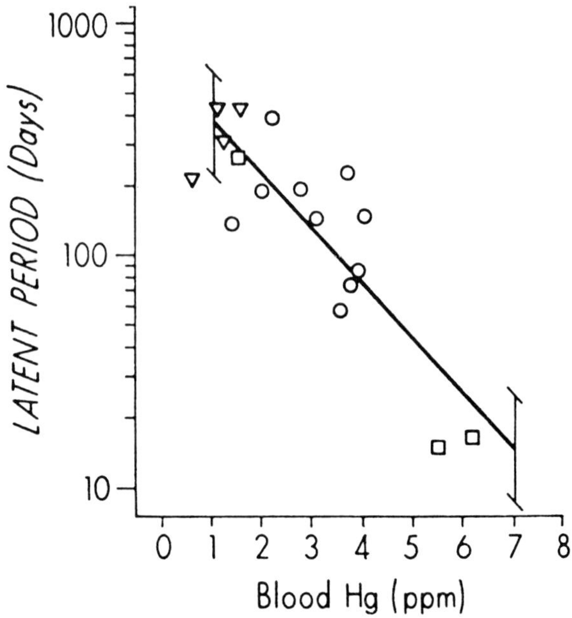

FIG. 9 The relationship between the latent period and the blood concentration of mercury in primates dosed with methyl mercury: the latent period is the time of onset of signs of methyl mercury poisoning (from Fig. 6 of Evans et al. 1977)

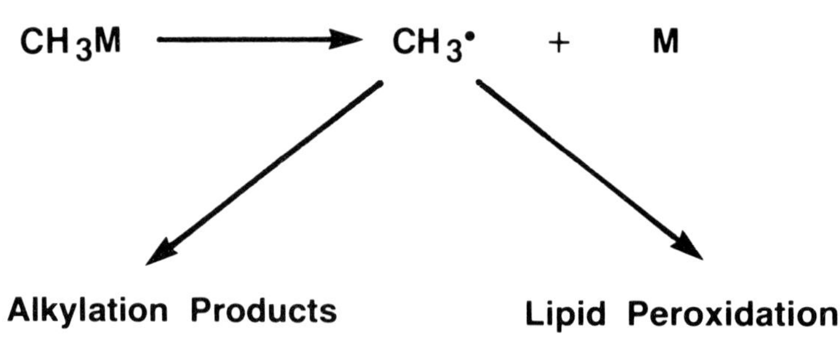

FIG. 10 The homolytic cleavage of methyl mercury resulting in alkylation or initiation of lipid peroxidation (for details see Ganther 1978).

methyl mercury can produce changes in membrane lipids in vivo.

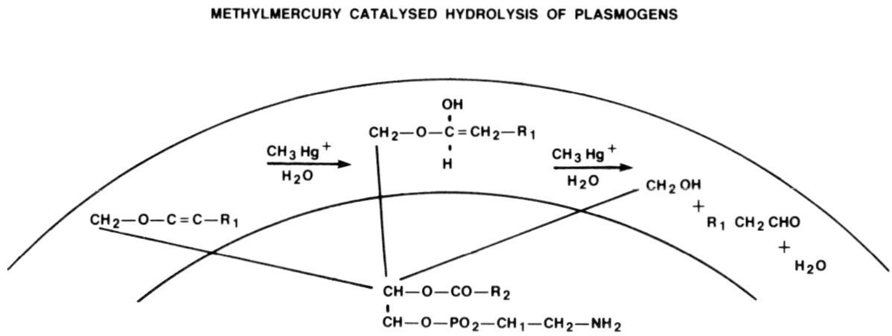

FIG. 11 The methyl mercury catalysed hydrolysis of a plasmogen (phosphatidyl ethanolamine) (adapted from Fig. 4 of Segall and Wood (1974)).

The above discussions with regard to both mercury vapor and methyl mercury are limited to effects on the mature nervous system. Actually, the most dramatic effects of methyl mercury are produced on the developing central nervous system. The early life cycle, particularly the prenatal stage in humans, is by far the most vulnerable to the effects of methyl mercury. The question arises as to why the developing central nervous system should be so specially sensitive, and to what extent the concept of a specific molecular target is helpful in understanding the toxic action of methyl mercury.

In fact, this area of research in mercury is at an exciting stage as there is a possibility that we can start talking about a specific molecular target. This putative target is the protein tubulin and the structures built from tubulin such as microtubules.

Methyl mercury produces a major deformity in the developing central nervous system. Prenatal exposure to methyl mercury produces microcephaly, that is to say an abnormally small head circumference. The brain is also being greatly disrupted. The normal layered structure of the cortex has been shown to be greatly disturbed in an infant suffering from prenatal mercury exposure (Choi et al. 1978). This disruption in the development of the brain in part may be due to an inhibition of the normal migration of neurons. For example, Choi et al. (1978) demonstrated that the Purkinje cells had failed to migrate to the outer surface of the cerebellum.

These two pathological effects of methyl mercury decreased head size and abnormal or deranged development of the brain might be explained by postulating the methyl mercury reacts with the protein tubulin and thereby destroys the microtubules in the cells. A considerable amount of investigation is now in progress to test this hypothesis. First of all, it has been shown in vitro that methyl mercury can indeed stop the assembly of microtubules from the protein tubulin (Sager et al. 1981) and depolymerise microtubules in vitro (Abe et al. 1975) and in fibroblast tissue culture (Sager et al. 1981).

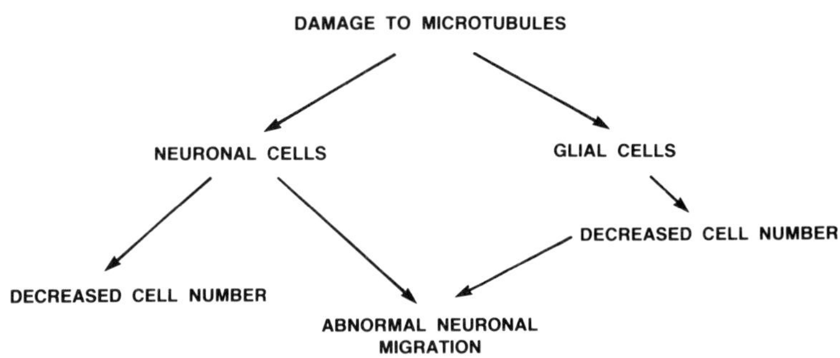

FIG. 12 Putative mechanism of prenatal brain damage.

Space will not allow a full discussion of all the detailed evidence but the hypothesis in its most general form is presented in Fig. 12. The interaction with tubulin causing the destruction of microtubules will result, first of all, in an inhibition of cell division. Microtubules are involved in the spindle apparatus necessary for the separation of the chromosomes. In fact, mitotic arrest does indeed occur in cells exposed to methyl mercury (Ramel 1968). More recently, late mitotic arrest has been demonstrated in the brains of neonatal mice by Sager et al. (1982). Microtubules are also necessary for the migration of neurons within the brain. Neuronal migration is also believed to be dependent upon the presence of support cells such as glial cells. Radial fibers from the glial cells act as a support and give direction for the migration of neuron cells to their ultimate destination in the mature brain

(Rakic 1972). The loss of these glial cells would in turn interfere with neuron migration.

3. CONCLUSION

In summary, research into the mechanism of action of methyl mercury on the developing CNS has reached an exciting stage. For the first time we have a molecular target to think about and one which offers a plausible explanation for the sensitivity of the developing central nervous system and for the type of effects seen, that is, failure of neurons to divide and failure of neurons to migrate.

There seems to be a variety of interactions of metals with molecular targets. The role of the putative molecular target may be very different for different metals. In some cases it may not be terribly important, for example, in the case of the selective accumulation of uranium in the proximal tubule of the kidney. For other metals, the molecular target plays a key role in explaining the toxicity of the metal such as in the reaction of trivalent arsenic with α-lipoic acid. There may be sequences of molecular targets. The carcinogenic action of chromium has been postulated to involve transport into the cell on a specific protein carrier for sulfate, its reduction to trivalent chromium by the enzyme P450, and the reaction of trivalent chromium with DNA. In a sense, all of these macromolecules are molecular targets. In some other cases the molecular target may already contain a metal, "an essential metal", which is displaced by the toxic metal. A molecular target might be an enzyme that normally adds a phosphate group to a substrate, and may be deceived into accepting an arsenate group. In addition to these main reactions, some others have been considered. It is not sure whether they play a role in the in vivo toxic action of metals, the oxidation reduction cycle of metals for instance; the possibility that the organometallic compounds might alkylate DNA or proteins; and the possibility that metals, whether inorganic metal or in organic form, might act as catalysts.

The value of the concept of the molecular target is not only in the possibility of explaining the signs and symptoms of poisoning but that it will stimulate more experimental work, such an example being the current work seeking to link the toxicity of methyl mercury to the developing central nervous system with the idea that tubulin is the molecular target.

ACKNOWLEDGEMENT

This paper is based on work performed under NIEHS Grants ES01247, ES01248 and NIGMS Grant GM25329.

REFERENCES

T. Abe, T. Haga and M. Kurokawa (1975) *Brain Res.*, 86, 504-508.

B.H. Choi, L.W. Lapham, L. Amin-Zaki and T. Saleem (1978) *J. Neuropath. Exper. Neurol.* 37, 719-733.

T.W. Clarkson, S. Halbach, L. Magos and Y. Sugata (1980) in *Molecular Basis of Environmental Toxicity* (Ed., R.S. Bhatnagher) Ann Arbor Science Pub., Ann Arbor, MI, pp. 419-427.

T.W. Clarkson and Z.A. Shaikh (1982) in *Nephrotoxicity Assessment and Pathogenesis* (Eds., P.H. Back, F.W. Bonner, J.W. Bridges and E.A. Lock) John Wiley and Sons, NY, pp. 263-279.

H.L. Evans, R.H. Garman and B. Weiss (1977) *Toxicol. Appl. Pharmacol.*, 41, 15-33.

H.E. Ganther (1978) *Environ. Health Perspect.*, 25, 71-76.

P.J. Gehring and P.B. Hammond (1967) *J. Pharmacol. Exp. Ther.*, 155, 187-201.

J.K. Grundt, E. Stensland and T.L.M. Syversen (1980) *J. Lipid. Res.*, 21, 162-168.

K.W. Jennette (1981) *Environ. Health Perspect.*, 40, 233-252.

F.W. Oehme (1972) *Clin. Toxicol.*, 5, 215-225.

P. Rakic (1972) *J. Comp. Neurol.*, 145, 61-84.

B. Ramazzini (1713) *Diseases of Workers*, Hafner Publ. Co., NY, pp. 34-43.

C. Ramel (1968) *Jpn. J. Exp. Med.*, 61, 1072-1076.

P.R. Sager, R.A. Doherty and J.B. Olmsted (1981) *The Toxicologist*, 1, 118-119.

P.R. Sager, R.A. Doherty and P.M. Rodier (1982) *Exp. Neurol.*, 77, 179-193.

H.J. Segall and J. M. Wood (1974) *Nature*, 248, 456-458.

D.C. Tosteson (1981) *Sci. Am.* 245, 164-174.

MAMMALIAN BIOTRANSFORMATION PROCESSES INVOLVING VARIOUS TOXIC METALLOIDS AND METALS

Paul Mushak

Department of Pathology, University of North Carolina at Chapel Hill, Chapel Hill, NC 27514 U.S.A.

1. INTRODUCTION

A number of toxic metalloids and metals are now known to undergo biotransformations such as oxidation/reduction and alkylation/dealkylation in humans and other mammals. While identification and characterization of these processes answer some questions concerning the toxicokinetics and toxicology of these substances, they raise additional questions about the mechanisms of toxic action. Most of the available information characterizing these transformations are of recent vintage since their elucidation has been dependent upon the availability of analytical methods which permit differentiation among the various metabolites generated.

In terms of both relevance to human toxicology and the amount of available information, biotransformation of the metalloids arsenic and selenium, the inorganic and organic forms of mercury, and the lower alkyl derivatives of tin (IV) and lead (IV) are of particular interest and are the subject of this review.

2. BIOTRANSFORMATION OF ARSENIC COMPOUNDS

Inorganic arsenicals in the tri- and pentavalent oxidation states are significant environmental pollutants and are associated with various acute and chronic adverse effects. With respect to biotransformation, we are primarily concerned with oxidation/reduction and biomethylation.

Studies of the biological transformation of inorganic arsenic date to the pioneering efforts of Challenger (1945; 1978) who first demonstrated that fungi can convert inorganic arsenic to the permethylated arsenical, trimethylarsine. In the interim, a number of microflora systems, anaerobic and aerobic, have been observed to carry out this reductive transformation (McBride et al., 1978).

That similar transformation occurs in mammalian systems was suggested by Braman and Foreback (1973), who employed speciation methodology to show that urinary arsenic in a group of human subjects consisted of a mixture of mainly dimethylarsinic acid (ca. 66 percent), monomethylarsonic acid (8 percent), arsenate (17 percent), and arsenite (8 percent). While this particular study did not permit any conclusions as to

Chemical Toxicology and Clinical Chemistry of Metals
ISBN 0.12.137520.X
Copyright © 1983 by IUPAC
All rights of reproduction in any form reserved

the source of the arsenicals, intake as such vs. transformation, subsequent investigations in humans where arsenic intake was clearly defined have confirmed that arsenic methylation occurs (Crecelius, 1977; Smith et al., 1977; Yamauchi and Yamamura, 1979; Tam et al., 1979a; Buchet et al., 1981a).

Biomethylation of inorganic arsenic as arsenate or arsenite has also been documented in a number of experimental animal species: dogs (Lakso and Peoples, 1975a; Tam et al., 1979b); cattle (Lakso and Peoples, 1975a); mice (Vahter, 1981); rats (Odanaka et al., 1980); and rabbits (Bertolero et al., 1981). By contrast, Vahter et al. (1982) have observed that the Marmoset monkey, alone among species studied so far, does not biomethylate inorganic arsenic.

From the available information, biomethylation of arsenic in mammals can be summarized as follows:

a) In humans and experimental animals, biomethylation of either tri- or pentavalent inorganic arsenic occurs to mainly yield dimethylarsinic acid and variable amounts of methylarsonic acid in a combined amount of 70-90 percent of the arsenic excreted in urine. The level of monomethyl arsenic is considerably higher in humans.

b) Reductive biomethylation in mammals apparently only proceeds to the step of dimethylarsinic acid formation or may proceed to formation of methylarsines which undergo oxidation before excretion, or quaternization may occur as in marine fauna.

c) The liver appears to be the major site of arsenic biomethylation based on both *in-vivo* (Rowland and Davies, 1982) and *in-vitro* data (Lakso and Peoples, 1975b; Shirachi et al., 1981; Lerman and Clarkson, 1983).

d) Since the methylated arsenicals, particularly dimethylarsinic acid, are much less toxic than their inorganic precursors and are rapidly excreted as such, biomethylation is viewed as a detoxification process. As a detoxification mechanism, biomethylation is only partially effective in the face of heavy inorganic arsenic exposure. One study suggests that methylation is relatively constant up to a level of ca. 500 ug/day, 5-day dosing (Buchet et al., 1981b).

e) Monomethylarsonic acid is only partially converted to dimethyl arsenic, which may be due to a difference in distribution *in-vivo*, since this form is apparently a precursor to dimethylarsinic acid.

The mechanism of biomethylation in mammalian systems has not been as well studied as in microbial systems, but available information suggests a similar pathway. In the pathway proposed for aerobic bacteria (Challenger, 1978; McBride et al., 1978):

$$\overset{V}{As}O_4^{3-} \xrightarrow[-O^{2-}]{2e} \overset{III}{As}O_3^{3-} \xrightarrow{CH_3^+} CH_3\overset{V}{As}O_3^{2-}$$

$$\xrightarrow[-O^{2-}]{2e} CH_3\overset{III}{As}O_2^{2-} \xrightarrow{CH_3^+} (CH_3)_2\overset{V}{As}O_2^{-} \xrightarrow[-O^{2-}]{2e}$$

$$(CH_3)_2\overset{III}{As}O^{-} \xrightarrow{CH_3^+} (CH_3)_3\overset{V}{As}O \xrightarrow[-O^{2-}]{2e} (CH_3)_3\overset{III}{As}$$

McBride et al. (1978) have determined that the methyl donor in aerobic methylation of arsenic by bacteria is S-adenosylmethionine, where the methyl group behaves as a carbonium ion and involves nucleophilic attack by arsenic precursors. This requires prior reduction of the arsenic precursor as seen in the above scheme. Shirachi et al. (1981) reported that arsenate is rapidly converted to dimethylarsenic by the supernatant fraction of rat liver, while monomethylarsonate undergoes methylation by both supernatant and mitochondrial fractions, suggesting the presence of two methyltransferases in rat liver. The isolated rat liver hepatocyte rapidly converts arsenite to dimethylarsinic acid, while little methylation occurs with arsenate, suggesting that arsenate is not transported as such across the cell membrane or must first be reduced to arsenite (Lerman and Clarkson, 1983).

Reduction of arsenate to arsenite, as required in the scheme given above, is consistent with the observations of Vahter (1981) that a) while both arsenate and arsenite undergo methylation in animals, the process is more rapid with arsenite; b) retention of arsenite is greater than arsenate, indicating that reduction of arsenate to arsenite competes with excretion of arsenate; and c) the retention of both forms of arsenic is dose-dependent.

Identification of arsenite formation with administration of arsenate to humans and animals has been reported, although the earlier literature supporting *in-vivo* reduction is inconclusive, owing to the analytical methods employed or ignorance of the presence of methylated forms. Recent studies confirm reduction in rats (Rowland and Davies, 1982) and humans (Yamauchi and Yamamura, 1979). Analysis of the forms of arsenic in either portal blood after dosing rats via the small intestine or venous blood after i.v. injection showed relatively rapid reduction of arsenate to arsenite, followed by linear rates of production of both monomethyl and dimethyl arsenic (Rowland and Davies, 1982). The authors ruled out artifactive conversion of one inorganic form to the other by adding valency-pure forms to samples carried through the analysis and finding that no reduction of arsenate to arsenite had occurred. In the study of Yamauchi and Yamamura (1979), a group of human subjects ingested a preparation validated as containing predominately arsenate and analysis of arsenical composition in urine revealed levels of arsenite, via reduction, which greatly exceeded the amounts ingested as minor component. The handling of the urine samples and the speciation techniques used indicated that such reduction was not an artifact of analysis.

3. BIOTRANSFORMATION OF SELENIUM

Chemical forms of selenium of interest in terms of mammalian biotransformation are the environmentally relevant selenate and selenite salts, elemental selenium and the selenoamino acids selenomethionine and selenocystine. Over the years, it has become evident that a common transformation pathway exists in mammals for metabolism of inorganic and certain organic selenium compounds. For example, exposure of laboratory animals to selenate, selenite or selenomethionine at varying dosage leads to excretion of selenium, as dimethylselenide, in expired air (Ganther, 1979; Bopp et al., 1982). Such data point to a reductive transformation, analogous to that for arsenic, rather than formation of oxidized metabolites seen in sulfur metabolism. This accords with the chemical properties of selenium as discussed by Williams (1979).

The significant features of *in vivo* transformation of selenium can be summarized as follows:

(a) At levels of exposure approximating physiological requirements, selenium given in the form of selenite, selenate, the seleno analogs of methionine and cystine, and seleniferous wheat is excreted in urine in the form of the trimethylselenonium ion and a second, uncharacterized seleno metabolite (Palmer et al., 1969, 1970; Nahapetian et al., 1983). With oral administration, the uncharacterized metabolite appears to be present in greater amount (Nahapetian et al., 1983). With increasing oral intake, the trimethylselenonium ion becomes the dominant metabolite.

(b) Still higher levels of selenium exposure to these forms lead to an increasingly greater fraction of selenium excreted through the lung as dimethylselenide (McConnell and Roth, 1966).

(c) Formation of trimethylselenonium ion and dimethyl selenide are held to be modes of detoxification, since these forms are much less toxic to the organism than their precursors.

(d) The relative amounts of trimethylselenonium ion or dimethylselenide formed varies with the form administered, at equivalent dosing, being highest with selenite, followed by selenate and then selenomethionine or selenocystine, suggesting that, in a common transformation pathway, metabolism of selenate to selenite and cleavage of selenium from the selenoamino acids is the rate-determining step.

(e) Recent data indicate that elemental selenium administered to experimental animals, via inhalation, gavage, nasal instillation or i.v. injection enters the organism as a transportable species with a subsequent tissue distribution similar to that for selenite, although the absorption rate for the latter is higher (Medinsky et al., 1981).

The mechanism of the reductive methylation pathway for metabolism of selenium compounds has been the subject of a number of studies and these have been reviewed (Diplock, 1976; Ganther, 1979; Magos and Webb, 1980; Bopp et al., 1982). Selenate is first reduced to selenite via the action of ATP, sulfurylase enzyme and glutathione (Dilworth and Bandurski, 1977; Ganther, 1979),

$$SeO_4^{2-} + ATP \xrightarrow[\text{Sulfurylase}]{} \text{Adenosine Selenophosphate} + PPi$$

$$Mg^{2+}$$

$$\xrightarrow[GSH]{} GSSeO_3^{2-} + AMP \xrightarrow[GSH]{} SeO_3^{2-} + GSSG$$

Selenite, assimilated as such or generated from selenate reduction, undergoes reduction to hydrogen selenide via the glutathione reductase pathway (Diplock, 1976; Ganther, 1979; Anundi et al., 1982):

$$SeO_3^{2-} + 4\,GSH \xrightarrow{2H^+} GS\text{-}Se\text{-}SG + GSSG + 3H_2O$$

$$GS\text{-}Se\text{-}SG \xrightarrow[-GSH]{NADPH} GS\text{-}Se\text{-}H \xrightarrow[-GSH]{NADPH} H_2Se$$

Formation of the intermediate glutathione selenotrisulfide appears to be nonenzymatic, while further conversion is under enzymatic control. This has been confirmed in the rat isolated hepatocyte by Anundi et al. (1982). Addition of selenite to the hepatocytes was associated with a decrease of intracellular reduced glutathione, an increase of the oxidized form, and decrease of NADPH levels. Pretreatment of the hepatocytes with the glutathione reductase inhibitor, 1,3-bis (2-chloroethyl) -1-nitrosourea, and addition of selenite resulted in markedly lowered levels of reduced glutathione.

Conversion of hydrogen selenide to methylated selenium metabolites involves at least two methyl transferases, the most active one in cytosol (kidney and liver) and a second in liver microsomes (Hsieh and Ganther, 1977). Microsomal methyltransferase activity is the more enhanced with increased selenium intake. The methyl donor is S-adenosylmethionine (Ganther, 1979). At moderate levels of selenium intake, methylating activity is adequate to allow formation of the trimethylselenonium ion but becomes less efficient at higher levels, leading to accumulation of dimethylselenide and excretion through the lungs.

The extreme sensitivity of the microsomal selenium methyltransferase activity to arsenite presumably accounts for reduced exhalation of dimethylselenide with co-administration of selenite and arsenite in experimental animals, while the concomitant enhancement of biliary clearance of selenium in the presence of arsenic, and vice versa, suggests formation of an arsenic-selenium complex.

The data of Medinsky et al. (1981) demonstrate that elemental selenium can be absorbed in the lung, upper respiratory and GI tracts and borne to tissues in a distribution pattern similar to that for selenite. Elemental selenium is polyatomic (an octameric ring) with a low vapor pressure, so that it is less diffusible than elemental mercury. On the other hand, elemental selenium is soluble in such organic solvents as carbon disulfide. Hence, it is not clear whether elemental selenium will diffuse into lipid material *in vivo* or must first be converted to a different valency state at the sites of absorption. It is possible that oxidation to a higher oxidation state analogous to the case with elemental mercury can occur via the primary complex of catalase, although degradation of the octameric ring can be envisioned as involving addition of 2 H atoms across Se-Se bonds, i.e., reduction, from e.g., glutathione.

Ontogenic dependency of selenium metabolism has not been well described. Ostadalova and coworkers (1982) have found that reductive methylation of selenite in young (10-day-old) vs. adult rats is considerably less on an equivalent dosing basis, as indexed by expiration rates of dimethylselenide and urinary excretion of trimethylselenonium ion. Since the time frame for these observations was relatively short, it is difficult to relate these differences to the known lower acute toxicity of selenite in young animals.

4. BIOTRANSFORMATION OF MERCURIALS

From the standpoint of mammalian biotransformations, forms of mercury which are of particular interest are a) elemental mercury vapor, b) the organomercurials phenylmercury(II), methoxyethyl mercury (II) and methyl mercury (II) salts, and c) mercuric ion. Since the transformation of these diverse forms of the element vary considerably, they are best discussed separately.

A. *In Vivo Oxidation of Elemental Mercury*

Inhalation of elemental mercury vapor by humans and experimental animals leads to rapid distribution to tissues, including the brain, because of its lipid solubility. It has long been known, however, that oxidation to the mercuric ion occurs, this form being responsible for the toxicity associated with mercury vapor inhalation (Clarkson, 1961; National Academy of Sciences, 1978). An early clue as to the mechanism of elemental mercury oxidation *in vivo* was the observation of Nielsen-Kudsk (1965) that ingestion of alcohol by human subjects in a study of the uptake of mercury vapor depressed pulmonary absorption of the element, indicating a linkage between ethanol metabolism and elemental mercury oxidation.

Over the past decade, a number of reports have appeared concerning the biochemical specifics of elemental mercury oxidation in humans and animals. These studies have been of several types: a) kinetic studies of mercury oxidation in erythrocytes; and b) studies of the effects of ethanol and the catalase inhibitor, 3-aminotriazole, on this oxidative process in various tissues.

These studies can be summarized as follows:

a) Oxidation of elemental mercury in erythrocytes and soft tissues of various mammalian species is mediated by the enzyme catalase, specifically the primary hydrogen peroxide complex functioning in the enzyme's peroxidative role (Halbach and Clarkson, 1978; Magos et al., 1978).

b) The apparent disparity between tissue catalase activity and corresponding mercury content in animals inhaling mercury vapor reflects a number of factors important to the oxidative transformation: (1) enzyme content of the tissue, (2) production of primary substrate, hydrogen peroxide, *in situ*, and (3) the entry rate of elemental mercury into the various tissues (Magos et al., 1978; Halback, 1981; Eide and Syversen, 1982).

c) While erythrocytes comprise a major site of elemental mercury oxidation, rapid diffusion of the element to various tissues can compete with oxidation in erythrocytes; subsequent oxidation to mercuric ion in various tissues then occurs (Halbach and Clarkson, 1978; Magos et al., 1978).

d) Ethanol inhibits oxidation of elemental mercury by catalase via substrate competition in the enzyme's peroxidative function (Cohen and Hochstein, 1964), while 3-aminotriazole is a potent inhibitor of catalase (Heim et al., 1956). In the presence of inhibitors, the level of mercury in liver rises, due to the fact that enzyme activity in liver never becomes rate-limiting; hence, the reduced level of oxidation in other tissues results in a higher delivery of elemental mercury to liver and a net higher amount of mercury converted to mercuric ion in this tissue.

e) Reduced oxidation of elemental mercury also occurs in acatalasemic mice, a mutant strain in which catalase content of tissues is greatly reduced.

As noted above, oxidation of zerovalent mercury occurs via the primary hydroperoxide complex (Compound II):

$$\text{E-Fe(III)-OH} + H_2O_2 \xrightarrow{-H_2O} \text{E-Fe}^{+}\text{(II)-OOH} \xrightarrow[-H_2O]{H+} \text{E-Fe}^{2+}\text{(IV)O}$$

$$(\text{E-Fe}^{2+}\text{(IV)O})\,(\text{Hg}^{\circ}) \longrightarrow \text{E-Fe}^{+}\text{(III)O-Hg}^{+} \longrightarrow \text{E-Fe(II)O-Hg}^{++}$$

$$\text{E-Fe(II)O-Hg}^{++} \xrightarrow{H+} \text{E-Fe(III)-OH} + \text{Hg}^{++}$$

The active complex is now held to be a complex with the ferryl iron (Compound II), by attachment of a proton and dehydration, producing an extremely reactive, highly oxidizing species in which the iron atom is formally tetravalent (Metzler, 1977). Stepwise transfer of two electrons from the initial complex of a mercury atom with Compound II can then occur. Protonation liberates mercuric ion and regenerates catalase.

B. Degradation of Organomercurials In Vivo

In vivo degradation of organomercurials in mammalian systems varies greatly as a function of the organic substituents; phenyl mercury and methoxyethyl mercury salts are rapidly and extensively degraded, while monomethyl mercury is more resistant to decomposition with variable dealkylation occurring across tissues.

The similarity of the behavior of phenylmercuric salts in terms of toxicokinetics and toxicity to the mercuric ion was first rationalized by Gage (1964), who observed that administration of the organomercurial to the rat resulted in levels of the mercurial in liver and kidney which were very low, ca. 3 and 10 percent respectively, of total mercury content. Hence, dearylation was almost complete, accounting for its similarity in behavior to mercuric ion.

Subsequently, Daniel and coworkers (1972) employed ^{14}C-labeled phenyl mercury to follow its metabolism in the rat. Generation of inorganic mercury and excretion of the latter in feces was accompanied by urinary excretion of labeled conjugates of phenol and quinol from the benzene liberated in degradation of the organomercurial. In parallel *in-vitro* studies, the soluble fraction of rat liver homogenates promoted dearylation, a process which was not NADH- or NADPH-linked, and produced benzene, which was determined as such by gas-liquid chromatography. These workers concluded that degradation was hydrolytic, rather than reductive, in nature. However, the data of Gage (1975) suggests reductive dearylation, since both ascorbate in the presence of copper (II) ion functioning aerobically and thiol-activated soluble proteins in rat liver homogenate promoted phenyl mercury degradation.

Administration of methoxyethyl mercury salts to experimental animals results in a mercury distribution pattern in tissues similar to that for mercuric and phenyl mercury salts (Ulfvarson, 1962), suggesting catabolism to inorganic mercury. The mechanism of this transformation in the rat has been studied by Daniel et al. (1971) using ^{14}C-labeled organomercurial. After s.c. administration of a single dose, the organomercurial was found to be rapidly degraded to ethylene and inorganic mercury. Taken with *in vitro* data showing chemical instability of the substance below pH 7.0, it was concluded that breakdown is nonenzymatic and occurs at sites such as muscle where pH would be low enough to promote chemical degradation. A simple chemical degradation of this organomercurial was also reported by Yonaha et al. (1975), who noted that oral administration to mice yielded a significant amount of inorganic mercury in stomach and small intestine, ca. 57 percent over the time frame studied.

Monomethylmercuric ion is relatively more resistant to *in vivo* degradation than those forms described earlier and this greater stability is considered a factor in differing toxicokinetics, e.g., entry into the brain, and toxicity, i.e., central and peripheral neurotoxicity. The available literature, however, indicates demethylation does occur, the extent of which is tissue-dependent.

Within mammalian systems, there appear to be two types of degradation processes involving methylmercury: a) demethylation by intestinal microflora, and b) variable demethylation within soft tissues. The end result of demethylation in the intestine is to decrease methyl mercury retention by interrupting enterohepatic circulation of the mecurial and increasing the fecal excretion of the inorganic mercury thus generated.

Norseth and Clarkson (1971), in their study of methyl mercury metabolism in the rat, concluded that intestinal flora were involved in demethylation of this agent. Enteric degradation of methyl mercury has been reported in the rat (Rowland et al., 1980), in mice (Nakamura et al., 1979; Seko et al., 1981), ruminant animals (Kozak and Forsberg, 1979), and human feces (Edwards and McBride, 1975). These investigations in mice and rats involved changes seen, vs. controls, in the catabolism rate in germ-free animals, animals treated with antibiotics, or after caecal resection.

That enteric bacteria constitute but one mechanism for *in vivo* breakdown of methylmercury is clear from available data. Lodgment of inorganic mercury in tissues after administration of methyl mercury does not reflect enteric degradation, per se, since the inorganic form is poorly absorbed from the GI tract. Secondly, reduction of enteric catabolism of the organomercurial by, for example, use of antibiotics (e.g., Rowland et al., 1980) is still associated with significant levels of inorganic mercury in liver and kidney.

With methyl mercury administration, the relative amount of inorganic vs. total mercury varies across tissues. Norseth and Clarkson (1970) observed that, in rats, highest relative proportions of inorganic mercury were in the kidney, liver and large intestine, while the corresponding values for brain never exceeded 4 percent. Comparable data have subsequently been reported for mice (Norseth, 1971), monkeys (Evans et al., 1977), guinea pigs (Iverson et al., 1974), and cats (Charbonneau et al., 1976). Kidney levels of inorganic mercury reflect both *in situ* transformation and movement from other organs. In this regard, Suzuki (1976) demonstrated *in vitro* degradation of methyl mercury by both kidney and liver slices.

The ontogenic dependency of methyl mercury degradation has been the subject of several studies. Landry (1981) noted that mice at 20 days of age demethylate the mercurial at a rate 5-fold greater than animals at 4 days of age. This is similar to what is seen in young rats (Konishi, 1978). This difference reflects a lower microorganism count (Konishi, 1978) and possibly less-developed transformation pathways in tissues of the newborn.

The mechanism of methyl mercury degradation in tissues and its relationship to its toxicity, specifically its neurotoxicity, is still open to debate. It does appear clear that demethylation does not involve the hepatic microsomal P-450 mediated monooxygenase system which does operate with other organometals (see next section). Nakayama (1976) saw no evidence *in vitro* of increased mercurial demethylation by microsomes from phenobarbital-treated rats. Chang and Desnoyers (1978) found that the mercurial was inhibitory for microsomal enzymes, with a rapid, transitory increase in enzyme activity with mercurial treatment being ascribed to early membrane damage.

Evidence continues to accumulate, however, that the degradation of methyl mercury is related to its toxicity via the formation of free radical intermediates which produce injury by interaction with lipid membranes, directly or indirectly.

a) Both selenium and vitamin E protect against methyl mercury toxicity (e.g., Ganther, 1980; Chang et al., 1977, 1978) as does the potent synthetic antioxidant, N,N' - diphenyl-p-phenylene diamine (Welsh, 1979).

b) Methyl mercury inhibits glutathione peroxidase in blood and other tissues (Chang and Suber, 1982; Yonaha et al., 1983). Selenium supplementation with mercurial administration prevents inhibition of the glutathione peroxidase and associated morphological changes in brain (Chang and Suber, 1982).

c) The recent studies of Yonaha et al. (1983) show that exposure of rats to methyl mercury resulted in greatly increased lipid peroxidation in liver and kidney, the rate of peroxidation paralleling the rate of depletion of glutathione peroxidase activity and levels of ascorbate. Vitamin E levels were not affected. As noted earlier, both liver and kidney demonstrate significant methyl mercury degradation. Since vitamin E levels are unchanged, hydrogen peroxide interaction with both mercurial and membranes may be the key.

d) Selenium is effective in protecting against methyl mercury toxicity at levels well below that which would involve direct, stoichiometric interaction (Ganther, 1980). This would accord with a protective role via increased glutathione peroxidase biosynthesis.

According to Ganther (1978; 1980), glutathione peroxidase protects against toxicity by preventing the interaction of hydrogen peroxide with methyl mercury and/or suppressing formation of radical intermediates while vitamin E serves as a radical scavenger. The synthetic antioxidant N,N'-diphenyl-p-phenylenediamine plays a similar protective function. Supporting the role of glutathione peroxidase is the observation of Konishi (1979) that hydrogen peroxide *in-vitro* readily degrades methyl mercury.

The question of the relationship of methyl mercury transformation to CNS toxicity via lipid peroxidation is of special interest. Norseth and Clarkson (1970) and other workers have noted that demethylation of the mercurial in brain is low. Measurement of simple ratios of inorganic to total mercury over time, however, could disguise a higher rate of transformation if the inorganic mercury formed is cleared from the nervous system at a higher rate than methyl mercury, i.e., the decay in mercury content of brain over time is mainly inorganic mercury formation and loss. In this regard, clearance half-time from brain of the mercuric form (inhaled as vapor elemental vapor and oxidized) was reported in humans to be 21 days (Hursh et al., 1976) vs. up to 70 days for the half-time of methyl mercury (Aberg et al., 1969).

Magos and Webb (1980) argue against the free-radical theory of Ganther by noting that ethyl mercury is degraded to a greater extent in brain without increase in neurotoxicity.

The key may well be methy/ethyl mercury inhibition of glutathione peroxidase activity, the increased hydrogen peroxide inducing lipid peroxidation and the formation of hydroxy radical from auto-catalysis. The radical then attacks RHg with degradation. Since radical substitution is favored for ethyl over methyl mercury, more degradation ensues for the former. The protective role for selenium and Vitamin E is preserved, and the point made by Magos and Webb is answered. The scheme is thus consistent will all the available elements of the debate.

A. *Reduction of Mercuric Ion to Elemental Mercury*

In vivo reduction of mercuric ion to elemental mercury occurs in mammalian systems, as first observed by Clarkson and Rothstein (1964), indicating the existence of an oxidation-reduction cycle. Reduction, at least in short-term studies appears to be a minor factor in excretion of mercury (Clarkson and Rothstein, 1964; Dunn et al., 1981a) although a long-term effect on distribution of mercury may be significant.

Given the discussion earlier on the effects of various factors on the oxidation of elemental vapor in mammalian systems, one would expect that such agents would affect the cycle by enhancing net reduction over re-oxidation. This clearly occurs with administration of ethanol to rats also treated with mercuric ion (Dunn et al., 1978) or use of acatalasemic mice and the catalase inhibitor, 3-aminotriazole (Sugata and Clarkson, 1979).

Reduction of mercuric ion reportedly occurs in mouse liver and kidney homogenates, with the highest reduction activity seen in the supernatant fraction of liver (Dunn et al., 1981b). Heat treatment markedly inhibited the transformation, suggesting an enzymatic process.

5. BIOTRANSFORMATION OF LOWER ALKYL DERIVATIVES OF LEAD AND TIN

Ethyl and methyl derivatives of lead (IV) and tin (IV) are of particular interest in this discussion given their status as environmental pollutants and the fact that the trialkyl analogs are potent neurotoxicants (Grandjean and Nielsen, 1978; World Health Organization, 1980). Since the lower alkyl derivatives of both lead and tin undergo similar transformations, they are best discussed together.

The tetraethyl- and tetramethyl derivatives of both lead (IV) and tin (IV) are as toxicologically active as their corresponding trialkyl compounds, due to their rather rapid conversion to the trialkyl form *in vivo* (Cremer, 1958, 1959; Bolanowska and Garczynski, 1968; Hayakawa, 1972).

Further transformation of the trialkyl lead and tin compounds occurs, although this is much slower.

Cremer (1958) concluded that triethyltin does not undergo further degradation, since this was the only metabolite claimed to be detected in animals treated with tetraethyl tin, using a two-step extraction/colorimetry procedure. This procedure would not rule out hydroxylated structures, however. Subsequently, Kimmel et al. (1977) reported that triethyltin undergoes hydroxylation *in-vitro* , using rat liver homogenates, with subsequent destannylation. Since Bridges (1967) noted that diethyltin is rapidly metabolized to monoethyltin with formation of ethane, the gist of these two studies is that transformation of tetraethyltin to at least the monoalkyl form can occur.

Similarly, triethyl lead undergoes slow conversion to products of dealkylation in rats (Bolanowska, 1968; Hayakawa, 1972), including inorganic lead. Ramstoeck et al. (1980a,b) have established that triethyl lead in rats is slowly dealkylated with the formation of ethane and ethylene. Generation of the hydrocarbons was enhanced in animals pretreated with phenobarbital. By 18 hours, decomposition with hydrocarbon formation amounted to ca. 3 percent in animals without barbiturate pre-treatment and ca. 6 percent with such treatment.

The mechanism of transformation of these agents appears to involve the hepatic microsomal P-450 mediated monooxygenase system (Casida et al., 1971; Kimmel et al., 1977; Ramstoeck et al., 1980a). With *in vitro* systems, hydroxylation of the ethyl group appears to occur to some extent prior to alkyl group cleavage (Kimmel et al., 1977), while the work of Ramstoeck et al. (1980a) indicates that the microsomal system generates ethyl radicals.

6. OVERVIEW

The net result of the various transformations of metalloid and metal examples described here is the alteration of intrinsic toxicity, although the direction of such change(s) is not always readily apparent. In the case of elemental mercury vapor, oxidation to mercuric ion clearly increases toxicity since the elemental form is quite inert. In the case of inorganic arsenicals or selenium compounds, reductive methylation would, on the surface, indicate detoxification since the methylated forms are of a lower order of toxicity. Transformation of phenylmercury and methoxyethyl mercury can be held to be beneficial if the absence of degradation would confer neurotoxic properties on these agents. As such, the inorganic mercury produced is toxic in its own right. Methyl mercury degradation poses a mixed interpretation. Enteric transformation is clearly beneficial, since the inorganic mercury formed is excreted rather than absorbed, thus intervening in the enterohepatic circulation of the organomercurial. The case of degradation in tissues is less clear. Since some hold that one mechanism of methylmercury toxicity is membrane damage via generation of methyl radicals in the course of degradation, then tissue transformation is adverse. It is to be hoped that this issue will continue to be studied.

As noted above, a number of the transformations described are held to be detoxifying processes. However, it can be argued that formation of a less toxic and/or rapidly excreted metabolite is but one facet of the whole metabolic process involved in such transformations.

The reductive methylation of arsenic and selenium and the oxidation of elemental mercury (and the reduction of mercuric ion) engage the metabolic machinery of the cell as substrates, which not only adds a new dimension to how toxicants such as the examples used impart their effects (compared to, say, binding to physiologically important sites such as sulfhydryl groups) but suggests that the increasing diversion of activity to transformation at increasing levels of exposure to the toxicants can only be adverse to native cellular processes, e.g., depletion of S-adenosylmethionine. At high levels of exposure to inorganic arsenic and selenium compounds, it is clear that biotransformation is overwhelmed and various acute and chronic effects become manifest. The apportioning of these adverse effects to untransformed toxicants or total usurpation of cellular processes in transformation *in-vivo* cannot be readily done, however.

Several studies are of interest in this regard. As noted earlier, the Marmoset monkey is unique among mammals in that arsenical reductive methylation does not occur. While this results in a relatively greater retention of arsenic, as arsenite, it would be of great interest to compare relative toxicity with species where transformation occurs. The study of Ostadalova and coworkers (1982) documents that the young rat does not show the level of reductive methylation seen in the adult. Whether this is related to the fact that young animals are less sensitive to the acute effects of selenite is an open question. Several studies with the isolated rat hepatocyte (Hogberg and Kristoferson, 1979; Anundi et al., 1982) make it clear that cellular lysis is a consequence of reductive methylation, due in part to a depletion of NADPH (Anundi et al., 1982).

REFERENCES

B. Aberg, L. Ekman, R. Falk, U. Greitz, G. Person and J.O. Snihs (1969) *Archives of Environmental Health*, 19, 478-484.

I. Anundi, J. Hogberg and A. Stahl (1982) *Archives of Toxicology*, 50, 113-123.

W. Bolanowska (1968) *British Journal of Industrial Medicine*, 25, 203-208.

W. Bolanowska and H. Garczynski (1968) *Medical Practice*, 19, 235-243.

F. Bertolero, E. Marafante, J. Edel-rade, R. Pietra and E. Sabbioni (1981) *Toxicology*, 20, 35-44.

B.A. Bopp, R.C. Sonders and J.W. Kesterson (1982) *Drug Metabolism Reviews*, 13, 271-318.

R.S. Braman and C.C. Foreback (1973) *Science*, 182, 1247-1249.

J.W. Bridges, D.S. Davies and R.T. Williams (1967) *Biochemical Journal*, 105, 1261-1266.

J.P. Buchet, R. Lauwerys and H. Roels (1981a) *International Archives of Occupational and Environmental Health*, 48, 71-79.

J.P. Buchet, R. Lauwerys and H. Roels (1981b) *International Archives of Occupational and Environmental Health*, 48, 111-118.

J.E. Casida, E.C. Kimmel, B. Holm and G. Widmark (1971) *Acta Chimica Scandinavica*, 25, 1497-1499.

F. Challenger (1945) *Chemical Reviews*, 36, 315-361.
F. Challenger (1978) in *Organometal and Organometalloids: Occurrence and Fate in the Environment* (Eds, F.E. Brinkman and J.M. Bellama), American Chemical Society, Washington, D.C. Pp. 1-22.
L.W. Chang and P.A. Desnoyers (1978) *Journal of Environmental Pathology and Toxicology*, 1, 569-579.
L.W. Chang and R. Suber (1982) *Bulletin of Environmental Contamination and Toxicology*, 29, 285-289.
L.W. Chang, A.W. Dudley, M.A. Dudley, H.E. Ganther and M.L. Sunde (1977) in *Neuro-Toxicology* (Eds, L. Roizin, M. Shiraki and N. Grcevic), Raven Press, New York, p. 225.
S.M. Charbonneau, G.K.H. Tam, F. Bryce and B. Collins (1978a) *Trace Substances in Environmental Health*, 12, 276-283.
S.M. Charbonneau, I.C. Munro, E.A. Nera, F.A.J. Armstrong, R.F. Willes, R. Bryce and R.F. Nelson (1978b) *Interim Reports of Toxicology*, 5, 337-349.
T.W. Clarkson, J. Gatzy and C. Dalton (1961) *University of Rochester AEC Report*, No. 582.
T. Clarkson and A. Rothstein (1964) *Health Physics*, 10, 1115-1121.
G. Cohen and P. Hochstein (1964) *Biochemistry*, 3, 895-900.
E.A. Crecelius (1977) *Environmental Health Perspectives*, 19, 147-150.
J.E. Cremer (1958) *Biochemical Journal*, 68, 685-692.
J.E. Cremer (1959) *British Journal of Industrial Medicine*, 16, 191-199.
J.W. Daniel, J.C. Gage and P.A. Lefevre (1971) *Biochemical Journal*, 121, 411-415.
J.W. Daniel, J.C. Gage and P.A. Lefevre (1972) *Biochemical Journal*, 129, 961-697.
G.L. Dilworth and R.S. Bandurski (1977) *Biochemical Journal*, 163, 521-529.
A.T. Diplock (1976) *CRC Critical Reviews in Toxicology*, 4, 271-329.
J.D. Dunn, T.W. Clarkson and L. Magos (1978) *British Journal of Industrial Medicine*, 35, 241-244.
J.D. Dunn, T.W. Clarkson and L. Magos (1981a) *Journal of Pharmacology and Experimental Therapeutics*, 216, 19-23.
J.D. Dunn, T.W. Clarkson and L. Magos (1981b) *Science*, 213, 1123-1125.
T. Edwards and B.C. McBride (1975) *Nature*, 253, 462-464.
I. Eide and T.L.M. Syverson (1982) *Acta Pharmacologica et Toxicologica*, 51, 371-376.
H.L. Evans, R.H. Garman and B. Weiss (1977) *Toxicology and Applied Pharmacology*, 41, 15-33.
J.C. Gage (1964) *British Journal of Industrial Medicine*, 21, 197-202.
J.C. Gage (1975) *Toxicology and Applied Pharmacology*, 32, 225-238.

H.E. Ganther (1978) *Environmental Health Perspectives*, 25, 71-76.
H.E. Ganther (1979) *Advances in Nutritional Research*, 2, 107-128.
H.E. Ganther (1980) *Annals of the New York Academy of Sciences*, 355, 212-226.
Phillippe Grandjean and Torben Nielsen (1979) *Residue Reviews*, 72, 97-148.
S. Halbach (1981) *Journal of Applied Toxicology*, 1, 303-307.
S. Halbach and T.W. Clarkson (1978) *Biochimica et Biophysica Acta*, 523, 522-531.
K. Hayakawa (1972) *Japanese Journal of Hygiene*, 26, 526-535.
W.G. Heim, D. Appleman and H.T. Pyfrom (1956) *American Journal of Physiology*, 180, 19-23.
H.S. Hsieh and H.E. Ganther (1977) *Journal of Nutrition*, 106, 1577-1582.
J.B. Hursh, T.W. Clarkson, M.E. Cherian, J.J. Vostal and R.V. Mallie (1976) *Archives of Environmental Health*, 31, 302-309.
F. Iverson, R.H. Downie, H.L. Trenholm and C. Paul (1974) *Toxicology and Applied Pharmacology*, 27, 60-69.
E.C. Kimmel, R.H. Fish and J.E. Casida (1977) *Journal of Agriculture and Food Chemistry*, 25(1), 1-8.
T. Konishi (1978) *Chudokoken Ho*, 11, 20-22.
T. Konishi (1979) *Chudokoken Ho*, 12 , 14-16.
Stan Kozak and C.W. Forsberg (1979) *Applied and Environmental Microbiology*, 38, 626-636.
J.U. Lakso and S.A. Peoples (1975a) *Journal of Agriculture and Food Chemistry*, 23, 674-676.
J.U. Lakso and S.A. Peoples (1975b) *APHA Academy of Pharmacological Sciences Abstracts*, 5, 159.
T.D. Landry (1981) *Dissertation Abstracts International*, 41, 2950B.
S.A. Lerman and T.W. Clarkson (1983, in press) *Fundamental and Applied Toxicology*.
B.C. McBride, H. Merilees, W.R. Cullen and W. Pickett (1978) in *Organometals and Organometalloids: Occurrence and Fate in The Environment* (Eds, F.E. Brinckman and J.M. Bellama), American Chemical Society, Washington, D.C. Pp. 94-115.
K.P. McConnell and D.M. Roth (1966) *Proceedings of the Society for Experimental Biology and Medicine*, 123, 919-921.
L. Magos and M. Webb (1980) *CRC Critical Reviews in Toxicology*, 8, 1-42.
L. Magos, S. Halbach and T.W. Clarkson (1978) *Biochemical Pharmacology*, 27, 1373-1377.
M.A. Medinsky, R.G. Cuddihy and R.O. McClellan (1981) *Journal of Toxicology and Environmental Health*, 8, 917-928.
D.E. Metzler (1977) *Biochemistry. The Chemical Reactions of Living Cells*, Academic Press, New York, p. 572.
A.T. Nahapetian, M. Janghorbani and V.R. Young (1983) *Journal of Nutrition*, 113, 401-411.

I. Nakamura, K. Hosokawa, H. Tamura and T. Miura (1977) *Bulletin of Enviornmental Contamination and Toxicology*, 17, 528-533.
National Academy of Sciences (1978) *Assessment of Mercury in the Environment*, Washington, D.C.
F. Nielsen-Kudsk (1965) *Acta Pharmacologica et Toxicologica*, 23, 263-274.
M. Nokayama (1976) *Kumamoto Medical Journal*, 29, 95-109.
T. Norseth (1971) *Acta Pharmacologica et Toxicologica*, 30, 172-176.
T. Norseth and T.W. Clarkson (1970) *Archives of Environmental Health*, 21, 717-727.
T. Norseth and T.W. Clarkson (1971) *Archives of Environmental Health*, 22, 568-577.
Y. Odanaka, O. Matano and S. Goto (1980) *Bulletin of Environmental Contamination and Toxicology*, 24, 452-459.
I. Ostadalove, A. Babicky and J. Kopoldova (1982) *Archives of Toxicology*, 49, 247-252.
I.S. Palmer, D.D. Fischer, A.W. Halverson and O.E. Olson (1969) *Biochimica Biophysica Acta*, 177, 336-342.
I.S. Palmer, R.P. Gunsalus, A.W. Halverson and O.E. Olson (1970) *Biochimica Biophysica Acta*, 208, 260-266.
E.R. Ramstoeck, W.G. Hoekstra and H.E. Ganther (1980a) *Chemical-Biological Interactions*, 32, 243-247.
E.R. Ramstoeck, W.G. Hoekstra and H.E. Ganther (1980b) *Toxicology and Applied Pharmacology*, 54, 251-257.
I.R. Rowland and M.J. Davies (1982) *Journal of Applied Toxicology*, 2, 294-299.
I.R. Rowland, M.J. Davies and J.G. Evans (1980) *Archives of Environmental Health*, 35, 155-160.
Y. Seko, T. Miura and M. Takahashi (1982) *Acta Pharmacologica et Toxicologica*, 50, 117-120.
D.Y. Shirachi, J.U. Lakso and L.J. Rose (1981) *Proceedings of the Western Pharmacological Society*, 24, 159-160.
T.J. Smith, E.A. Crecelius and J.C. Reading (1977) *Environmental Health Perspectives*, 19, 89-93.
Yasuo Sugata and T.W. Clarkson (1979) *Biochemical Pharmacology*, 28, 3474-3476.
T. Suzuki (1976) *Tokuku Journal of Experimental Medicine*, 120, 361-363.
K.H. Tam, S.M. Charbonneau, E. Bryce and G. LaCroix (1979) *Toxicology and Applied Pharmacology*, 50, 319-322.
U. Ulfvarson (1962) *International Archiv fur Gewerbepathologie*, 19, 412.
M. Vahter (1981) *Environmental Research*, 25, 286-293.
M. Vahter, E. Marafante, A. Lindgren and L. Dencker (1982) *Archives of Toxicology*, 51, 65-77.
S.O. Welsh (1979) *Journal of Nutrition*, 109, 1673-1681.

R.J.P. Williams (1978) in *New Trends in Bio-Inorganic Chemistry* (Eds, R.J.P. Williams, and J.R.R.F. de Silva), Academic Press, New York, pp. 253-260.

World Health Organization (1980) *Tin and Organotin Compounds: A Preliminary Review*, United Nations Environment Programme and the World Health Organization, Geneva, Environmental Health Criteria 15, 109 pp.

H. Yamauchi and Y. Yamamura (1979) *Japanese Journal of Industrial Health*, 21, 47-54.

R.K. Yip and L.W. Chang (1982) *Environmental Research*, 28, 84-95.

M. Yonaha, S. Ishikura and M. Uchiyama (1975) *Chemical and Pharmaceutical Bulletin*, 23, 1726-1732.

M. Yonaha, M. Seito and M. Sagai (1983) *Life Sciences*, 32, 1507-1514.

INTERACTIONS BETWEEN LEAD AND CALCIUM IN ISOLATED BONE CELL POPULATIONS

John F. Rosen

Department of Pediatrics, Albert Einstein College of Medicine, Montefiore Medical Center, Bronx, NY 10467 U.S.A.

1. INTRODUCTION

Bone is the major reservoir of lead in humans (Barry, 1975), yet little information is available concerning the metabolism of lead in bone. Metabolic studies using radiotracers of lead in adults have indicated that short and long-lived reservoirs of lead in bone are sources of significant amounts of lead in whole blood and soft tissues (Rabinowitz *et al.*, 1976). The concentration of lead in long bones and in tooth dentine correlate closely with the presence of lead nephropathy in adults and impairment of school performance in children, respectively (Rosen, 1983a). Further definition of lead metabolism in bone appears necessary to understand clinical and biochemical toxicity of lead at critical soft tissue sites.

To understand the complex interactions between calcium, phosphate, lead-chelating agents and calcium-regulating hormones, we have reported studies of lead transport in bone organ culture (Rosen and Markowitz, 1980). We have described and defined, in part, a rapidly exchangeable bone lead compartment that is regulated, to an extent, by the same ions and hormones that control bone mineral metabolism. Specifically, the lead-chelating action of $CaNa_2EDTA$ (EDTA) is enhanced by low medium calcium concentrations, parathyroid hormone (PTH) and 1,25-dihydroxyvitamin D_3. These actions of EDTA were inhibited by increasing medium concentrations of calcium, phosphate and calcitonin. To characterize this subcompartment of exchangeable bone lead at the cellular level, bone cell populations, derived from mouse calvaria, were enriched for osteoclasts (OC) and osteoblasts (OB) by sequential collagenase digestion (Rosen, 1983b).

2. MATERIALS AND METHODS

From this digestion of 75-100 calvaria, 6 separate bone cell populations were cultured and grew to confluency by 6-7 days; then, individual cell populations were subcultured in multiwells

Chemical Toxicology and Clinical Chemistry of Metals
ISBN 0.12.137520.X
Copyright © 1983 by IUPAC
All rights of reproduction in any form reserved

to which fresh medium, hormones and tracers were added. Isolated bone cell populations were characterized as OC or as OB according to specific biochemical parameters: parathyroid hormone stimulates acid phosphatase activity and hyaluronate synthesis in OC cells; PTH inhibits alkaline phosphatase activity and the decarboxylation of citrate in OB cells (Wong et al., 1978; Rosen, 1983b).

3. RESULTS AND DISCUSSION

The time-dependent uptake of lead, particularly in OC cells, was characterized by an initial period of rapid uptake, which most likely represents initial binding to plasma membranes and influx into the cell interior. This is followed by a second, more prolonged phase of accumulation in OC cells, that is completed after 15-18 hours of incubation. By this time, OC cells reached an equilibration state in which the amount of lead uptake into the cell interior and release are about equal. In contrast, OB cells evidence a plateau soon after incubation with ^{210}Pb; and there is little, if any, additional uptake by these cells.

OB cells evidence very little increase in lead accumulation as media lead concentrations were increased from 6.5 to 65 μM during an incubation period of 15 hours. In contrast, the ability of OC cells to accumulate tracer was relatively unlimited.

To ascertain the amount of incorporated lead that was exchangeable, cells were pre-loaded with ^{210}Pb for 24 hours, washed and suspended in medium without tracer. The amount of radioactivity remaining in the cells, compared to time = 0, was measured. The overall release of tracer was greatest from OC but a significant amount of release was demonstrable for both cell types. These in vitro data indicate that a significant fraction of bone cell lead, like bone cell calcium, is readily exchangeable. It is apparent that OC cells are the predominant cell type involved in the in vitro metabolism of bone lead after exposure to low concentrations of the metal. In this regard, during experimental lead intoxication in vivo, the prevalence of lead-protein complexes in osteoclasts reflects, in degree, the concentration of lead in bone (Hsu et al., 1973).

OC cells became the focus of experiments to characterize possible interactions between cellular lead and calcium metabolism. Similar membrane channels for lead and calcium have been proposed in other cell systems. The response of OC cells to physiological concentrations of PTH was examined in media that contained a relatively low concentration of lead - 6.5 μM. As reported previously, a rapid rise and subsequent fall in OC cell calcium uptake was observed. During this response to PTH, there was also an increase in ^{210}Pb uptake - during a time course that was initially similar to that of calcium; thereafter, ^{210}Pb accumulation persisted (Fig. 1). It is reasonable to suggest that PTH increases OC membrane permeability for lead as for calcium. Though lead influx appears to be closely coupled to that of calcium, initially at least, we have not defined in this study the intracellular localization or subsequent metabolism of lead after PTH treatment. These studies are being carried out using desaturation techniques that will enable us to characterize lead compartments within OC cells.

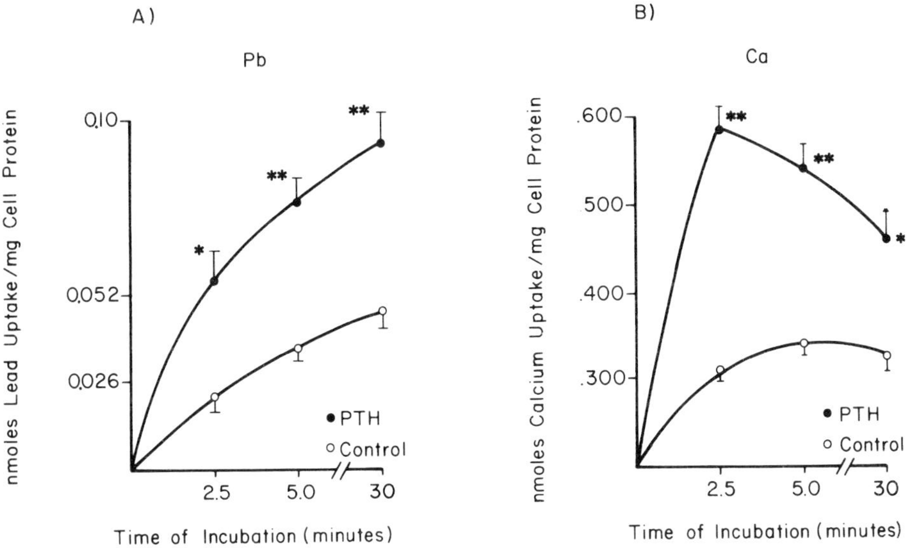

FIG. 1 Effects of PTH on lead and calcium uptake in OC cells. In the same experiment, fresh medium containing PTH (100 ng/ml), 6.5 μM ^{210}Pb and ^{45}Ca (3.0 $\mu Ci/ml$) were added simultaneously. After various incubation periods, cells were collected and cell lead and calcium uptake was measured. Values are mean ±SE. * = $p < .01$, PTH-treated vs. control cells; ** = $p < .001$.

We have shown further that the uptake of ^{210}Pb in OC cells was responsive to graded concentrations of PTH; it is unlikely that these PTH-induced increases in cellular lead are the result of non-specific protein-cell interactions.

The possibility was also considered that lead ion might modify cellular calcium metabolism, as reported in hepatocyte cultures (Pounds *et al.*, 1982). Increasing concentrations of medium lead, at or above 26 μM, in the absence of PTH, induced marked exaggeration or overamplification of calcium uptake by OC cells. Compared to PTH, there was a marked overamplification of calcium uptake at medium lead concentrations above 26 μM. These summarized data are in marked contrast to those presented previously, where physiological responses, in cell calcium uptake, were observed to a naturally occurring hormonal peptide (Rosen, 1983a, 1983b). It will be of interest to determine if OC cells, exposed to lead at medium concentration >26 μM and thereby pre-loaded with calcium, will respond normally to PTH and other humoral stimuli.

4. CONCLUSIONS

In conclusion, the results of these studies (Rosen, 1983a, 1983b) indicate that: 1) Uptake of ^{210}Pb by OB and OC cells is rapid. Quantitatively, OC cells are the predominant cell type active in the metabolism of lead in this monolayer system. OB cells evidence relatively little avidity for lead, compared to OC cells. 2) The uptake of ^{210}Pb by OC is almost linear, as medium lead concentrations increase. 3) Significant release of label occurs within short time periods. 4) PTH effects an increase in ^{210}Pb and ^{45}Ca uptake. Pb accumulation continues, as calcium uptake decreases. 5) Increasing media concentrations of lead, above 26 μM, markedly exaggerate calcium uptake - far above that produced by PTH.

It is tempting to speculate that: 1) Mediated incorporation of lead occurs in bone cells which is likely to involve changes in membrane permeability produced by hormonal stimuli, such as PTH. 2) Modulations in cellular calcium metabolism induced by lead at relatively low concentration may be an early and discrete toxic effect of lead common to different cell types. 3) Such pertubations in cell calcium metabolism have the potential of disturbing stimulus-response cascades of many hormone-target cell systems that depend on calcium as a second messenger.

5. ACKNOWLEDGEMENT

This study was supported by NIH Grant #ES01060-08.

6. REFERENCES

P.S.I. Barry (1975) *Brit. J. Industr. Med.* 32:119-1139.

F.S. Hsu, L. Krook, J.N. Shively and J.R. Duncan (1973) *Science* 181:447-448.

J.G. Pounds, R. Wright, D. Morrison and D.A. Casciano (1982) *Toxicol. Appl. Pharmacol.* 63:389-401.

M.B. Rabinowitz, G.W. Wetherill and J.D. Kopple (1976) *J. Clin. Invest.* 58:260-270.

J.F. Rosen and M.E. Markowitz (1980) *Pediatr. Res.* 14:330-335.

J.F. Rosen (1983a) in *Dietary and Eivironmental Exposure To Lead* (Ed. K.R. Mahaffey), Elsevier Biomedical, New York, Amsterdam, and (1983b) *Toxicol. Appl. Pharm.* In press.

G.L. Wong, G.N. Kent, K.Y. Ku and D.V. Cohn (1978) *Endocrinol.* 103:2274-2282.

CELLULAR METABOLISM OF ZINC: A KINETIC ANALYSIS IN CULTURED RAT HEPATOCYTES

Joel G. Pounds and David R. Morrison

Food and Drug Administration
National Center for Toxicological Research
Jefferson, AR 72079, U.S.A.
and
Division of Interdisciplinary Toxicology
University of Arkansas for Medical Sciences
Little Rock, AR 72205, U.S.A.

1. INTRODUCTION

The zinc ion has multiple essential functions including the maintenance of enzyme configuration or catalytic function and stabilization of membranes. Lead intoxication decreased tissue zinc levels, and oral zinc supplementation ameloriated some of the toxic effects of ingested lead (Petering, 1980). In vivo and in vitro zinc reversed the inhibition of delta-amino levulinic acid dehydratase (Finelli et al., 1975; Haeger-Aronsen et al., 1976). The use of primary cell cultures provides an experimental tool for investigating the organotypic homeostasis, metabolism, function or toxicity, and interactions of essential and toxic elements. The steady state homeostasis of metal salts can be analyzed mathematically by kinetic analysis of radioisotope washout curves and the cellular metabolism of the element described by compartmental analysis in terms of pool sizes, rate constants, fluxes, and halftimes of exchange. This approach is particularly useful in that it provides a quantitative, noninvasive, integrated measure of whole cell metabolism. The pH, ionic, substrate gradients and compartmentation of cell function are maintained, providing an integrated analysis of the multitude of kinetic processes which occur in the intact cell (isotopic exchange, diffusion, active transport, etc.). Since the liver has a central role in the short-term homeostatic regulation and storage of zinc for systemic use, the present studies were performed to provide a kinetic model of cellular zinc metabolism for evaluating lead-zinc interactions.

Chemical Toxicology and Clinical Chemistry of Metals
ISBN 0.12.137520.X
Copyright © 1983 by IUPAC
All rights of reproduction in any form reserved

2. METHODS

Hepatocytes were obtained from male Sprague-Dawley rats (250-350 g) by in situ retrograde perfusion of collagenase, and cultured in Williams' medium E (WE). The kinetic analysis was conducted in a manner similar to previous experiments using lead-210 (Pounds, et al., 1982). Briefly, following adaptation to culture (2-4 hr) the cells were labeled with zinc-65 (1 µCi/ml) in serum-free WE supplemented with 2 mU/ml insulin, 2 mM glutamine, 20 µg/ml gentamicin, and 10 µM Zn as zinc sulfate. After 20 hours, the labeling medium was removed and the cells quickly washed 4 times with sucrose-HEPES buffer. Fresh WE without the radionuclide was added to each culture and the steady-state efflux of zinc-65 was followed for 210 minutes. The fraction of the radioactivity remaining in the cells with time for each culture was fit to a three-term, sum-of-exponential equation and the kinetic parameters derived from the coefficients and exponents by general compartmental modeling procedures.

3. RESULTS AND DISCUSSION

The goodness of fit of the data to a sum of exponential equation, as determined from the sum of squares of residuals, was significantly improved by assuming a three-term, as compared to a two-term exponential equation. The fit was not further improved by a four-term exponential equation. This information indicates that hepatocytes contain three kinetically distinct pools of zinc.

To identify the presence of kinetic compartments which were associated with the glycocalyx and extracellular face of the plasma membrane, cultures were labeled for 20 hr in 10 µM zinc (zinc-65). Triplicate cultures were washed once with 0.5 mM EGTA in sucrose-HEPES buffer at pH 7.0, or in buffer alone, for 30 sec. The cells were washed three additional times with sucrose-HEPES. The kinetic pool S_1 was reduced in size indicating the association of a small amount of zinc (<0.2 nmol/mg cell protein) with the extracellular surface of the hepatocyte (Table 1).

To correlate zinc thionein, an integral component of hepatocellular zinc metabolism, cells were labeled with zinc-65 in medium containing 10 µM Zn^{2+} with or without dexamethasone (1 µM). The dexamethasone treatment increased the size of the most slowly exchanging zinc compartment, S_3, by 60% of the control data (Table 2). This observation indicates that zinc bound to metallothionein, a major storage protein for zinc in the liver, is associated with the slowest exchanging pool, S_3.

TABLE 1

Effect of EGTA Rinse on the Kinetic Distribution of Zinc-65 in Cultured Rat Hepatocytes

Pool Size (nmol/mg protein)	Sucrose-HEPES	EGTA (0.5 mM)
S_1	0.282 ± .012	*0.095 ± .009
S_2	0.220 ± .017	0.237 ± .007
S_3	0.661 ± .057	0.635 ± .024
S_{Total}	1.164 ± .083	*0.966 ± .029

Rat hepatocytes were incubated in serum-free WE containing 10 μM zinc (1 μCi zinc-65/ml) for 20 hr. The cultures were washed with sucrose-HEPES with or without EGTA to chelate extracellular zinc. The data were obtained from kinetic analysis of zinc-65 washout curves as described in METHODS and represent the mean ± SEM of triplicate cultures from a representative experiment. *$p<0.05$ by Students' t test.

TABLE 2

Effect of Dexamethasone on the Kinetic Distribution of Zinc-65 in Cultured Rat Hepatocytes

Pool Size (nmol/mg protein)	Control	Dexamethasone (1 μM)
S_1	0.251 ± .016	0.266 ± .019
S_2	0.240 ± .014	0.271 ± .029
S_3	0.603 ± .049	*0.988 ± .105
S_{Total}	1.14 ± .092	*1.525 ± .116

Rat hepatocytes were incubated in serum-free WE containing 10 μM zinc (1 μCi zinc-65/ml) alone or with 1 μM dexamethasone for 20 hrs. The data were obtained from kinetic analysis of zinc-65 washout curves as described in METHODS and represent the mean ± SEM of triplicate cultures from a representative experiment. *$p<0.05$ by Students' t test.

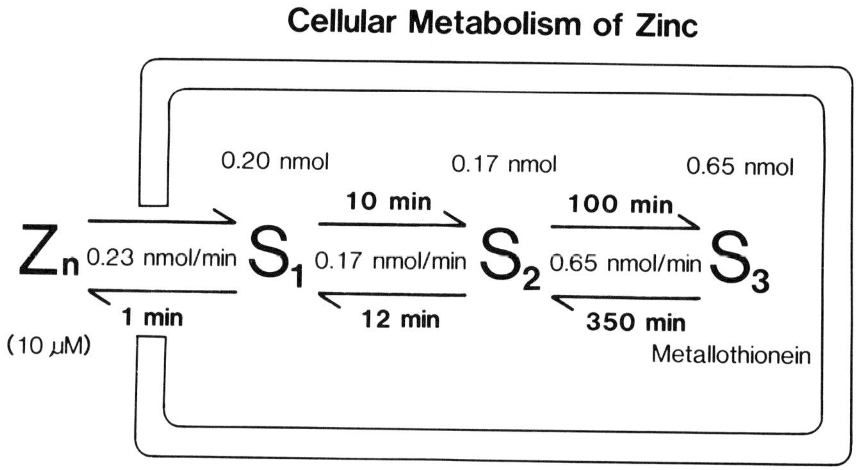

Fig. 1. Kinetic model of the steady-state metabolism of zinc in the cultured rat hepatocyte. All data are normalized to 1 mg cell protein. Data represent the mean of experiments in serum-free Williams' medium E containing 10 μM zinc (1 μCi zinc-65/ml).

Figure 1 summmarizes the kinetic data into an anatomic model of the steady-state cellular metabolism of zinc in cultured rat hepatocytes. At least part of S_1 appears to be associated with the plasma membrane while S_3, which was increased in size by the glucocorticoid, dexamethasone was presumed to include that cellular zinc associated with metallothionein. The three kinetic compartments can alternatively be interpreted as physical phase transitions. The kinetic pool S_1 would represent zinc in solution as zinc ions, soluble salts or chelates, S_2 would represent the low affinity zinc-enzyme complexes, and S_3, zinc metalloenzymes, including metallothionein. This approach should provide useful means to evaluate the role of the cell microenvironment, e.g., vitamins, hormones, and toxic elements in the hepatocellular homeostasis of zinc.

4. REFERENCES

R.J. Cousins (1979) Nutr. Rev., 37, 97-103.

V.N. Finelli, D.S. Lauder, M.A. Karaffa and H.F. Petering (1975) Biochem. Biophys. Res. Comm., 65, 303-311.

B. Haeger-Aronsen, M. Abdulla and A. Schutz (1976) Arch. Environ. Hlth., 31, 215-220.

H.G. Petering (1980) Ann. N.Y. Acad. Sci., 51, 298-308.

J.G. Pounds, R. Wright and R.L. Kodell (1982) Toxicol. Appl. Pharmacol., 66, 88-101.

EFFECT OF METHYL MERCURIC CHLORIDE AT NANOMOLAR CONCENTRATIONS ON DRUG METABOLISM IN ISOLATED RAT HEPATOCYTES

A. I. Ally, J. G. Pounds* and K. R. Reuhl

Ecotoxicology Group, National Research Council of Canada, Ottawa K1A 0R6, Canada
**National Center for Toxicological Research, FDA, HHS, Jefferson, Arkansas 72079, U.S.A.*

1. INTRODUCTION

Caffeine (1,3,7 trimethylxanthine) is extensively metabolized *in vivo* and *in vitro* by N-demethylation. There is a growing awareness that the toxicity of many compounds may be mediated via an alteration in the mixed function oxidase system (MFO), the enzyme activity most often involved being the cytochrome P450. The metabolism of theophylline (1,3 dimethylxanthine) has become a useful probe for the study of human MFO *in vivo* (Jenne, 1982).

Methyl mercury (MMC) is an environmental contaminant which achieved worldwide prominence as a result of the incident at Minamata Bay, Japan. MMC toxicity has been associated with changes in the activity of enzymes such as Na-K ATPase. It has been reported that inorganic mercury inhibited rat hepatic MFO system (Alvares *et al.*, 1972). Conflicting data showing an induction of cytochrome P450 and NADPH cytochrome C reductase has been reported by Pekkanen and Salmien (1972) after a single dose of methyl mercury in mice.

2. METHODS

Hepatocyte isolation and culture: Hepatocytes were isolated from male Sprague-Dawley rats by *in situ* perfusion of collagenase and cultured as previously described (Pounds *et al.*, 1982). Briefly adapted to culture for 2-4 hours and then treated with 0, 2.0 ηM (0.5 ηg/ml), or 4.0 ηM (1.0 ηg/ml) methyl mercuric chloride and caffeine (5 μg/ml) or theophylline (5μg/ml). All experiments were conducted in Williams' medium E supplemented to contain 10% fetal bovine, 2 mU/ml insulin, 2 mU/ml L-glutamine, and 20 μg/ml gentamicin sulfate. After 5 hours of treatment, the culture medium was collected, deproteinized with TCA (2.5% final), filtered (0.45 μM) and stored at -70°C until analysis.

Chemical Toxicology and Clinical Chemistry of Metals
ISBN 0.12.137520.X
Copyright © 1983 by IUPAC
All rights of reproduction in any form reserved

HPLC analysis: Caffeine, theophylline, and their metabolites were separated by HPLC using a Merck C-18 radial compression column. The methylxanthines were separated using a tertiary solvent system: 5 ηM sodium acetate, methanol, acetonitrile, tetrahydrofuran, acetic acid (91.5: 3.2:2.8:1.7:1; v/v/v/v/v; pH 5) at a flow rate of 1.5 ml/min and monitored at 268.5 ηM. The retention times were (min): 1,3-dimethyluric acid (5.60), theobromine (6.06), paraxanthine (8.57), theophylline (9.15), caffeine (15.16).

3. RESULTS AND DISCUSSION

We found that cultured rat hepatocytes interconvert theophylline and caffeine. The methylation of the theophylline 7 nitrogen to give caffeine was prevented by the presence of 4.0 ηM methyl mercury in the cell culture medium. When used as the substrate, caffeine was demethylated to yield theophylline. There was no detectable oxidation of theophylline to 1,3-dimethyluric acid in this study. In the presence of 4.0 ηM methyl mercury, the demethylation of caffeine was blocked.

The metabolism of caffeine and theophylline *in vivo* has been well characterized and proceeds via a combination demethylation and oxidation. Caffeine can be demethylated in positions 1, 3 or 7 giving rise to theobromine (3,7 dimethylxanthine), paraxanthine (1,7 dimethylxanthine) or theophylline (1,3 dimethylxanthine). The dimethylxanthines found depend not only on the demethylation but also on subsequent metabolism by alternative metabolic pathways. There are species differences in the profile of dimethylxanthines produced *in vivo* from caffeine. The rat metabolizes caffeine to theophylline (7.7%), theobromine (13%) and paraxanthine (22.5%) whereas man demethylates caffeine to theophylline (6.5%), theobromine (9%) and paraxanthine (38%). These metabolites are found primarily in the blood with minor quantities being excreted in the urine. Tu *et al.*, (1982) found that all the caffeine metabolites produced by primary rat hepatocytes were in the medium fraction. Thus in studies using hepatocytes to study methylxanthine metabolism the culture medium can be analyzed directly for these compounds.

The mechanism for the inhibitory action of methyl mercury is unknown. Theophylline elimination is unaffected by phenobarbital in man (Piafsky *et al.*, 1977), however metabolism is enhanced by treatment with polycyclic aromatic compounds (Lohman and Miech, 1976; Betlach and Tozer, 1980) and by tobacco smoking in man (Powell *et al.*, 1977; Jusko *et al.*, 1978) suggesting that cytochrome P448 is involved in methylxanthine biotransformation. Aldridge *et al.* (1979) suggested that both cytochrome P450 and P448 are involved in methylxanthine metabolism. If these are the MFOs which metabolize the methylxanthines then methyl mercury inhibits both cytochrome P450 and

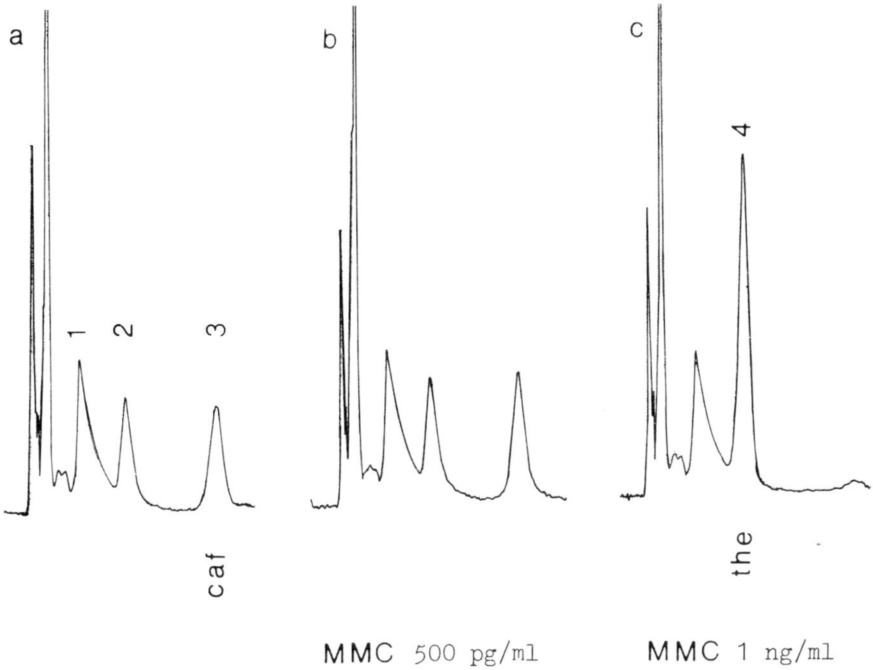

FIG. 1. HPLC analysis of rat hepatocyte methylation of theophylline and the effect of methyl mercury. Panel a: control, peaks 1 and 2 are endogenous to the culture media, peak 3 is the caffeine produced through methylation of the 7 position nitrogen of theophylline. Panel b: in the presence of methyl mercury 500 pg/ml (2.0 ηM). Panel c: in the presence of 1 ng/ml methyl mercury (4.0 ηM), peak 4 is the theophylline superimposed on the endogenous coeluting substance in the media.

P448. Such an inhibition would account for the inhibition of the demethylation of caffeine to theophylline in this study. The enzymatic processes responsible for methylation of theophylline to produce caffeine are presently unknown. It is however known that S-adenosylmethionine is the methyl donor during the *in vitro* methylation of theophylline in human fetal liver homogenate. It is therefore possible that methyl mercury could inhibit both the MFO and the methyl transferase to produce the inhibition seen in this study.

4. REFERENCES

Aldridge, A.A., Aranda, J.V. and Niems, A.H. (1979). *Clin. Pharm. Ther.* 25, 447-452.

Alvares, A.P., Leigh, S., Cohn, J. and Kappas, A. (1972). *J. Exp. Med.*, 135, 1406-1409.

Betlach, C.J., Tozer, T.M. (1980). *Drug Metab. Dispos.*, 8, 271-273.

Jenne, J.W. (1982). *Chest*, 81, 529-530.

Jusko, W.J., Schenlag, J.J., Clark, J.H., Gardener, M. and Yurchak, A.M. (1978). *Clin. Pharmacol. Ther.*, 24, 406-410.

Lohman, S.M., Miech, R.P. (1976). *J. Pharmacol. Exp. Ther.*, 186, 213-225.

Pekkanen, T.J. and Salmien, K. (1973). *Acta Pharmacol. et Toxicol.*, 32, 289-293.

Piafsky, K.M., Sitar, D.S. and Ogilvie, R.I. (1977). *Clin. Pharmacol. Ther.*, 22, 336-340.

Pounds, J.G., Wright, R. and Kodell, R.L. (1982). *Toxicol. Appl. Pharmacol.*, 66, 88-101.

Powell, J.R., Thiercelin, J.F., Vozeh, S., Sansom, L. and Riegelman, S. (1977). *Am. Rev. Resp. Dis.*, 116, 17-23.

Tu, A.S., Robertson, R.S., Callahan, M.M. and Thayer, P.S. (1982). *Mutation Res.*, 95, 475-492.

UPTAKE OF CADMIUM AND COPPER IN ISOLATED RAT HEPATOCYTES

W.S. Din and J.M. Frazier

Division of Environemtnal Toxicology,
School of Hygiene and Public Health,
The Johns Hopkins University,
Baltimore, MD 21205 U.S.A.

1. INTRODUCTION

Cadmium is known to cause both acute and chronic toxicity in humans. When laboratory animals were exposed to cadmium intraveneously, more than 90% of the injected dose was cleared from the plasma within 15 minutes and 50-70% of the dose initially accumulated in the liver (Frazier, 1980). Uptake of cadmium into the liver plays an important role in the systemic distribution and therefore, toxicity of the heavy metal. However, the mechanism of this uptake process remains unknown. Frazier et. al. (1976) have studied the kinetics of hepatic cadmium transport in isolated perfused liver and suggested that cadmium uptake into the liver occurred by a combination of simple diffusion and a carrier mediated process. Stacey et. al. (1980) have further studied the cadmium uptake in isolated rat hepatocytes. In their study, cadmium uptake exhibited bi-phasic uptake kinetics over the dose range of 25 to 800 μM. However, these concentrations were quite high and carrier-mediated transport may have been masked. Furthermore, subtle toxicity due to cadmium may occur. The objective of this study was to investigate the kinetics of cadmium uptake at low concentrations and its metabolism after a 15 minute-pulse exposure in isolated hepatocytes. In addition, cadmium kinetics were compared to that of an essential metal, copper. Copper was selected since excess copper is also taken up by the liver, but is readily excreted via the bile, or incorporated into ceruloplasmin. It was of interest to compare the metabolism of these metals in the same system, in order to gain some insights into the differences in vivo.

2. METHODS

(a) Hepatocyte Preparation

Male Wistar rats (200-300 g) from Charles River (Wilmington,

Chemical Toxicology and Clinical Chemistry of Metals
ISBN 0.12.137520.X
Copyright © 1983 by IUPAC
All rights of reproduction in any form reserved

Mass.) were used for hepatocyte preparation, according to the method of Seglen (1976), with slight modifications. Viability was assessed by both the trypan blue exclusion test and leakage of the cytosolic enzyme, lactate dehydrogenase. Only preparations which have viability greater than 90% were used for experiments.

(b) Uptake of Metal

Hepatocyte preparations in synthetic RPMI media 1640 (Gibco Lab., N.Y.), supplemented with 10% dialyzed bovine serum albumin and insulin (0.09 I.U./ml), were exposed either to 3 to 18 μM stable Cd^{2+} or to 1.6 to 16 μM Cu^{2+}, labelled with a constant specific activity of Cu^{67}. After a 15 minute-incubation period, cells were harvested, washed(2X), and homogenized to determine the total metal uptake.

(c) Cytosolic Distribution of Metal and Induction of Metallothionein

After the 15 minute-exposure period, cells were harvested, washed and resuspended in fresh media supplemented with ^{35}S-cysteine. At 0.5, 1, 3 and 6 hours after the commencement of incubation, cells were harvested and cytosol was prepared by centrifugation of the homogenate at 30,000 X g for 20 minutes, followed by 100,000 X g for 1 hour. Cytosolic distribution of the metals and induction of metallothionein (MT) were determined by analyzing the Sephadex G-75 column profile of the cytosol.

3. RESULTS

Both cadmium and copper uptake increase as the metal concentration in the media increases (Fig. 1). At the concentrations studied, both uptake mechanisms are saturable. For cadmium, 19 nmoles were accumulated per 3×10^7 cells at the lowest dose, and 34 nmoles were accumulated per 3×10^7 cells at the highest dose. For copper, 4.0 nmoles were accumulated at the lowest dose and 12.5 nmoles at the highest dose. At equal molar concentrations of the two metals, cadmium uptake is 3.5 times greater than copper uptake. Presence of preinduced zinc-MT in hepatocytes had no effect on the uptake kinetics of either metal.

Cells were harvested at 0.5, 1, 3 and 6 hours after a 15 minute-pulse exposure to 4.5 μM of cadmium. Cytosol was prepared and chromatographed on Sephadex G-75 and MT quantitated (Fig. 2). In hepatocytes exposed to 4.5 μM cadmium, there was a lag period of 30 minutes before MT content increased from 1 to 6 hour (Fig. 3). The induction of MT shifts the binding of cadmium from high molecular weight protein to the MT pool.

No synthesis of MT could be detected at the concentrations of copper studied.

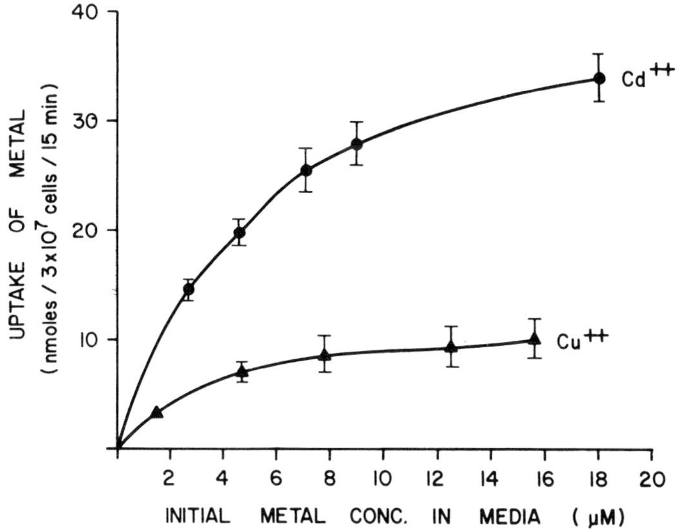

FIG. 1 Uptake of cadmium and copper in isolated rat hepatocytes. The data are expressed as nmole of metal accumulated per 15 minutes of exposure, normalized to 3 X 10^7 cells. The cell number was verified by quantitative determination of the DNA content in each sample (Burton, 1962). Each data point represents mean ± S.D. of four hepatocyte preparations.

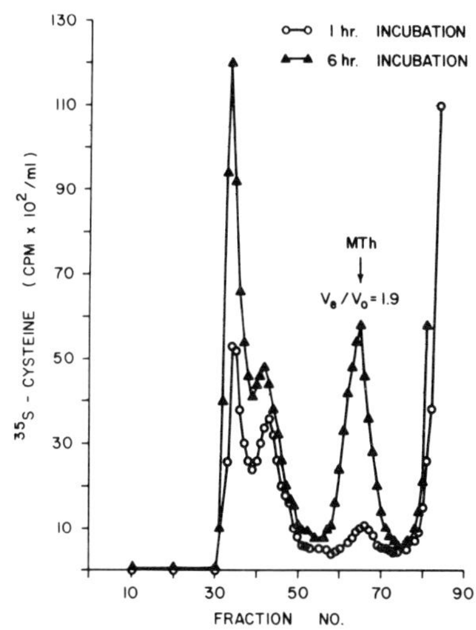

FIG. 2 Cytosolic distribution of ^{35}S-cysteine in hepatocytes which were pulsed-exposed to 4.5 μM of cadmium for 15 minutes and fractionated on Sephadex G-75 column.

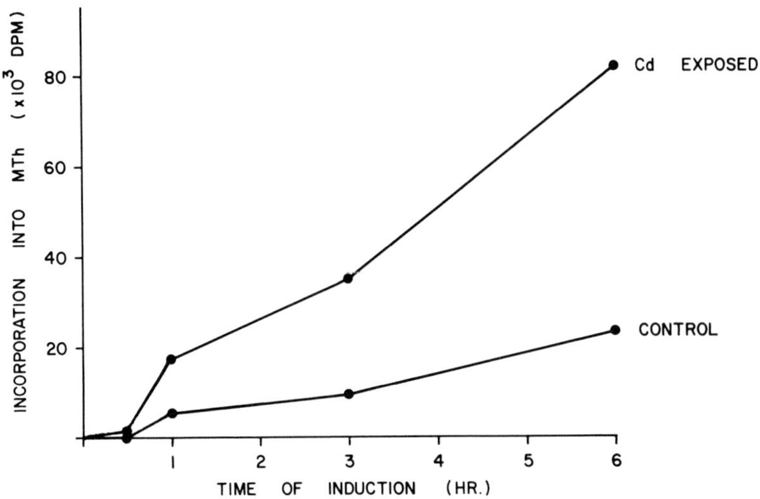

FIG. 3 Time course of net ^{35}S-cysteine incorporation in cytosolic MT after hepatocytes were pulse-exposed to 4.5 uM of cadmium for 15 minutes.

4. CONCLUSIONS

Both cadmium and copper exhibit saturable kinetics for their uptakes at the concentrations studied. Presence of preinduced zinc-MT in hepatocytes had no effect on the uptake kinetics of either metal. These data suggest that uptake of metal is limited at the membrane transport level at physiological concentration. If simple diffusion were the dominant mechanism, presence of excess binding capacity in the cytosol would have acted as a sink for the metal in the cell and enchanced uptake. Once cadmium enters the cells, there is a subcellular redistribution within 6 hours, in correlation with MT induction. Synthesis of MT shifts the binding of cadmium from high molecular weight proteins to MT pool. No MT synthesis can be detected at the concentrations of copper studied. It suggested that at the cellular level, cadmium may be a significantly stronger MT inducer than copper. In order to validate this hypothesis, the effect of these metals on protein synthesis must be investigated to rule out the possible toxic effect of the metals.

5. REFERENCES

K. Burton (1962) Biochem. J., 62, 315-322.

J.M. Frazier and B.S. Kinsley (1976) Tox. Appl. Pharmacol., 38, 583-593.

J.M. Frazier (1980) J. Tox. Env. Health, 6, 503-518.

P.O. Seglen (1976) Methods Cell Biol., 13, 29-83.

N.H. Stacey and C.D. Klaassen (1980) Tox. Appl. Pharmacol., 55, 448-455.

EFFECTS OF NICKEL CHLORIDE AND DIETHYLDITHIOCARBAMATE ON METALLOTHIONEIN IN RAT LIVER AND KIDNEY

F. William Sunderman Jr. and Carol B. Fraser

Departments of Laboratory Medicine and Pharmacology
University of Connecticut School of Medicine
263 Farmington Ave., Farmington, CT 06032 U.S.A.

1. INTRODUCTION

The goals of this study were to elucidate the dose-effect relationships for $NiCl_2$-induction of metallothionein (MT) in rat liver and kidney, and to determine if $NiCl_2$-induction of MT is blocked by actinomycin D, cycloheximide, or sodium diethyldithiocarbamate (DDC), a chelating agent for Ni[II]. This extended abstract is a condensation of a report by Sunderman and Fraser (1983).

2. MATERIALS AND METHODS

The test substances were ultra-pure nickel chloride ($NiCl_2 \cdot 3H_2O$); actinomycin D: cycloheximide; and sodium diethyldithiocarbamate, recrystallized according to Baselt et al (1977). The experimental animals were 90 male rats of the Fischer-344 strain (200 to 300 g) housed in polypropylene cages and fed Purina rat chow and water ad libitum. There were 5 to 8 rats in each experimental group. Control rats were treated with NaCl vehicle solution (0.15 mol/litre, 0.3 to 0.5 ml/rat) by the same routes and at the same times that test rats were treated with the test substances. Food was removed from the cages 17 h before the rats were killed by inhalation of diethyl ether. The livers and kidneys were perfused with NaCl solution (0.15 mol/litre, 4°C), and homogenized in sucrose solution (0.25 mol/litre, 4°C) by four strokes of a motor-driven teflon-glass tissue homogenizer (1 g of liver or 0.5 g of kidney to 4 ml of sucrose solution). The homogenates were centrifuged at 18,000 g for 20 min at 4°C; the supernatant cytosol was used for MT assay by the Cd-saturation method of Onosaka and Cherian (1982). Analysis of Cd in the final MT-extract was performed by flame atomic absorption spectrophotometry at 228.8 nm, using a dual-channel atomic absorption spectrometer (Jarrell-Ash model 810) with background compensation at 226.6 nm non-resonance line

Chemical Toxicology and Clinical Chemistry of Metals
ISBN 0.12.137520.X
Copyright © 1983 by IUPAC
All rights of reproduction in any form reserved

TABLE 1

Measurements of Metallothionein (MT) Concentrations in Rat Liver (L) and Kidney (K)

Group	Treatments: compound, route, number of injections, dose, and h between last injection and sacrifice	No of rats	MT conc.[a] (μg/g tissue) Mean ± SD
A	Controls, vehicle sc, 1x, -6.5 h	7	L 49 ± 15 K 136 ± 6
B	Controls, vehicle sc, 1x, -17 h	6	L 39 ± 10 K 120 ± 8
C	Controls, vehicle ip, 4x[a], -72 h	5	L 46 ± 9 K 118 ± 7
D	$NiCl_2$, sc, 1x, 0.75 mM/kg, -6.5 h	7	L 129 ± 49[c] K 191 ± 24[c]
E	$NiCl_2$, sc, 1x, 0.25 mM/kg, -17 h	6	L 156 ± 38[c] K 239 ± 49[c]
F	$NiCl_2$, sc, 1x, 0.50 mM/kg, -17 h	6	L 226 ± 23[c] K 246 ± 32[c]
G	$NiCl_2$, sc, 1x, 0.75 mM/kg, -17 h	7	L 318 ± 58[c] K 276 ± 75[c]
H	DDC, im, ix, 1.33 mM/kg, -17.2h	5	L 296 ± 54[c] K 113 ± 28
I	$NiCl_2$, sc, 1x, 0.25 mM/kg, -17 h & DDC, im, 1x, 1.33 mM/kg, -17.2	7	L 356 ± 37[c] K 208 ± 19[c]
J	Actinomycin D, ip, 3x[b]	6	L 59 ± 24 K 166 ± 62
K	$NiCl_2$, sc, 1x, 0.25 mM/kg -17 h, & Actinomycin D, ip, 3x[b]	8	L 143 ± 59[c] K 285 ± 66[c]
L	Cycloheximide, ip, 1x, 2 mg/kg, -18 h	5	L 49 ± 21 K 131 ± 12
M	$NiCl_2$, sc, 1x, 0.25 mM/kg -17 h & Cycloheximide, ip, 1x, 2 mg/kg, -18 h	5	L 92 ± 43[cd] K 170 ± 14[cd]
N	$NiCl_2$, ip, 4x[c], 0.1 mM/kg, -72 h	5	L 66 ± 17[e] K 175 ± 40[e]
O	$CdCl_2$, ip, 4[a], 0.1 mM/kg, -72 h	5	L 752 ± 226[c] K 428 ± 169[c]

[a]Four daily injections; sacrifice 72 h after last injection
[b]Actinomycin D, total dose = 1 mg/kg, (0.5 mg/kg -18 h, 0.25 mg/kg at -16 and -14 h)
[c]$p < 0.001$ versus corresponding controls
[d]$p < 0.05$ versus corresponding values for $NiCl_2$ alone (Group E)
[e]$p < 0.05$ versus corresponding controls

of Cd. Based upon 100 consecutive duplicate analyses of MT concentrations in livers and kidneys from 18 control rats and 32 rats that received injections of $NiCl_2$ at various dosages, the within-run coefficient of variation of MT-analysis was 9.8% (mean MT concentration: 90 μg/g).

3. RESULTS

Administration of $NiCl_2$ to rats by three treatment schedules caused significant induction of MT in rat liver and kidney (Table 1). Dose-effect relationships were observed for $NiCl_2$-induction of MT in liver and kidney of rats killed 17 h after sc injection of $NiCl_2$. At $NiCl_2$ dosages of 0.25, 0.50 and 0.75 mmol/kg, hepatic MT concentrations increased 4.0-fold, 5.8-fold, and 8.2 fold respectively, and renal MT-concentrations increased 2.0-fold, 2.1-fold, and 2.3-fold respectively. Administration of actinomycin D or cycloheximide to control rats did not significantly affect MT concentrations in liver or kidney; $NiCl_2$-induction of MT in liver and kidney was not blocked by actinomycin D, an inhibitor of RNA synthesis, but was prevented by cycloheximide, an inhibitor of protein synthesis. Administration of (DDC) to control rats increased MT concentration 7.6-fold in liver, but did not affect MT concentration in kidney. $NiCl_2$-induction of MT in liver and kidney was not prevented by DDC. In control experiments, MT concentrations in liver and kidney were not affected by <u>in vitro</u> addition of DDC (0.2 mmol/litre) to the tissue homogenization medium.

4. DISCUSSION

Although Vasak et al. (1981) showed that Ni[II] binds to apothionein in vitro, Webb (1972), Sabbioni and Marafante (1975), and Suzuki and Yoshikawa (1976) detected little or no nickel in MT-fractions obtained by gel-filtration chromatography of hepatic cytosol from Ni[II]-treated rats. On the other hand, the present study and previous reports by Suzuki and Yoshikawa (1976), Eaton et al. (1980), and Chmielnicka et al. (1982) demonstrate increased MT concentrations in liver and kidney of $NiCl_2$-treated rats. Three possible mechanisms for Ni-induction of MT merit consideration: (a) Ni may induce MT synthesis indirectly by enhancing uptake of Zn, Cu, or other endogenous metals into liver and kidney, (b) Ni may induce MT synthesis indirectly by hormonal stimulation (e.g., epinephrine, corticosteroids, or glucagon): and (c) Ni may induce MT synthesis directly by stimulating translation of MT-mRNA by liver and kidney ribosomes. Experiments to test these hypotheses are underway in our laboratory, as well as attempts to explain the unexpected discovery that DDC, by itself, causes 7.6-fold increase of MT concentration in rat liver.

5. SUMMARY

Effects of $NiCl_2$ and sodium diethyldithiocarbamate (DDC) upon metallothionein (MT) concentrations were studied in liver and kidney of male Fischer rats. After injection of $NiCl_2$ (0.75 mmol/kg, sc), hepatic MT concentration increased 2.6-fold at 6.5 h and 8.2-fold at 17 h; renal MT concentration increased 1.4-fold at 6.5 h and 2.3-fold at 17 h. Dose-related increases of MT concentrations were observed in liver and kidney of rats killed 17 h after injection of $NiCl_2$ (0.25 to 0.75 mmol/kg, sc). Repeated administration of $NiCl_2$ (0.1 mmol/kg, ip) on 4 successive days, with sacrifice 3 days after the last treatment, increased MT concentrations 1.4-fold in liver and kidney, whereas $CdCl_2$-treatment at the same dosage schedule increased MT concentration 16-fold in liver and 3.3-fold in kidney. $NiCl_2$-induction of MT in liver and kidney was not prevented by actinomycin D (1 mg/kg, ip), but was inhibited by cycloheximide (2 mg/kg, ip). DDC, given alone (1.33 mmol/kg, im), 17 h before death, increased MT concentration 7.6-fold in liver but did not affect MT concentration in kidney; administration of DDC prior to injection of $NiCl_2$ did not inhibit $NiCl_2$-induction of MT.

6. ACKNOWLEDGMENTS

Supported by U.S. Department of Energy Grant EV-03140 and National Institute of Environmental Health Sciences Grant ES-01337.

7. REFERENCES

R.C. Baselt, F.W. Sunderman Jr., J.M. Mitchell and E. Horak (1977) Res. Commun. Chem. Pathol. Pharmacol., 18, 677-688.

J. Chmielnickca, J.A. Szymanska and J. Tyfa (1982) Environ. Res., 27, 216-221.

D.L. Eaton, N.H. Stacey, K.L. Wong and C.D., Klaassen (1980) Toxicol. Appl. Pharmacol., 55, 393-402.

S. Onosaka and M.G. Cherian (1982) Toxicol. Appl. Pharmacol., 63, 270-274.

E. Sabbioni and E. Marafante (1975) Environ. Physiol. Biochem., 5, 132-141.

F.W. Sunderman Jr. and C.B. Fraser (1983) Ann. Clin. Lab. Sci. (in press).

Y. Suzuki and H. Yoshikawa (1976) Industr. Health, 14, 25-31.

M. Vasak, J.H.R. Kagi, B., Holmquist and B.L. Vallee (1981) Biochemistry, 20, 6659-6664.

M. Webb (1972) Biochem. Pharmacol., 21, 2751-2765.

THE METABOLIC FATE OF CADMIUM BOUND TO METALLOTHIONEIN AND POLYMERS OF METALLOTHIONEIN

Douglas M. Templeton and M. George Cherian

Department of Pathology, Health Sciences Building,
University of Western Ontario
London, Canada N6A 5C1

1. INTRODUCTION

Different forms of Cd show markedly different tissue distributions and toxicity. It is well known that injection of Cd^{2+} salts into experimental animals results in major deposition of Cd in the liver, since Cd loosely bound to plasma proteins is readily taken up by the hepatocyte. Cadmium administered as its metallothionein (MT) complex is tightly bound to the protein ($\log \beta_3 = 25.5$), and the Cd-MT is filtered intact by the glomerulus. Deposition of Cd in the proximal tubular cells results (1,2). Low molecular weight Cd-thiolate chelation complexes show an intermediate distribution (3), and renal Cd deposition correlates with the stability constants of the complexes which Cd forms with co-administered chelators (4). We have prepared two polymeric forms of MT, with the aim of altering its filtration characteristics. These novel Cd-containing metalloproteins are useful in determining the factors influencing the renal handling of protein-bound metal.

2. MATERIALS AND METHODS

Rat liver MT-II was isolated from animals exposed to high levels of Cd, as described previously(5), and equilibrated against $^{109}CdCl_2$ of known specific activity. This MT-II was polymerized with either glutaraldehyde (GA-MT) or dimethylsuberimidate (DMS-MT) as described(6,7). The single major polymeric species were separated by Sephadex G-75 gel chromatography.

Male Sprague Dawley rats (150-200 g) were given Cd(7 ± 1 µg/kg) iv through a jugular catheter placed under light anaesthesia with a ketamine (60 mg/kg) / xylazine (15 mg/kg) mixture. Cadmium was labelled with ^{109}Cd, and given as either $CdCl_2$, Cd-MT, GA-MT, or DMS-MT. Animals were sacrificed by exsanguination at 3 or 24 hrs. and liver, kidney, spleen, pancreas, duodenum, and blood saved for determination of Cd as ^{109}Cd. Kupffer cell enrichment was achieved by the method of Page and Garvey (8), and

Chemical Toxicology and Clinical Chemistry of Metals
ISBN 0.12.137520.X
Copyright © 1983 by IUPAC
All rights of reproduction in any form reserved

subcellular fractionation of tissue homogenates was performed by differential centrifugation.

3. RESULTS

The physical properties of the GA-MT and DMS-MT polymers are summarized in Table I (5,6).

TABLE I

Physicochemical Properties of MT Polymers

	MT-II	GA-MT	DMS-MT
MW (by electrophoresis)	6,250	58,900	49,000
Frictional ratio (f/f_0)	1.3	1.06	1.04
Stokes' radius (Å)	16.0	27.6	25.4
Isoelectric point	4.6	5.2	4.6
Cysteine/monomer[a]	17	16	17
Lysine/monomer	9	2	5
-SH (by pOH-$HgBzO^-$)	18	18	18
ratio-SH/(Cd + Zn)	3.0	4.3	3.0
Cd/monomer	5.2	4.2	5.2
Zn/monomer	0.8	0	0.8
Cu/monomer	0.2	-	-

[a]By amino acid analysis following performic acid oxidation.

Following administration of all forms of Cd, most of the dose is recovered from kidney and liver at 3 hrs. (Table II and Fig. 1). Whereas most Cd is deposited in the liver following $CdCl_2$ administration, Cd-MT results in major deposition of Cd in the kidney. The polymers result in an intermediate distribution pattern, as shown clearly by the liver:kidney ratios of recovered dose (Table II).

TABLE II

Recovery of -Cd in Liver and Kidney

	% dose in liver+kidney[a]	Ratio of Cd, Liver:Kidney[a]	% of liver Cd in Kupffer cells[b]
$CdCl_2$	100 ± 1	38 ± 6	1.6 , 1.6
Cd-MT	95 ± 4	0.37 ± 0.03	- -
Cd-GA-MT	90 ± 4	17 ± 9	14 , 16
Cd-DMS-MT	93 ± 3	2.6 ± 0.7	17 , 17

[a]% recovered ^{109}Cd, mean ± S.D. (n=5)
[b]Two separate experiments

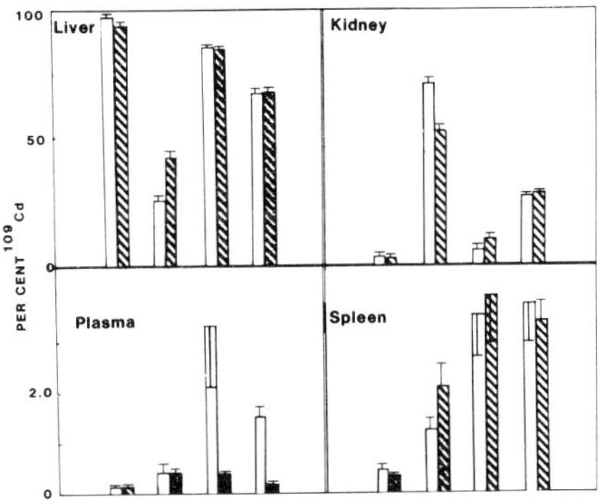

FIG. 1 Tissue distribution of Cd following iv administration of various forms. Values are expressed as % of ^{109}Cd recovered from all tissues, as described in Methods. Animals were sacrificed at 3 hrs. (solid bars) or 24 hrs. (dashed bars) after injection. Each bar is Mean ± S.D. (n=5), and from right to left in each panel represent $CdCl_2$, Cd-MT, GA-MT and DMS-MT respectively.

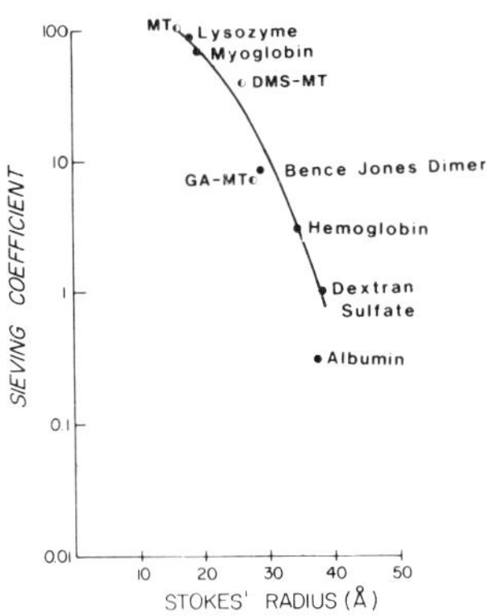

FIG. 2 Relationship of the glomerular sieving coefficient to molecular size for a variety of proteins. Values are based on the data of Pesce *et al.* (10), -●-, or the present work, -◐-.

The polymers result in a prolonged circulation time of Cd, which remains in plasma at 3 hrs. and is taken up to a significant extent by the spleen (Fig. 1). This may in part account for the increased antigenicity of glutaraldehyde treated MT(9). Because of the prolonged circulation time, the reticuloendothelial system plays a part in clearance of these proteins, and increased Cd is found in Kupffer cells.

Subcellular fractionation of renal homogenates shows that 60% of renal Cd remains in the cytosol at 3 hrs., regardless of the form of Cd administered, and no other differences in subcellular distribution occur.

4. DISCUSSION

Cadmium bound to either polymer of MT remains in circulation longer than any other soluble form of Cd. These polymers bind Cd tightly, as does MT, but show retarded glomerular filtration. Our results are compared with the data of Pesce *et al.* (10) regarding the glomerular sieving coefficient and molecular radius of proteins (Fig. 2). All MT species are acidic proteins (Table I) and highly negatively charged at the pH of the plasma filtrate. Consequently, filtration differences are determined primarily by molecular size. If the sieving coefficient is assumed to be proportional to the renal deposition of Cd measured here, and a value of 100 is used for MT (10), then these MT species may be added to Fig. 2. They fit the curve well, supporting the supposition that protein size is a major determinant of renal Cd deposition. The polymers fall in a critical range for filtration, ie. in the region of steepest descent of the curve in Fig. 2, and thus are useful for probing the mechanism of renal clearance. As a consequence of the retarded plasma clearance of Cd bound to the polymers, the reticuloendothelial system also plays a part in removing these forms of Cd from the blood.

5. REFERENCES

1. J. J. Vostal and M. G. Cherian (1974) *Fed. Proc. Fed. Amer. Soc. Exp. Biol.*, 33, 519.
2. E. C. Foulkes (1978) *Toxicol. Appl. Pharmacol.*, 45, 505-512.
3. M. G. Cherian (1979) *Environ. Health Perspect.*, 28, 127-130.
4. K. T. Suzuki and Y. Yamamura (1979) *Biochem. Pharmacol.*, 28, 3643-3649.
5. M. G. Cherian and Z. A. Shaikh (1975) *Biochem. Biophys. Res. Commun.*, 65, 863-869.
6. D. M. Templeton and M. G. Cherian, Submitted.
7. D. M. Templeton and M. G. Cherian (1983) *Environ. Health Perspect.*, in press.
8. D. T. Page and J. S. Garvey (1979) *J. Immunol. Meth.*, 27, 159-173.
9. R. J. Vander Mallie and J. S. Garvey (1978) *Immunochem.*, 15, 857-868.
10. A. J. Pesce, D. H. Clyne, V. E. Pollak, S. K. Kant, E. C. Foulkes and W. M. Selenke (1980) *Clin. Biochem.*, 13, 209-215.

BINDING OF MERCURY TO METALLOTHIONEIN

Martin J. Stillman, Annie Y.C. Law and Jadwiga A. Szymanska

Department of Chemistry, University of Western Ontario
London, Ontario, CANADA N6A 5B7

1. INTRODUCTION

Metallothionein (MT), a low molecular weight and cysteine-rich protein, is isolated from a wide range of organisms (Kagi & Nordberg, 1979), following induction by a variety of different metal ions, e.g. Cd^{2+}, Zn^{2+}, Cu^{2+}, Bi^{3+}, Hg^{2+} and Ag^{+}. The most thoroughly studied metallothioneins so far are hepatic MTs that contain a mixture of Cd and Zn ions. Recently, metallothioneins containing other metal ions, in particular Hg^{2+}, have been examined more closely. A recent report describes the optical characterization of Cu,Hg-MT isolated from rat kidney (Szymanska et al., 1983a) following induction using $HgCl_2$ (Zelazowski & Piotrowski, 1980). In vitro studies have shown that Hg^{2+} can readily displace Cd^{2+} and Zn^{2+} present in MT (Vasak et al., 1981; Sokolowski et al., 1974; Szymanska et al., 1983b); however, little detail is known of the binding of Hg^{2+} to Cd,Zn-MT (Szymanska et al., 1983b). In this paper, we report the changes that take place in the absorption, circular dichroism (CD) and magnetic circular dichroism (MCD) spectra during the in vitro titration of Cd,Zn-MT with $HgCl_2$. The Hg-dimercapto-propanol (BAL) complex has also been prepared and the UV and MCD spectra obtained.

2. EXPERIMENTAL

Cd,Zn-MT isoform 1 was isolated from the livers of rats exposed to $CdCl_2$ (Szymanska et al., 1983b). The protein was purified according to Zelazowski et al. (1980). Spectroscopic techniques are described in a earlier report (Szymanska et al., 1983a).

3.RESULTS AND DISCUSSION

Fig. 1 shows the absorption, CD and MCD spectra recorded as Hg^{2+} was added to rat liver Cd,Zn-MT 1. Lines 1-5 in all three

Chemical Toxicology and Clinical Chemistry of Metals
ISBN 0.12.137520.X
Copyright © 1983 by IUPAC
All rights of reproduction in any form reserved

sets of spectra, represent the first stage of what appears to be a two stage binding process. In the absorption spectrum there is a general increase in the 250 nm shoulder. Similar results have been observed when Cd^{2+} is added to Cd,Zn-MT 1 solutions (Law et al., 1983). This change is consistent with the replacement of Zn^{2+} by Hg^{2+}, as line 5 represents the point at which the Zn^{2+} and Hg^{2+} concentrations in the protein solution are equivalent (as shown by AAS determinations). This suggests that the apparent increase in the 250 nm absorption is, in fact, a wavelength

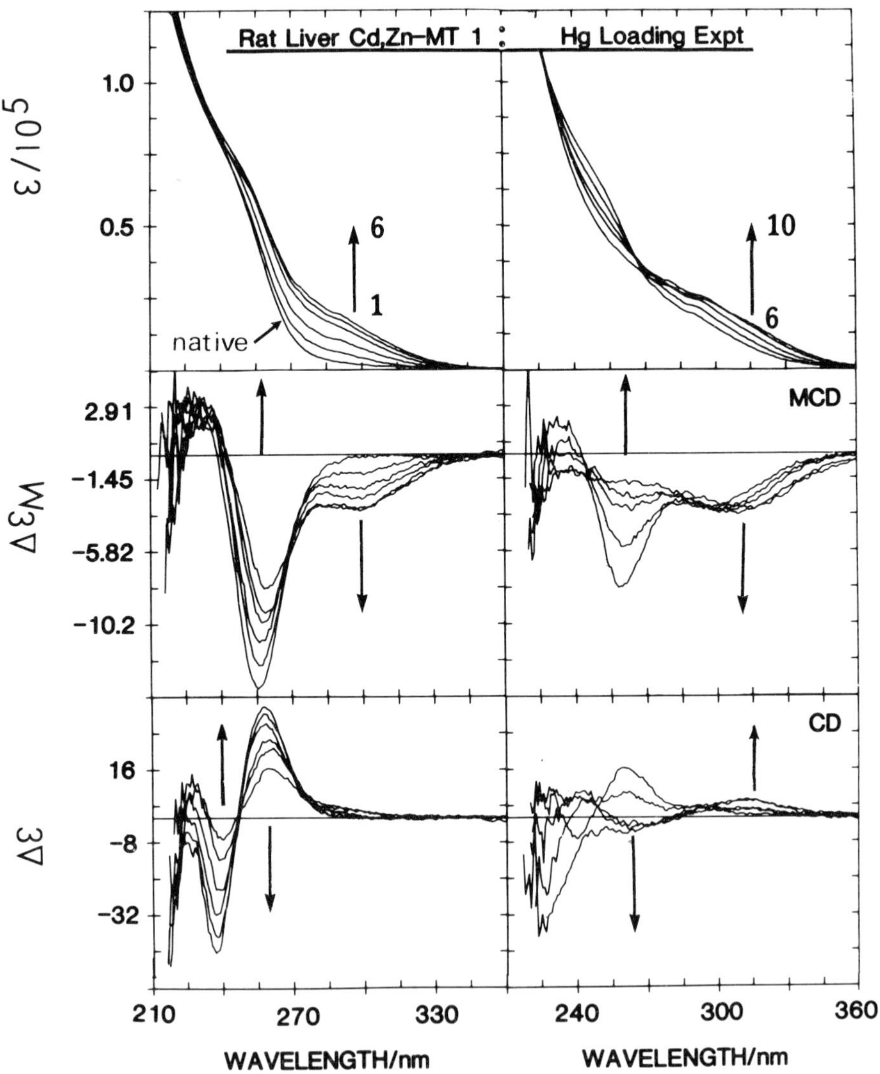

Fig. 1. Titration of rat liver Cd,Zn-MT 1 with $HgCl_2$.

shift in the S → Cd charge transfer band when the Zn^{2+} is displaced from the binding sites. Addition of Hg^{2+} beyond this concentration now removes the bound Cd^{2+} ions. The absorption shows an isosbestic change with an increase in the broad band near 300 nm and a gradual decrease in the shoulder at 250 nm which indicates that Cd^{2+} has dissociated. The MCD spectral changes observed during the titration also reflect the two stages in the binding of Hg^{2+}. With the displacement of Zn^{2+} by Hg^{2+} (lines 1-5), the MCD band decreases in intensity, and the band centre shifts to a longer wavelength. This can be rationalized by the fact that the unsymmetrical MCD band is an envelope containing both Cd- and Zn-related transitions. As the Zn^{2+} is removed from the protein, a broad negative band centred around 290 nm increases in intensity. This band is assigned to a S→Hg charge transfer transition (Gunter et al., 1975). The absence of a derivative-shaped MCD band as observed in the tetrahalo complexes of mercury, as well as in the Hg-BAL model compound, may well suggest that the Hg^{2+} bound in the protein is not in a tetrahedral environment. Line 9 represents a saturated Hg-MT. The CD spectrum shows three bands: 315nm(+), 250nm(-) and 240nm(+).

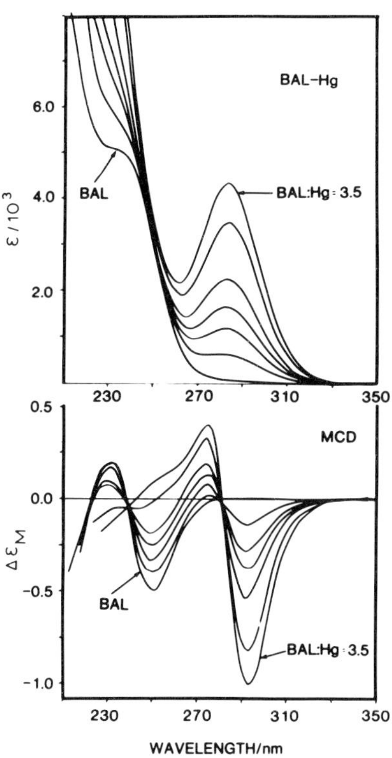

FIG. 2. A titration of a solution of dimercaptopropanol (BAL) in a Na_2CO_3/$NaHCO_3$ buffer at pH 10 with $Hg(NO_3)_2$. The molar ratios for BAL:Hg are 27.7, 13.8, 9.2, 6.9, 4.6, and 3.5.

This CD spectrum is quite different from the one obtained by other workers (Vasak & Kagi, 1983), where the HgMT was prepared by adding Hg^{2+} to apoMT at low pH. It is not unreasonable to expect that the Hg^{2+} may be bound to the protein in a slightly different manner for the two different preparative routes. Further addition of Hg^{2+} beyond this concentration was found to diminish the 300 nm absorption band as well as to change the CD spectrum. This suggests that as excess Hg^{2+} is added, the mode of binding changes; we suggest that the coordination number decreases allowing the additional Hg^{2+} to be bound to the thiolate groups in the protein. A similar set of Hg^{2+} titrations curve was obtained with horse kidney Cd,Zn-MT 1 (not shown).

Fig. 2 shows a titration curve recorded for the addition of Hg^{2+} to a BAL solution. The MCD spectrum shows a well defined derivative-shaped band similar to those obtained for the tetrahalo complexes of Hg (Gunter et al., 1975), which can be assigned as an MCD A term arising from the S→Hg charge transfer transition.

4. ACKNOWLEDGEMENTS

We thank NSERC of Canada for financial support and the Centre for Chemical Physics for a Visiting Fellowship (to JAS).

5. REFERENCES

W. Bernhard, M. Good, M. Vasak and J.H.R. Kagi (1983) Inorg. Chim. Acta, 79, 154-155.

J.D. Gunter, A.F. Schreiner and R.S. Evans (1975) Inorg. Chem., 14, 1589-1592.

J.H.R. Kagi and M. Nordberg, (1979) Experientia Supp., 34, Birkhauser Verlag, Basel.

A.Y.C. Law and M.J. Stillman (1980) Biochem. Biophys. Res. Commun., 102, 397-402.

A.Y.C. Law, M.G. Cherian and M.J. Stillman (1983) Biochim. Biophys. Acta, in press.

G. Sokolowski, W. Pilz and U. Weser (1974) FEBS Lett., 48, 222-225.

J.A. Szymanska, A.J. Zelazowski and M.J. Stillman (1983a) Biochem. Biophys. Res. Commun., in press.

J.A. Szymanska, A.Y.C. Law, A.J. Zelazowski and M.J. Stillman (1983b) Inorg. Chim. Acta, 79, 123-124.

M. Vasak and J.H.R. Kagi (1983) in Metal Ions in Biological Systems (Ed, H. Sigel) M. Dekker, vol. 15, pp 213-273.

M. Vasak, J.H.R. Kagi and H.A.O. Hill (1981) Biochemistry, 20, 2852-2856.

A.J. Zelazowski, J.A. Szymanska and H. Witas (1980) Prep. Biochem., 10, 495-505.

A.J. Zelazowski and J. K. Piotrowski (1980) Biochim. Biophys. Acta, 625, 89-99.

SPECTROSCOPY OF METALLOTHIONEINS: CIRCULAR DICHROISM, MAGNETIC CIRCULAR DICHROISM AND EMISSION STUDIES.

Annie Y. C. Law, Jadwiga A. Szymanska and Martin J. Stillman

Department of Chemistry, University of Western Ontario, London, Ontario, CANADA N6A 5B7

1. INTRODUCTION

Metallothionein (MT) is a sulfur-rich, low molecular weight protein that can bind several atoms per molecule of a remarkably wide variety of metals (Webb, 1979; Vasak & Kagi, 1983). Mammalian proteins appear to be able to bind up to seven metal atoms in vivo; in vitro binding experiments have shown that the protein can bind more than seven ions but whether the coordination geometry now resembles that in the native protein is not clear. The characteristic features of the optical spectra are derived from the lack of aromatic amino acids and the presence of S→M charge transfer bands in the 220 nm to 400 nm wavelength region (Weser & Rupp, 1979). Circular dichroism (CD) and magnetic circular dichroism (MCD) spectra provide sensitive indicators of the conformation of the coordinating groups in the protein and the geometry of the metal binding site, respectively. ^{113}Cd nmr studies have established that metal binding takes place, in hepatic mammalian MT, in a three-metal and a four-metal cluster (Boulanger et al., 1983).

We describe in this paper the optical spectra for native Cd,Zn-MT and Cd,Cu-MT, a Cd-dimercaptopropanol (BAL) model compound and a Cd,Cu-MT formed in vitro, and the spectral changes that occur with changes in the pH and with metal loading.

2. EXPERIMENTAL

The metallothionein proteins were isolated from the livers (Cd,Zn-MT) and kidneys (Cd,Cu-MT) of rats which had been exposed to $CdCl_2$. Protein preparations and the spectroscopic techniques used have been described previously (Law & Stillman, 1981; Law et al., 1983a,b; Stillman & Szymanska, 1983).

3. RESULTS AND DISCUSSION

The absorption spectrum of Cd,Zn-MT (Fig. 1) is charac-

Chemical Toxicology and Clinical Chemistry of Metals
ISBN 0.12.137520.X
Copyright © 1983 by IUPAC
All rights of reproduction in any form reserved

terized by a poorly resolved shoulder at 250 nm on the edge of an intense, protein-derived band; at low pH the 250 nm absorption disappears leaving the spectrum of the metal-free MT. The CD and MCD spectra for Cd,Zn-MT show much more detail than the corresponding absorption spectrum. A derivative-shaped CD band is characteristic of Cd^{2+} binding. The CD spectra (Fig. 1) are a superposition of both S→Cd and S→Zn bands, with the dominant features being due to the S→Cd transitions. The MCD spectrum exhibits a distorted derivative-shaped envelope. For such compounds, where the metal binding site geometry is tetrahedral, we expect to observe a Faraday A term. Fig. 2 shows the MCD spectrum of the Cd-BAL complex; similar MCD spectra have been reported for complexes of Hg^{2+} (Stillman et al., 1983). Both acidification and Cd-loading of Cd,Zn-MT lead to the preferential displacement of Zn^{2+}. At pH 4 (Fig. 1), we observe that the 228 nm band intensity in the CD spectrum is diminished with only slight changes at 238 nm and 260 nm, which suggests that a Zn-related band is present at 228 nm. Changes in the MCD spectrum show more detail: as Zn^{2+} is displaced, the main band becomes more symmetric and the crossover point moves about 4 nm to the red. This suggests that there also are S→Zn bands which distort

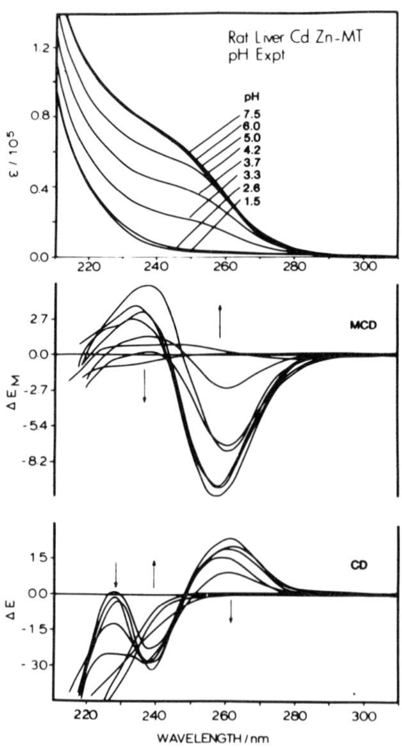

FIG 1. Spectra of rat liver Cd,Zn-MT.

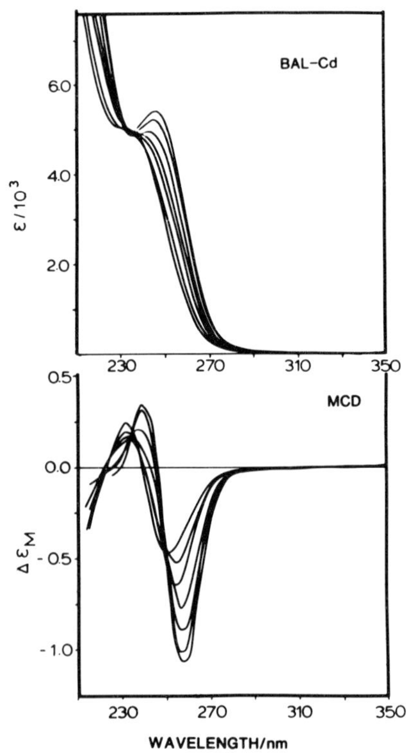

FIG 2. Spectra of the Cd-BAL complex.

the symmetric A term expected for Cd^{2+} in a tetrahedral site.

Fig. 3 shows the spectral changes that occur when $CdCl_2$ is added to a sample of Cd,Zn-MT. Although there are reproducible effects observed in the absorption spectrum, much more detail is found in the CD spectrum. Initially, intensity at 228 nm increases as the Zn^{2+} is displaced by the Cd^{2+}; at the same time the positive 258 nm band shifts to 255 nm. At this stage, the protein contains the same number of cadmium atoms as the sum of zinc and cadmium in the native sample; however, it is still possible to add more Cd^{2+} to the protein. Now the 228 nm band intensity collapses, while there is an increase in the new band at 255 nm. Because the increases at 255 nm are related isosbestically, we suggest that only a single species is being formed. The dramatic changes in intensity of the 228 nm band suggest that this may in fact be a marker band for Cd^{2+} binding in a cluster; it may arise from the interaction between groups on amino acid residues other than cysteine that take part in weak binding to the Cd^{2+}. As the metal loading increases so the coordination number for Cd^{2+} binding must decrease from the four proposed for the bridging model (Boulanger et al., 1983).

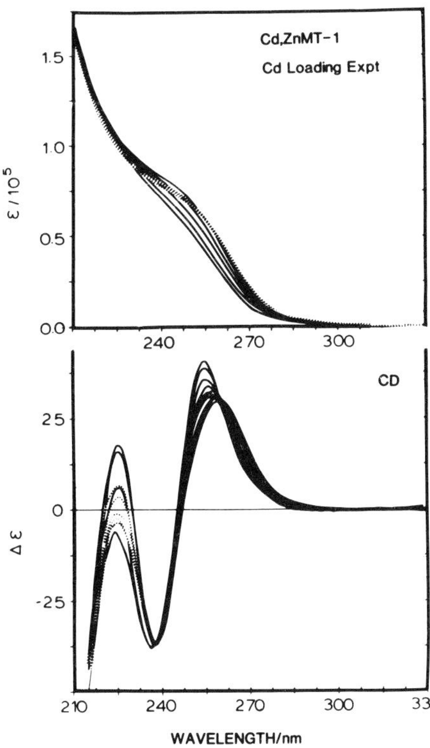

FIG 3. Absorption and CD spectra of Cd,Zn-MT.

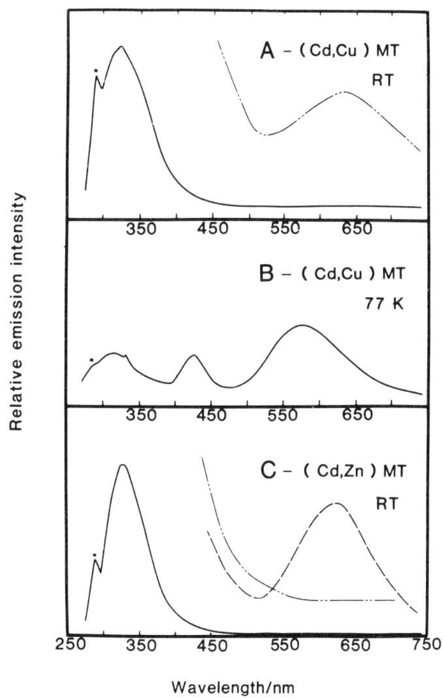

FIG 4. Emission spectra of Cd,Cu-MT and Cd,Zn-MT.

Renal Cd,Cu-MT is a native Cd- and Cu-containing protein. Absorption, CD and MCD spectra for Cd,Cu-MT closely resemble those of the Cd,Zn-MT. This implies that spectral intensity from the Cu-thiolate groups is considerably lower than that observed for the Cd-thiolates. However, unlike the Cd,Zn-MT protein, Cd,Cu-MT is luminescent in the 500-700 nm region (Fig. 4A,B). A weak band observed near 610 nm in aqueous solutions at room temperature intensifies considerably at 77 K (Fig. 4B). Excitation spectra recorded by monitoring the 610 nm emission intensity, show a band at 270 nm suggesting that the Cu-thiolate groups contribute significantly to the intensity observed at 610 nm. Addition of Cu^+ to a Cd,Zn-MT solution results in the development of a band at 610 nm (Fig. 4C) which also has an excitation spectrum with a band near 270 nm. Thus the emission band at 610 nm is an indicator for Cu^+ binding to metallothioneins. The weak band near 340 nm (its intensity is comparable with that of the Raman band of the solvent) is found in the spectra of all samples of MT that we have studied (rat, horse and rabbit), as well as in the spectra of complexes of BAL and cysteine. The intensity appears unrelated to the metal content of the protein. The phosphorescence band at 430 nm is also observed in the spectrum of the glycerol solvent blank.

4. ACKNOWLEDGEMENTS

The authors gratefully acknowledge support from NSERC of Canada through the Strategic Grants program. MJS is a member of the Centre for Chemical Physics at the UWO.

5. REFERENCES

Y. Boulanger, C.M. Goodman, C.P. Forte, S.W. Fesik and I.M. Armitage (1983) Proc. Natl. Acad. Sci. USA, 80, 1501-1505.

A.Y.C. Law and M.J. Stillman (1981) Biochem. Biophys. Res. Commun., 102, 397-402.

A.Y.C. Law, M.G. Cherian and M.J. Stillman (1983a) Biochim. Biophys. Acta, in press.

A.Y.C. Law, J.A. Szymanska and M.J. Stillman (1983b) Inorg. Chim. Acta, 79, 114-115.

M.J. Stillman, A.Y.C. Law and J.A. Szymanska (1983) These Proceedings, p .

M.J. Stillman and J.A. Szymanska (1983), submitted to Biophysical Chemistry.

M. Vasak and J.H.R. Kagi (1983) in Metal Ions in Biological Systems (Ed, H. Sigel) M. Dekker, vol.15, pp 213-273.

M. Webb (1979) in The Chemistry, Biochemistry and Biology of Cadmium (Ed, M. Webb) Elsevier, pp 195-266.

U. Weser and H. Rupp (1979) in The Chemistry, Biochemistry and Biology of Cadmium (Ed, M. Webb) Elsevier, pp 267-283.

METAL EXCRETORY PROTEINS: POSSIBLE ROLE OF A RENAL METAL SEQUESTERING PROTEIN IN METAL EXCRETION

Bibudhendra Sarkar and A.W. Abdulwajid

Research Institute,
The Hospital for Sick Children, Toronto, M5G 1X8;
and Department of Biochemistry, University of Toronto,
Toronto, M5S 1A8, Ontario, Canada

1. INTRODUCTION

The mechanism by which toxic metals like nickel, cadmium and mercury are removed from the body is not known. Most of the studies have been focussed on the transportation, utilization and effects of these metals on different organs. Very little is known about the process of detoxification and elimination of these metals in a situation of excessive intake. In our continuing studies of the transport, storage and excretion of metals, we have been studying the state of nickel in rat kidney and its urinary excretion. Our previous studies have shown that kidney is the target organ for the accumulation and excretion of nickel following a single intraperitoneal administration of $^{63}NiCl_2$ in rats (Sarkar, 1980; Sarkar, 1983). We have also shown that nickel binds to a soluble low molecular weight glycoprotein in kidney, which has been isolated, purified and partially characterized (Abdulwajid and Sarkar, 1983). In the present study, we have preliminary evidence that a similar protein or a fragment of the same protein does exist in the urine, which not only binds nickel but other metals like copper, zinc, cadmium and manganese.

2. TECHNIQUES and FINDINGS

Male Sprague-Dawley rats (Camm, USA), weighing 200-250 g, were used for these experiments. Radioactive $^{63}NiCl_2$ (Amersham) with specific activity 11.7 mCi/mg was used as a tracer. Rats were administered intraperitoneally a dose of 102 μg Ni per kg body weight as $NiCl_2$. They were kept in metabolic cages and the 24 hr urine was collected in plastic cylinders. This urine was used for the fractionation of metal-binding components. The urine was fractionated on a Sephadex G-75 column equilibrated with 0.001 M Tris-HCl buffer pH 8.6. The fractions eluted

Chemical Toxicology and Clinical Chemistry of Metals
ISBN 0.12.137520.X
Copyright © 1983 by IUPAC
All rights of reproduction in any form reserved

were determined for ^{63}Ni radioactivity and the optical density read at both 280 and 254 nm. The liquid scintillation counting of ^{63}Ni was carried out using 0.5 ml of eluent with 10 ml of aquasol (New England Nuclear) in a Beckman LS8100 scintillation counter.

The Sephadex G-75 elution profile of urine is shown in Fig.1. It shows three peaks, which absorb highly at 280 and 254 nm. The first two peaks showed the presence of radioactivity, out of which the second peak has about 96% of total radioactivity. Figure 2 shows the elution profile of protein and radioactivity of rat kidney supernatant containing $^{63}NiCl_2$ on the same column. It appears from these figures that both kidney and urine probably contain a protein of similar molecular size (V_e/V_o = 3.2-3.5) which binds nickel.

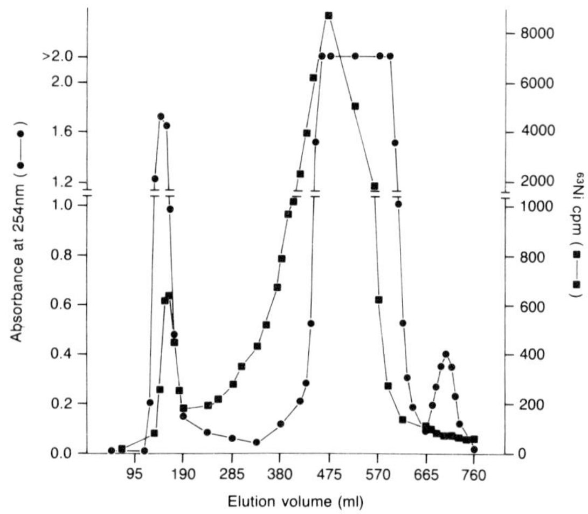

FIG. 1 Gel filtration profile of urine on Sephadex G-75 column (100 x 3 cm), 11 ml urine (~289 mg of proteins) was loaded. Elution buffer, 0.001 M Tris-HCl (pH 8.6); rate 20 ml/hr.

The second peak was lyophilized and on polyacrylamide gel electrophoresis, it showed the presence of 6 to 7 protein bands. This peak was further fractionated by high pressure liquid chromatography (HPLC) of Waters Associates, Inc., on protein analysis columns I-60 and I-125 using Tris-HCl buffer pH 7.2 (0.025 M). The results show the presence of nine peaks, which absorb at both 280 and 254 nm (Fig. 3). Out of these, only three protein peaks have most of the radioactivity.

The urine of control rats was equilibrated overnight with $^{63}NiCl_2$ and fractionated on the same Sephadex G-75 column. A similar elution profile was seen as that of urine obtained from rat administered with ^{63}Ni, indicating the presence of similar proteins. Equilibration of urine of control rats with other

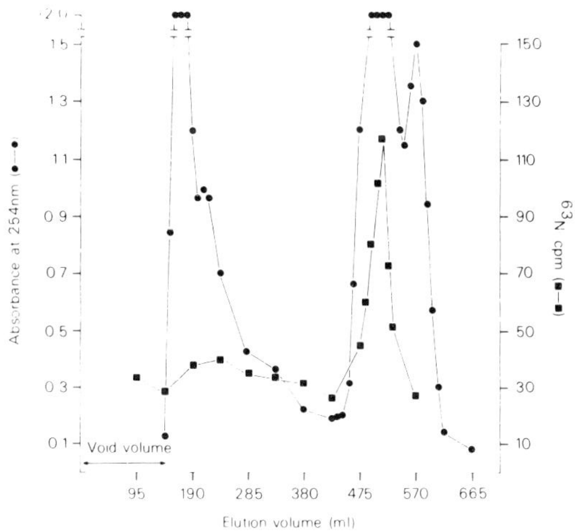

FIG. 2 Gel filtration profile of kidney cytosol on Sephadex G-75 column (100 x 3 cm); 20 ml supernatant (~200 mg of proteins) was loaded. Elution buffer, 0.001 M Tris-HCl (pH 8.6); rate, 20 ml/hr (from Abdulwajid and Sarkar, 1983).

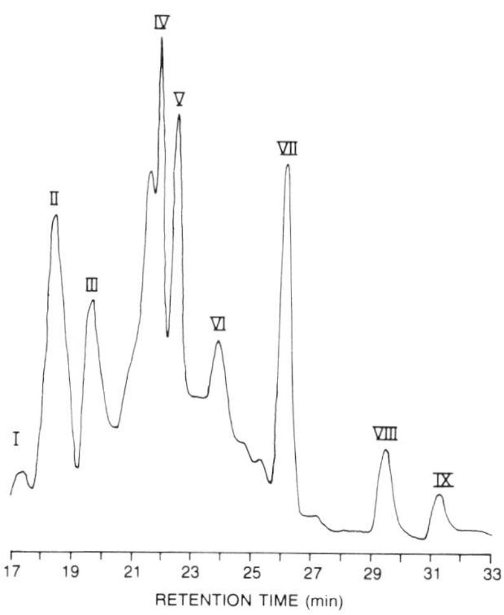

FIG. 3 High pressure liquid chromatography profile of the second radioactive peak obtained from Sephadex G-75 fractionation on protein analysis column I-60 and I-125. About 1 mg of protein was loaded. Elution buffer 0.025 M Tris-HCl pH 7.2; rate 1 ml/min; chart speed, 1 cm/min.

metals like ^{67}Cu, ^{65}Zn, ^{109}Cd and ^{54}Mn also showed a similar Sephadex G-75 elution profile and further fractionation of the radioactive peak by HPLC showed similar results as those obtained with nickel.

3. CONCLUSIONS

Although the urinary metal-binding proteins have not been fully characterized, it appears that there are proteins in urine which do bind nickel and other metals. Our preliminary results tend to suggest that the urinary metal-binding protein is either a protein similar to that which we isolated from kidney, or a fragment of it after proteolysis. It is not known if the carbohydrate part of the protein is still retained. In the light of these results, it seems possible that these proteins, which bind not only nickel but other metals as well, may constitute an important vehicle for the excretion of these metals in the event of an excessive intake. The characterization of these proteins is in progress.

4. ACKNOWLEDGEMENT

The research was supported by grants from the Medical Research Council and Natural Sciences and Engineering Research Council of Canada.

5. REFERENCES

A.W. Abdulwajid and B. Sarkar (1983) *Proc. Natl. Acad. Sci. USA*, in press.
B. Sarkar (1980) in *Nickel Toxicology* (Eds, S.S. Brown and F.W. Sunderman, Jr.), Academic Press, London, pp. 81-84.
B. Sarkar (1983) *Chem. Scripta*, 21, 103-110.

AN IN-VITRO EVALUATION OF THE CLEARANCE OF DESFERAL-CHELATED IRON DURING HEMODIALYSIS

Roger L. Boeckx,[1] Jose T. Salcedo,[2] Glenn H. Bock[2] and Naomi L.C. Luban[1]

Departments of Clinical Laboratories[1] and Nephrology[2], Children's Hospital National Medical Center and the Department of Child Health and Development, George Washington University School of Medicine and Health Sciences, Washington, DC, 20010 U.S.A.

1. INTRODUCTION

Anemia is a common complication of chronic renal failure (CRF). The causes of this anemia may include blood loss during dialysis procedures, shortened red cell survival due to uremia, loss of folate and other water soluble vitamins during dialysis, chronic gastrointestinal blood loss, dietary iron and protein deficiency, chronic infection, hypertension, erythropoietin deficiency or the presence of a serum inhibitor to erythropoietin. In an attempt to optimize red cell mass and growth in children with CRF, iron supplementation and frequent red cell transfusions are used. These multiple transfusions may result in an iron-over-load syndrome in some patients. This transfusion related hemochromatosis results in iron deposition in the heart, liver, pancreas and other organs which can ultimately result in death.

Chelation therapy with Desferal^R (desferrioxamine mesylate) is the only effective therapy for iron-overload. However, the Desferal-iron complex is predominately excreted in the urine, precluding its use in anephric patients with CRF. A few single case reports of Desferal therapy during hemodialysis have been reported (Baker, 1976; McCarthy, et. al., 1982), but no detailed study of the conditions under which optimal removal of iron can be accomplished has been performed. We evaluated the in-vitro clearance of the Desferal-iron complex by conventional hemodialysis systems using a mock hemodialysis system.

2. PROCEDURE

$Fe^{59}Cl_3$ (approximately 30 mCi/mg) was obtained from New England Nuclear. Desferal was a kind gift of Ciba Pharmaceutical

Chemical Toxicology and Clinical Chemistry of Metals
ISBN 0.12.137520.X
Copyright © 1983 by IUPAC
All rights of reproduction in any form reserved

Co. For each dialysis run, 1500 mL of fresh heparinized human blood was treated with 5 mL of $Fe^{59}Cl_3$ or Desferal-Fe^{59} complex. The Desferal-Fe^{59} complex solution contained 36.5 mmol/L $FeCl_3$ in saline. The $Fe^{59}Cl_3$ solution (control) contained the same amounts of $FeCl_3$ and $Fe^{59}Cl_3$ but no Desferal. The treated blood contained 6.8 μg of added iron per mL and had a specific activity of 0.13 μCi/mL. Treated blood was placed in a 2L transfer pack which served as the blood reservoir of the mock hemodialysis system shown in Figure 1. This system uses a Drake Willock hemodialysis machine and a polyacrylonitrile copolymer artificial kidney (Hospal 607).

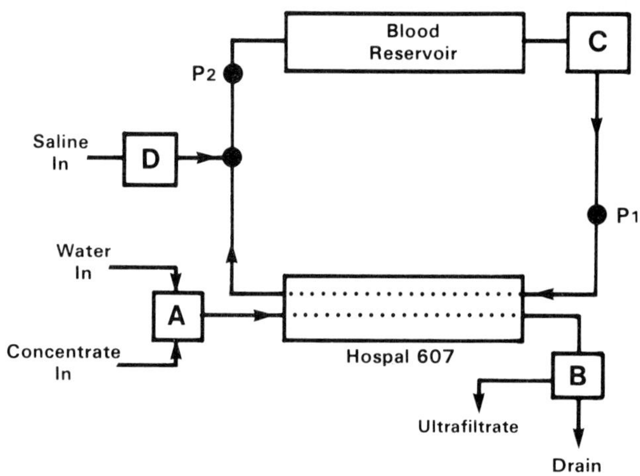

Figure 1. Schematic diagram of the mock dialysis system. A- Proportioning pump, B- Effluent pump, C- Blood pump, D-Saline pump, P_1- Arterial sampling port, P_2- Venous sampling port.

Dialysis was continued for 210 minutes at a total blood flow of 50 mL/min, a dialysate flow of 500 mL/min and an ultra filtration rate of 330 mL/h. The blood reservoir volume was maintained by isovolemic saline replacement. At 20 minute intervals throughout the dialysis period, 0.5 mL aliquots of blood were removed from the "arterial" and "venous" sides of the kidney, diluted to 5.0 mL with 1% Triton X-100 and counted in a Searle automatic gamma counting system optimized for Fe^{59}. The results are shown in Figure 2.

Clearances were calculated using the formula shown below:

$$C = Qi\,\frac{Fv - Fa}{Fa} + Qf\,\frac{Fa}{Fv}$$

where C is the clearance (mL/min), Qi is the blood flow into the kidney, Qf is the ultrafiltration rate (the difference between the blood flow into and out of the kidney; approximately 5.5 mL/min), and Fa and Fv are the amounts of added iron remaining (μg/mL) on the arterial and venous sides of the system respectively. The clearance (C) is corrected for the surface area of the membrane to yield corrected clearances ($mL/min/m^2$).

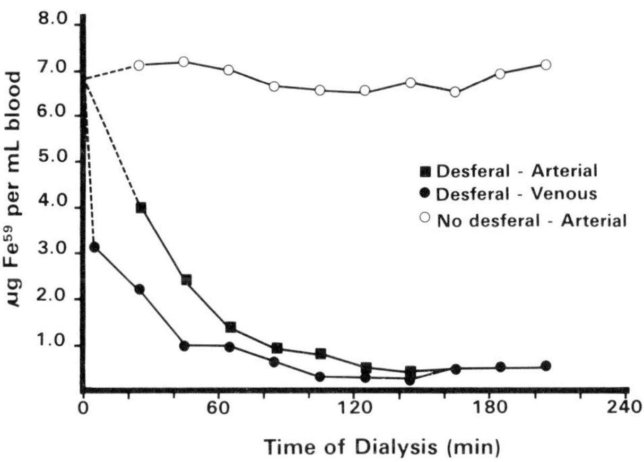

Figure 2. Fe^{59} remaining at various times during analysis.

The corrected clearances for added Desferal-iron and $FeCl_3$ were 37.1 ± 3.3 $ml/min/m^2$ and 8.4 ± 0.1 $mL/min/m^2$ (mean ± SEM) respectively. Following 3 hours of dialysis, less than 10% of the added Desferal-iron remained in the blood, while 90% of the added $FeCl_3$ remained in the absence of Desferal.

3. CONCLUSION

Our in-vitro studies show that Desferal-iron is effectively cleared by conventional hemodialysis systems suggesting that this drug can be used in the treatment of iron-overload in anephric patients.

4. REFERENCES

L. R. I. Baker, M. D. Barnett, B. Brozonic, W. R. Cattell, P. Ackrill, J. McAllister, and C. Nimon (1976) Clin. Nephrol., 6, 326-328.

J. T. McCarthy, C. R. Libertin, J. C. Mitchell III, and V. F. Fairbanks (1982) Mayo Clinic Proc., 57, 439-441.

EFFECT OF MANGANESE ON CEREBRAL RNA POLYMERASE AND FREE RIBOSOMAL PROTEIN SYNTHESIS: A POSSIBLE MECHANISM OF THE RETARDATION IN LEARNING AND MEMORY

S. Magour, H. Maser and I. Steffen

Institute of Toxicology and Biochemistry,
Gesellschaft fur Strahlen- und Umweltforschung,
D-8042 Neuherberg-Munich, F.R.G.

1. INTRODUCTION

Manganese (Mn) is present in varying concentrations in human and animal nutrition and is thought to be of minor toxicity. However, long-term exposure to high levels of Mn induce severe neurological disorders (Locura manganica), similar in many respects to those of Parkinson's disease (Mena, 1980). It has also been reported that daily oral intake of Mn (1000 ppm) for 30 days impaired learning and memory of growing rats (Murthy et al., 1981). The biochemical mechanism of this effect is not known. At present there is increasing evidence implicating a role for cerebral RNA and proteins in learning and memory according to which acquired informations is taken up and stored in RNA and protein-like material(s) in the brain. Therefore it was assumed that Mn may also inhibit RNA and protein synthesis of immature rat brain.

2. METHODS

Two groups of three weeks old female Wistar rats received Mn (either 55 or 1000 ppm) in their drinking water for four weeks. Cerebral nuclei, free ribosomes (which are thought to synthesize proteins destined for intracellular use) and pH 5 enzyme fractions were isolated (Steele et al., 1975; Magour et al., 1983) every three days. RNA polymerase (Yanaghara, 1974) and ribosomal protein synthesis (Campagnani and Harris, 1977) were determined <u>in vitro</u>, using ^{14}C guanosine triphosphate (^{14}C-GTP) and ^{14}C-labelled reconstituted protein hydrolysate as substrates, respectively.

3. RESULTS AND DISCUSSION

During daily oral intake of 55 ppm Mn no alteration of food and water intake or body weight gain were observed. However, free ribosomal protein synthesis was inhibited up to 30% between weeks two and three but returned toward the control level

Chemical Toxicology and Clinical Chemistry of Metals
ISBN 0.12.137520.X
Copyright © 1983 by IUPAC
All rights of reproduction in any form reserved

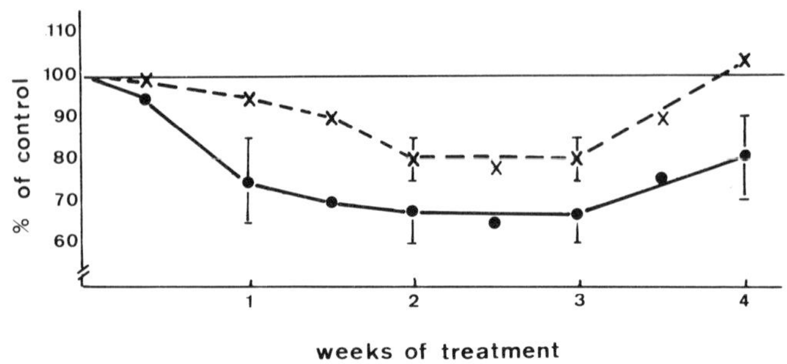

FIG. 1 Effect of daily intake of 55 ppm manganese in drinking water on free ribosomal protein synthesis (●—●) and RNA content of the pH 5 enzyme fraction (x—x) of immature rat brain. Data are expressed as percent changes from the control ± S.D.

by week four of treatment (Fig. 1). From cross incubation experiments with free ribosomes and pH 5 enzyme fractions from control and Mn-treated rats we previously observed that this inhibition was due to alteration of the pH 5 enzyme fraction rather than to changes in the polysomal activity. Figure 1 also shows that t-RNA content in the pH 5 fraction was reduced up to 20% during the first three weeks and also returned to the control level after 4 weeks. The adaptation phenomena shown in Figure 1 was not observed at the rather high dose of 1000 ppm Mn. The reason for the reduction of t-RNA content is not known but could be due to an inhibition of RNA synthesis and/or stimulation of RNase activity in the brain. Therefore, RNA synthesis was determined (in vitro) during daily oral intake of Mn. Under these experimental conditions we did not observe any significant change in RNA synthesis even at the highest level of 1000 ppm Mn in drinking water. Since Mn was not detectable in any of the isolated subcellular fractions, it was assumed that this element may not be firmly bound in the cells and may have been washed out during the isolation procedure. Therefore it was necessary to conduct experiments with manganese added to the reaction mixture in vitro.

As shown in Fig. 2, manganese (10^{-6}-10^{-3}M) inhibited the overall RNA synthesis up to 52%. This inhibition could be reversed by EDTA when the chelator was added to the reaction mixture after the enzyme had been preincubated with Mn for 10 min.

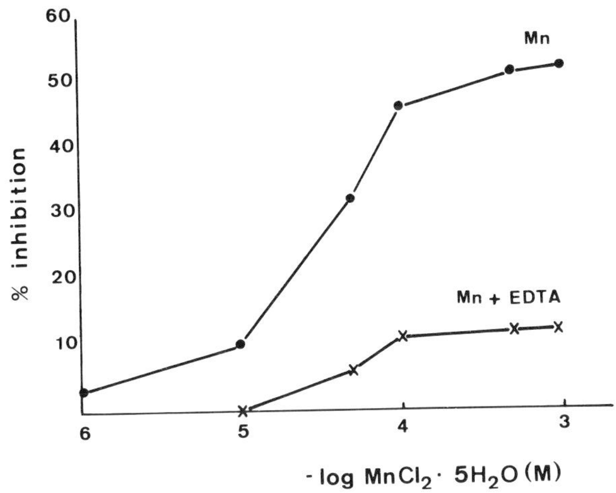

FIG. 2 The percent inhibition of RNA synthesis by manganese *in vitro* before (•-•) and after (x-x) the addition of EDTA.

The reversibility by EDTA would suggest that the inhibition of RNA synthesis caused by Mn is not due to an irreversible destruction of the enzyme. However, the inhibition by Mn added *in vitro* may be due to: 1) A reversible binding of Mn to the active site of the enzyme. 2) chelation with the phosphate groups of the nucleotides thus blocking the polymerization to RNA. 3) By competition with magnesium ions which are essential to RNA synthesis. In order to examine these possibilities, we performed kinetic experiments either by varying the substrate (nucleoside triphosphates, NTP) or the magnesium concentrations in the reaction mixture.

As shown in Table 1, Mn inhibited the nuclear RNA synthesis by 43-47% at all substrate concentrations used (0.04-0.20 mM) suggesting that the inhibition is at least not of a competitive manner with respect to the substrates. Due to the linearity of the GTP incorporation data, it was not possible to determine precisely the type of inhibition according to Lineweaver-Burk. In another set of experiments the magnesium concentrations in the reaction mixture were varied from 0.05-8.0 mM. The data presented in Table 2 indicate that manganese stimulated RNA synthesis (171% to 33% over controls) at the low magnesium concentrations (0.05-0.30 mM), and strongly inhibited the enzyme activity when the magnesium concentration was elevated from 1.0-8.0 mM.

TABLE 1

Effect of Manganese (0.1 mM) on RNA Synthesis at Different Substrate (NTP) Concentrations

Total NTP concentration (mM)	GTP incorp. dpm · 10^{-3}/mg DNA		% Inhibition
	Control	Manganese	
0.04	45 ± 8	24 ± 5	47
0.06	70 ± 9	39 ± 8	44
0.08	97 ± 11	57 ± 8	41
0.10	131 ± 15	69 ± 9	47
0.15	203 ± 30	111 ± 15	45
0.20	273 ± 35	155 ± 25	43

TABLE 2

Effect of Manganese (0.1 mM) on RNA Synthesis at Different Magnesium Concentrations

Magnesium concentration (mM)	GTP incorp. dpm · 10^{-3}/mg DNA		% Changes
	Control	Manganese	
0.05	14 ± 2	38 ± 5	+ 171
0.10	18 ± 4	41 ± 8	+ 128
0.15	21 ± 4	40 ± 8	+ 90
0.20	23 ± 2	39 ± 8	+ 70
0.30	30 ± 6	40 ± 6	+ 33
0.50	43 ± 3	46 ± 8	+ 7
1.00	70 ± 8	52 ± 12	- 26
2.00	97 ± 10	56 ± 15	- 42
4.00	115 ± 19	63 ± 15	- 45
8.00	116 ± 20	63 ± 15	- 46

TABLE 3

Effect of Manganese (0.1mM) on RNA Synthesis at Low Magnesium Concentration (0.1 mM) in the Presence and Absence of α-Amanitin (2 μg/ml)

Addition	^{14}C-GTP incorporated (dpm · 10^{-3}/mg DNA)	% Changes
Saline	17 ± 2	-
Manganese	39 ± 5	+ 129
α-amanitin + Manganese	28 ± 4	+ 65

At present we do not have experimental data to explain this biphasic effect of Mn but certain points should be discussed. It is generally known that there are at least two RNA polymerases I and II, which are magnesium and manganese dependent, respectively. Although we did not control the ionic strength of the incubation mixture, it is tempting to speculate that Mn may have differentially affected the activities of the two enzyme systems. At high magnesium concentrations RNA polymerase I may be operating and may be inhibited by Mn. However, when the magnesium concentration was kept low, the activity of RNA polymerase II may be detectable and may be stimulated with Mn. In an attempt to elucidate this point further, we determined the effect of Mn on RNA synthesis at low magnesium concentration (0.1 mM) in the presence and absence of α-amanitin (a specific inhibitor of RNA polymerase II). If Mn stimulated RNA polymerase II, it may not do so in the presence of α-amanitin.

As shown in Table 3, the stimulation by Mn was strongly suppressed in the presence of α-amanitin, suggesting that Mn may have mostly affected polymerase II under these *in vitro* conditions. However, at high magnesium-concentration (8.0 mM), Mn either with or without α-amanitin inhibited RNA synthesis by 47% and 41% respectively (Table 4), indicating that Mn inhibits RNA polymerase I. Accordingly Mn was ineffective in the presence of Actinomycin D. Certainly the results concerning the polymerases I and II are preliminary in nature and further investigations *in vivo* and *in vitro* are needed to clarify this point and to give more information on the type of inhibition of RNA synthesis.

TABLE 4

Effect of Manganese (0.1 mM) on Cerebral RNA Synthesis at High Magnesium-Concentration (8.0 mM) in the Presence and Absence of -Amanitin (2 μg/ml) and Actinomycin D (20 μg/ml)

Addition	^{14}C-GTP incorporated (dpm · 10^{-3} DNA)	% Changes
Saline	122 ± 10	-
Manganese	71 ± 9	- 41
α-Amanitin + Manganese	64 ± 8	- 47
Actinomycin D	7 ± 3	- 94
Actinomycin D + Manganese	6 ± 2	- 95

4. CONCLUSION

These results, and our previous findings (Magour et al., 1983) suggest that Mn when present in high concentrations in brain cells, may inhibit RNA polymerase I (which is thought to be involved, at least in part, in t-RNA synthesis), thus leading to the inhibition of ribosomal protein synthesis. It is concluded that manganese impairs learning and memory of chronically treated young rats partly through inhibition of cerebral RNA and protein synthesis. Accordingly, the Mn-treated rats may not be able to store the learned information in macromolecules.

5. REFERENCES

A.T. Campagnoni and J.R. Harris (1977) J. Neurochem. 28, 589-596.

S. Magour, H. Maser, and I. Steffen (1983) Acta Pharmacol. Toxicol. (in press).

I. Mena (1980) Manganese in Metals in the Environment (H.A. Waldron, Ed.) Academic Press, London, pp.191-220.

R.C. Murthy, S. Lal, D.K. Saxena, G.S. Shukla, M. Mohd. Ali and S.V. Chandra (1981) Chem.-Biol. Interactions 37, 299-308.

W.I. Steele, and T. Johannesson (1975) Acta Pharmacol. Toxicol. 37, 265-273.

T. Yanagihara (1974) J. Neurochem. 22, 113-117.

ENHANCEMENT OF NUCLEOLAR RNA SYNTHESIS BY CHROMIUM(III) IN REGENERATING RAT LIVER

Shoji Okada, Hideo Tsukada, and Hiroshi Ohba

Department of Environmental Biochemistry
Shizuoka College of Pharmacy
Shizuoka 422, Japan

1. INTRODUCTION

Work on intracellular chromium (Cr) has indicated that Cr accumulates in the nuclei mainly in the form of Cr(III) even when Cr(VI) is administered. Although it remains unknown whether the nuclear accumulation of Cr results in the physiological or toxicological action, it is possible to assume that nuclear Cr may regulate or alter the gene functions. Our previous findings support this possibility; Cr(III), not Cr(VI), enhanced cell-free RNA synthesis directed by DNA or chromatin by increasing the number of initiation sites [Okada *et al.* 1981,1982a], and also enhanced hepatic RNA synthesis in mice [Okada *et al.* (1983)]. The present study using regenerating rat liver indicates that Cr(III) administration enhanced specifically the nucleolar RNA synthesis followed by a less synchronized DNA synthesis. Furthermore, we isolated from the nucleoli a Cr-containing protein which has enhancing activity on nucleolar RNA synthesis.

2. METHODS AND RESULTS

(a) *Accumulation of Cr and Enhancement of RNA Synthesis in Regenerating Liver of Rats Treated with Cr (III)*

Male Wistar rats (110-120 g) were i.p. dosed with $CrCl_3$ (5 mg Cr/kg) containing $^{51}CrCl_3$ and, 24 hr later, partially (*ca.*68%) hepatectomized. The accumulation of Cr in the liver cells proceeded in the course of regeneration up to 6 hr. In particular, nuclear and nucleolar Cr increased remarkably up to 6 hr. On the other hand, RNA synthesis stimulated by partial hepatectomy was further and significantly enhanced at least for 12 hr. It is noteworthy that this enhancement occurred particularly in nucleolar RNA synthesis (Fig. 1). The enhanced nucleolar RNA synthesis was also observed at a lower dose, 0.5 mg Cr/kg, while not at 0.05 mgCr/Kg.

Chemical Toxicology and Clinical Chemistry of Metals
ISBN 0.12.137520.X
Copyright © 1983 by IUPAC
All rights of reproduction in any form reserved

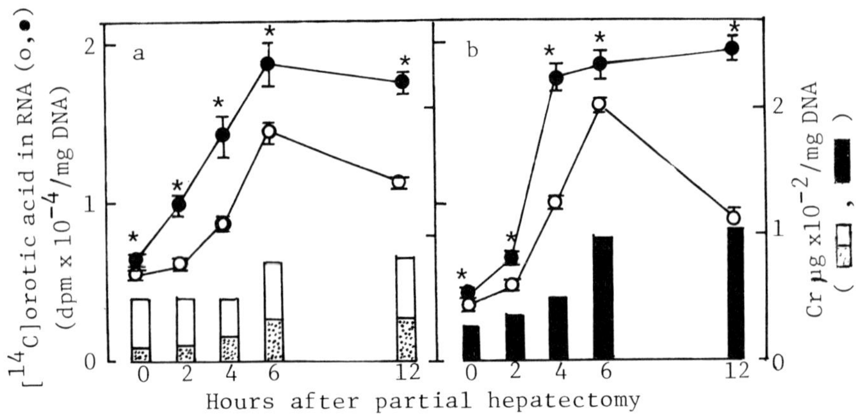

FIG. 1 Cr-accumulation and RNA synthesis in the whole cells (a) and nucleoli (b) of regenerating rat liver. Cr-accumulation: ▭ in whole cells, ▨ in nuclei, ■ in nucleoli. ● $CrCl_3$-dosed (5 mg Cr/kg), o control. mean ± S.E. for 3 rats. *$p < 0.05$ to each control.

(b) *Processing of Nucleolar RNA in Regenerating Liver of Cr(III)-Treated Rats*

The hepatic RNA of Cr-dosed (5 mg Cr/kg) rats was pulse-labeled with [^{14}C]orotic acid 4 hr after the partial hepatectomy and chased with actinomycin D (0.55 mg/kg, i.p.) (Fig. 2). The radioactivity in nucleolar RNA of Cr-dosed rats, initially much higher than that of control, decreased with the chase time and appeared in rRNA, the processing product of nucleolar pre-rRNA, similarly to, but higher than the control. Furthermore, the analyses of sucrose density gradient centrifugation profiles and nucleotide compositions indicated that the nucleolar

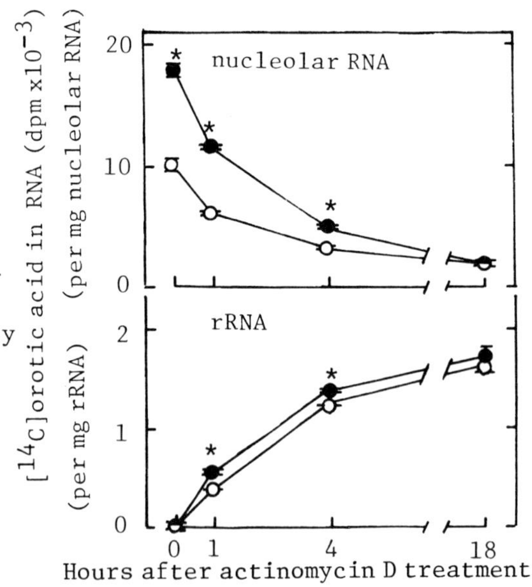

FIG. 2 Fate of pulse-labeled nucleolar RNA of $CrCl_3$-dosed (5 mg Cr/kg) (●) and control (o) regenerating liver. mean ± S.E. for 3 rats. *$p < 0.05$ relative to each control.

RNA of Cr-dosed regenerating liver was normally processed into rRNA.

(c) *Effect of Cr (III)-Administration on DNA Synthesis in Regenerating Liver*

It is well-known that the rapid increase in nucleolar RNA synthesis in the early stage of liver regeneration triggers the synchronized DNA synthesis. In Cr-dosed (5 mg Cr/kg) rats the DNA synthesis started a little earlier and continued longer, i.e., it was less synchronized than in control rats (Fig. 3) (Okada et al. 1982b) .

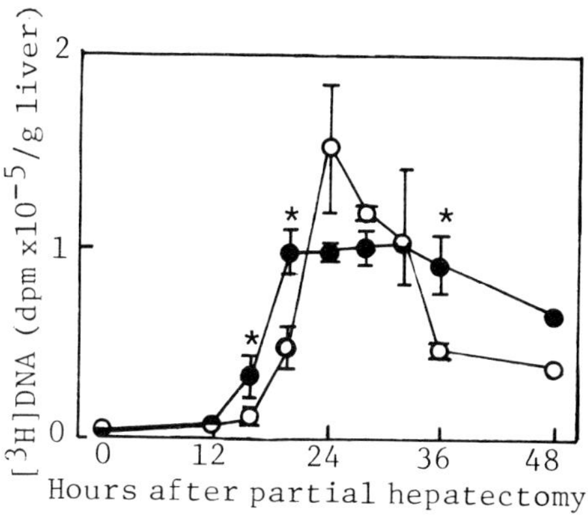

FIG. 3 DNA synthesis in regenerating liver of $CrCl_3$-dosed (5 mg Cr/kg) (●) and control (o) rats. mean ± S.E. for 3-4 rats except at 0, 12 and 48 hr.(one rat for each). *$p < 0.05$ to each control.

(d) *Isolation and Properties of Enhancing Factor for Nucleolar RNA Synthesis*

To see whether the enhancement of nucleolar RNA synthesis in Cr-dosed regenerating liver arose from the activation of nucleolar chromatin itself or from the occurrence of RNA synthesis-enhancing factor in the nucleoli, nucleolar chromatin was fractionated by the method of Muramatsu and Onishi (1977) from which EDTA-treatment was omitted. The nucleoli isolated from Cr-dosed (5 mg Cr/kg) regenerating liver had a significantly higher activity of *in vitro* RNA synthesis catalyzed by *E. coli* RNA polymerase than those from control, whereas the isolated nucleolar chromatin showed the same level of activity as control. During the isolation procedure, 0.15 M NaCl extraction of nucleoli lowered the activity of nucleoli of Cr-dosed rats to the same level as that of control, suggesting that the RNA synthesis-enhancing factor was removed by this extraction. The presence of this factor in the 0.15 M NaCl-extract was confirmed by recombining this extract with the residual nucleoli, which resulted in reactivation of the nucleoli.

This enhancing factor was isolated and purified from the 0.15 M NaCl-extract by successive chromatography with Sephadex G-75, DEAE-cellulose, CM-cellulose, and Sephadex G-100 or high performance liquid chromatography using TSK G-3000SW column. Thus purified, this factor is a monomeric protein of ca. 70000 molecular weight and was found to contain 5-6 atoms of Cr which were not effectively removed by further gel filtration (Table 1).

TABLE 1

Properties of Enhancing Factor for Nucleolar RNA Synthesis
(Cr-containing 70K Nucleolar Protein)

Molecular weight	
by SDS-PAGE with and without 2-mercaptoethanol	*ca.* 70000
by HPLC with and without 2-mercaptoethanol	*ca.* 70000
λ_{max}	278 nm
Cr/molecule (determined by using ^{51}Cr)	5.8 atoms
pKI	5.2
Stability	
60 °C, 5 min	stable
90 °C, 5 min	unstable
Amount giving maximum enhancement (per μg DNA in nucleoli)	0.8 μg

3. CONCLUSIONS

In the regenerating liver of $CrCl_3$-dosed (5 mg Cr/kg) rats, nucleolar RNA synthesis was enhanced specifically and DNA synthesis was less synchronized in comparison with control regenerating liver. This nucleolar RNA seemed to be normal in character and in its processing into rRNA. The factor enhancing nucleolar RNA synthesis was isolated from the nucleoli and estimated to be 70K protein containing Cr; its occurrence and role in or effect on the nuclear gene functions must be of particular interest.

4. REFERENCES

M. Muramatsu and T. Onishi (1977) in *Methods in Cell Biology* (Ed. D.M. Prescott), Academic Press, New York, vol. 17, pp. 141-161.

S. Okada, H. Ohba and M. Taniyama (1981) *J. Inorg. Biochem.*, 15, 223-231.

S. Okada, M. Taniyama and H. Ohba (1982a) *J. Inorg. Biochem.*, 17, 41-49.

S. Okada, M. Suzuki, H. Tsukada and H. Ohba (1982b) *Chem. Pharm. Bull.*, 30, 3439-3441.

S. Okada, M. Suzuki and H. Ohba (1983) *J. Inorg. Biochem.*, in the press.

GENOTOXICITY OF STANNOUS CHLORIDE IN MAMMALIAN CELL TESTS

S.H.H. Swierenga and J.R. McLean

Health and Welfare Canada
Drug Research Bureau, Ottawa, K1A 0L2

and Radiation Protection Bureau
Ottawa, Canada, K1A 1C1

1. INTRODUCTION

The accumulation of tin compounds in the environments of industrialized societies has been recently documented, Hodge et al. (1979). The long term consequences to human populations from this chronic exposure however have not yet been determined. One novel but little known use of inorganic tin (II) is in injectible radiopharmaceuticals for diagnostic purposes, where tin functions as a reducing agent for the tracer, resulting in a relatively high tin (II) exposure. Recently, it has been shown that inorganic tin can be biomethylated to more toxic compounds, Ridley et al. (1977) and that tin (II) can damage the DNA of cells in vitro, McLean et al. (1983).

In this report we show that tin (II) can: in some radiopharmaceuticals damage the DNA of human WBC; inhibit DNA synthesis in replicating cells; enhance the level of unscheduled DNA synthesis (UDS) in methylmethanesulfonate (MMS)-treated hepatocytes, and induce mutations in mammalian cells in vitro.

2. MATERIALS AND METHODS

Tin (II) chloride, tartrate or fluoride, or tin (IV) chloride, were dissolved in dilute HCl immediately before use. Tin containing radiopharmaceuticals were obtained as nonradioactive "kits" and reconstituted before use with isotonic saline.

Human WBC were isolated from freshly collected whole blood and resuspended in a balanced salt solution, Birnboim and Jevak (1981).

Tin (II)-113 dichloride (NEN, 18.2 mCi/mg) was used to measure WBC binding of tin (II) at 37°C, Strauss et al. (1977).

Fluorimetric analysis of DNA unwinding (FADU) was determined with human WBC, Birnboim and Jevak (1981).

A rat liver cell line T51B, Swierenga et al. (1978), was used to determine effect of tin (II) on entry of synchronized cells into S-phase (arrested by lowering medium calcium concentration to 0.02 mM and stimulated to initiate S-phase by calcium readdition) and to determine the ability of tin (II) to induce 8-azaguanine resistant mutants,

Chemical Toxicology and Clinical Chemistry of Metals
ISBN 0.12.137520.X
Copyright © 1983 by IUPAC
All rights of reproduction in any form reserved

according to the method of Tong et al., 1980. DNA repair assays were performed with freshly isolated hepatocytes, as described by Williams, 1980.

3. RESULTS AND DISCUSSION

Tin (II), but not tin (IV), was rapidly taken up by human WBC and caused DNA strand breaks as measured by the FADU technique (Table 1). It has been shown previously by X-ray fluorescence analysis that only tin II was incorporated into nuclei of intact cells. (McLean et al., manuscript in preparation). EDTA, a polar metal chelator that does not cross cell membranes, prevented tin (II) from binding to cells and from producing DNA damage. In contrast, tin (II) in the presence of the related but less polar chelator CDTA, was able to bind cells and produced extensive DNA damage (Table 1). Tin (II)-phytate (mesoinositol), a liver scanning agent with 2 or 3 tin (II) binding sites, increased both the cell-bound tin (II) and the level of DNA damage (Table 1). Tin (II)-fluoride colloid-another liver scanning agent - and tin (II)-methylene diphosphonate (MDP), a bone scanning agent, were also bound by cells and produced DNA/damage in vitro (Table 1). Since tin (II) has a high affinity for hard bases (such as oxygen), any tin (II) able to reach the cell nucleus could produce DNA strand breaks by a reductive cleavage of the sugar-phosphate back bone. Alternatively, tin (II) may reduce the oxygen dissolved in the medium by a one electron transfer to the highly reactive and DNA damaging superoxide anion.

TABLE 1 DNA DAMAGE IN WBC EXPOSED TO TIN (II) - CONTAINING RADIOPHARMACEUTICALS

Treatment[1]	Tin (II) Binding (f moles/cell)	Strand Breaks per Genome[2]
Tin (II) Chloride	7.66	7458
Tin (II) CDTA	1.15	4658
Tin (II) Phytate	16.75	7235
Tin (II) Fluoride (Colloid)	3.42	2094
Tin (II) EDTA	0.02	119
Tin (II) MDP	0.18	1865

[1] 50 μM tin (II), 30 min. at 37°C; [2] assayed by FADU, Genome = $3X10^{12}$D

Tin (II) inhibited the rate at which Ca(II) re-initiated DNA synthesis in cells previously arrested in the G_1 phase of their cell cycle by a Ca(II) deprivation (Table 2). This inhibition may be due to an inactivation of the DNA template, as occurs after the exposure of cells to low doses of ionizing radiation.

TABLE 2 EFFECT OF STANNOUS CHLORIDE ON INITIATION OF DNA SYNTHESIS IN SYNCHRONIZED CULTURES OF T51B RAT LIVER CELLS

Treatment[1]	Per Cent Labeled Cells[2]
HOH	12.9 ± 0.9
10 μM $SnCl_2$	5.4 ± 1.1
$CaCl_2$	45.0 ± 3.5
$CaCl_2$ + 10 μM $SnCl_2$	33.3 ± 1.6
$CaCl_2$ + 25 μM $SnCl_2$	25.9 ± 2.7

[1] test compounds added to cells arrested in G_1 by a 24 hr medium calcium deprivation.

[2] measured autoradiographically 18 hr after addition of test compounds.

Tin (II) did not induce UDS in freshly isolated rat hepatocytes but did enhance the UDS in MMS exposed cells (Table 3). Tin (II) could be methylated in situ by MMS to a more damaging organotin or, alternatively, the transfer of methyl groups to DNA could be catalyzed by the presence of tin (II). Both mechanisms would increase the levels of DNA damage and UDS in the cells.

TABLE 3 EFFECT OF STANNOUS CHLORIDE ON DNA REPAIR INDUCED IN RAT HEPATOCYTES BY METHYLMETHANE-SULFONATE (MMS)

Treatment	Net Grains Per Nucleus	Percent of Cells in Repair
untreated	2.7 ± 4.7	0
MMS 20 μM	19.1 ± 11.0	42
$SnCl_2$ 0.5-10 μM	4.5 ± 6.4	0
MMS + 1 μM $SnCl_2$	29.8 ± 12.2	61
MMS + 5 μM	32.4 ± 10.8	84
MMS + 10 μM	49.0 ± 9.3	95
MMS + 25 μM	28.6 ± 7.9	51

Tin (II) induced mutations in T51B cells, in a random manner, that was not related to concentration, or the presence of freshly isolated hepatocytes as an added activation system (Table 4). The presence of the hepatocytes however, increased the consistency of this response. The inconsistency of the mutation response may be due to the labile nature of tin (II) and the as yet undefined complex changes which it undergoes in solution. The presence of hepatocytes, for examples, could introduce endogenous alkylating agents (e.g. S-adenosylmethionine) into the system which could interact with tin (II) to produce the mutation-inducing lesions in DNA.

TABLE 4 INCIDENCE OF 8-AZAGUANINE-RESISTANT MUTANTS IN T51B RAT LIVER CELLS

Treatment	Mutants per 10^6 survivors Experiment 1	Experiment 2	Experiment 3
HOH	138 ± 32	8 ± 15	47 ± 18
HOH ± hep[1]	17 ± 29	15 ± 17	60 ± 42
DMBA		444 ± 85	536 ± 198
DMBA + hep[1]	1271 ± 311	972 ± 45	789 ± 156
5 μM $SnCl_2$			29 ± 23
5 μM $SnCl_2$ + hep[1]			324 ± 146
10 μM $SnCl_2$			17 ± 19
10 μM $SnCl_2$ + hep[1]		15 ± 17	
25 μM $SnCl_2$	363 ± 238	184 ± 81	18 ± 20
25 μM $SnCl_2$ + hep[1]	1154 ± 239	8 ± 16	21 ± 43

[1] freshly isolated rat hepatocytes added as activating system.

4. REFERENCES

Birnboim, H.C. and Jevak, J., (1981) Cancer Res. 41, 1889-1893.

Hodge, V.F., Seidel, S.L. and Goldberg, E.D. (1979) Anal. Chem. 51, 1256-1259.

McLean, J.R.N., Blakey, D.H., Douglas, G.R. and Kaplan, J.G. (1983) Mutat. Res. Lett. 119, 195-201.

Ridley, W.P., Dizikes, L.J. and Wood, J.M. (1977) Science 197, 329-332.

Strauss, R.R., Sheehan, J.M. and Kashket, E.R. (1977) Immunol. 118, 1328-1334.

Swierenga, S.H.H., Whitfield, J.F., and Karasaki, S., Proc. Natl. Acad. Sci. USA 75, 6069-6072, (1978).

Tong, C. and Williams, G.M. Mut. Res. 74, 1-9, (1980).

Williams, G.M., Cancer Letters, 4, 69-75 (1978).

CLINICAL AND EXPERIMENTAL STUDIES

IATROGENIC METAL POISONING IN MAN - ALUMINIUM

Michael R. Wills and John Savory

Departments of Pathology, Internal Medicine and Biochemistry
University of Virginia Medical Center
Charlottesville, Virginia 22908 U.S.A.

1. INTRODUCTION

The word iatrogenic is derived from the Greek words 'iatric', a healer or physician, and 'genesis', the origin, creation or generation. According to Webster's Dictionary, the word means "induced by a physician" and is "used chiefly of imagined ailments induced in a patient by autosuggestion based on a physician's words or actions during examination". The more common clinical usage of the word iatrogenic is to describe an adverse reaction in a patient that has occurred as a result of treatment prescribed by a physician. The excess or lack of a number of trace and heavy metals has been recognized for many years to be associated with clinical disturbances. The established list of metals of importance in this context includes antimony, arsenic, bismuth, cadmium, chromium, copper, cobalt, fluorine, gold, lead, maganese, mercury, nickel, selenium, silicon, silver, thallium and zinc. Of the metals listed, the only ones that have been associated with toxic phenomena and may be considered to be truly iatrogenic are bismuth and gold. Bismuth was used in the past for the treatment of syphilis and for some intestinal disorders; gold is still used in the treatment of rheumatoid arthritis. Aluminium can now be included in the list. In the past decade aluminium, has been recognized as a toxic metal and the associated clinical toxic phenomena may truly be classified as iatrogenic poisoning.

In reviewing the events of the past decade and the recognition of aluminium as a toxic metal, it is of interest that only nine years ago in an extensive

Chemical Toxicology and Clinical Chemistry of Metals
ISBN 0-12-137520-X
Copyright © 1983 by IUPAC
All rights of reproduction in any form reserved

review (Sorenson et al., 1974) it was concluded that "there is still no need for concern by the public or producers of aluminium or its products concerning hazards to human health derived from well established and extensively used products".

2. ALUMINIUM TOXICITY - NORMAL RENAL FUNCTION

It has been recognized for many years that the inhalation of dust heavily contaminated with aluminium, in industrial situations, can cause a severe and sometimes fatal pulmonary fibrosis. The latter is usually seen as an interstitial fibrosis of the upper lobes. It has also been proposed that aluminium may be neurotoxic and play a role in the pathogenesis of Alzheimer's disease. These proposals were based on observations, in experimental animals, that the intracerebral or intrathecal injection of aluminium salts induced neurofibrillary degeneration. Increased brain aluminium concentrations in patients with senile and presenile dementia of the Alzheimer type, comparable to those used in the experimental animals, added support to these proposals. Aluminium-induced neurofibrillary degeneration has, however, some ultrastructural differences from the lesions found in Alzheimer's disease. The potential role, if any, of aluminium as a specific neurotoxic agent in the pathogenesis of Alzheimer's disease and other neurological disorders awaits clarification, as do the mechanisms involved in the neurotoxicity of aluminium in experimental animals. Even if an increased tissue load of aluminium is incriminated as of importance in the pathogenesis of Alzheimer's disease, it cannot be classified as of iatrogenic origin.

3. ALUMINIUM TOXICITY - CHRONIC RENAL FAILURE

Although the toxicological importance of aluminium, if any, in subjects with normal renal function in non-industrial situations, remains to be defined, the toxicological importance of aluminium has been established in patients with chronic renal failure undergoing haemodialysis and on treatment with aluminium-containing oral medications, and dialysis. In this group of patients the tissue accumulation of aluminium causes an encephalopathy; it accounts for one component of dialysis osteodystrophy and may be responsible for some of the other complications of the long-term treatment of end-stage chronic renal failure. The importance of the recognition of the iatrogenic aluminium-induced complications of the treatment of chronic renal failure is that they can be prevented by the iatric!

(a) Hyperaluminemia

Aluminium salts are extensively used in the therapeutic management of the hyperphosphatemia which occurs in patients with chronic renal failure. Aluminium is absorbed from the gastrointestinal tract and in normal subjects (Figure 1) there is an increase in serum concentration after oral administration, followed by excretion in the urine (Mauras et al., 1982). In patients with impaired renal function due to end-stage chronic renal failure on

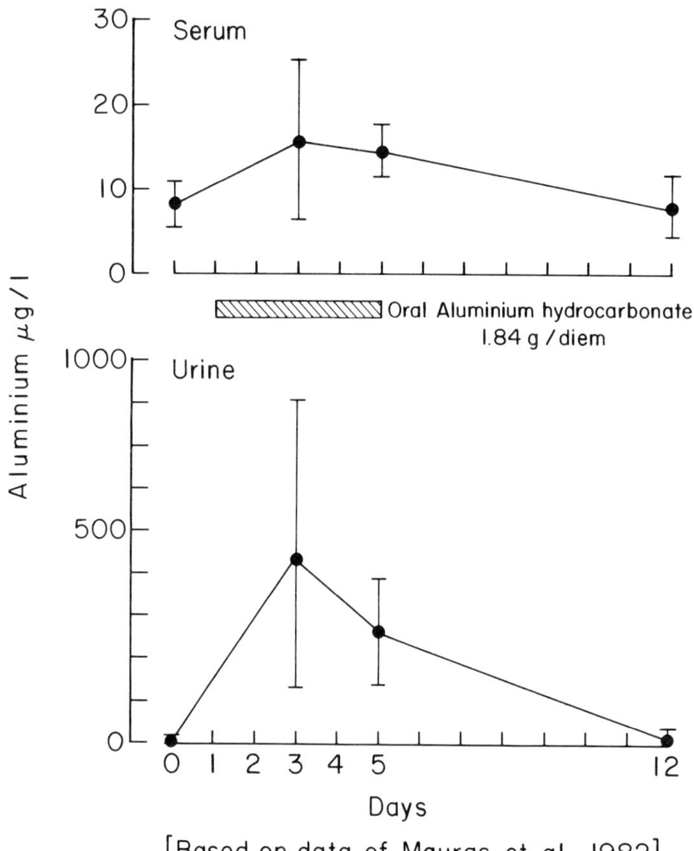

Fig. 1. Changes in serum and urine aluminium concentration and excretion following oral administration in 10 normal subjects.

long-term treatment there is hyperaluminemia with an accumulation of aluminium in a variety of tissues. The sources of the increased serum and tissue content are recognized to be two-fold by intestinal absorption, following oral administration of aluminium salts, and across the dialysis membrane from the dialysate. The aluminium content of the latter is dependent on the content in the water used to prepare the dialysate. Aluminium is present in high concentrations in some domestic tap water supplies. The source of the aluminium is either of natural origin or from the use of aluminium salts as a flocculant during the water purification process. In the hyperaluminemia of chronic renal failure, it has been proposed that parathyroid hormone may play a contributory role by increasing intestinal absorption and may also affect tissue distribution (Mayor et al., 1981). The precise interrelationships between aluminium and calcium homeostasis, and parathyroid hormone secretion await clarification.

Marked increases in serum aluminium concentration in some patients with end-stage chronic renal failure were first reported by Berlyne and his colleagues in 1970. The values were increased in some patients on oral treatment with aluminium-cycle resins and aluminium hydroxide, and in some patients on dialysis treatment who were not taking oral aluminium salts but who had been "fortuitously exposed to a dialysate with a relatively high aluminium content". They proposed that the increased values in these patients were "probably accounted for by considerable intestinal absorption of aluminium ions from the intestine and impairment of the normal efficient renal mechanism of excretion of aluminium". Berlyne and his colleagues stated that the toxic effects of hyperaluminemia were unknown in man. They subsequently reported the effects of aluminium poisoning, in uremic and non-uremic rats, after modest doses of oral and parenteral aluminium salts (Berlyne et al., 1972). In these animals aluminium poisoning was associated with periorbital bleeding, lethargy, anorexia and death. These clinical findings were associated with significant increases in the serum concentration and the aluminium contents of liver, heart, striated muscle, brain and bone tissues. Berlyne and his colleagues (1972) proposed that, because of the massive deposition in these various tissues, it was possible that aluminium had a toxic role and possibly accounted for some of the features of the clinical syndrome of uremia.

(b) Dialysis Encephalopathy

In 1972 the occurrence was reported of a new progressive fatal neurological syndrome in patients with chronic renal failure on long-term intermittent hemodialysis treatment (Alfrey et al., 1972). In these patients, the first recognizable symptom was a speech disorder, followed by the development of dementia, convulsions and myoclonus. Alfrey and his colleagues proposed initially, in 1972, that the syndrome was possibly a metabolic disorder due to the accumulation of trace metals; they subsequently proposed that it was due to aluminium intoxication (Alfrey et al., 1976). The proposal was based on their findings of increased aluminium content in brain, muscle and bone tissues in affected patients. The syndrome has been called either dialysis encephalopathy or dialysis dementia, and unless preventive measures are taken, it occurs in patients with end-stage chronic renal failure after three to seven years on treatment. The source of the increased tissue aluminium in patients with the dialysis encephalopathy syndrome, was originally considered to be derived from the oral aluminium hydroxide which had been administered therapeutically to control serum phosphate concentration. A positive aluminium balance had been reported previously in patients with chronic renal failure on hemodialysis treatment who were taking oral aluminium containing phosphate-binding gels (Clarkson et al., 1972). The bulk of the available evidence is now consistent with the view that the source of the increased brain tissue aluminium in patients with dialysis encephalopathy can be from either the dialysate or the gastrointestinal tract.

Aluminium is now generally accepted as the major toxic factor in the etiology of the dialysis encephalopathy syndrome (Sideman and Manor, 1982). It is also recognized that this syndrome appears to have a common etiology with one of the osteomalacic components of dialysis osteodystrophy (Parkinson et al., 1981). Hyperaluminemia and dialysis encephalopathy may occur not only in patients on hemodialysis treatment but also in those on peritoneal dialysis and in some patients who have not been dialyzed. Non-dialyzed patients with the syndrome have been confined to children with renal failure on oral treatment with aluminium hydroxide. It has been proposed by other workers that the dialysis encephalopathy syndrome has a multifactorial etiology (Prior et al., 1982) but even if the syndrome has a multifactorial etiology, there is no doubt that aluminium plays a major role. Inhibition by aluminium of dihydropteridine reductase is one potential mechanism for its neurotoxicity. The inhibition of this enzyme would result in a reduction in brain tetrahydrobiopterin, tyrosine and neurotransmitter concentrations. The neurotoxicity of aluminium may also involve alterations in the major postsynaptic enzymes of cholinergic neurotransmission. Although the mechanism remains to be defined, there is now a considerable amount of evidence that the accumulation of aluminium in brain tissue is associated with toxic sequelae.

(c) Dialysis Osteodystrophy

Bone pain is a major problem among patients with chronic renal failure on long-term intermittent hemodialysis treatment. The bone pain in these patients is due to metabolic bone disease, a direct consequence of chronic renal failure. The term "dialysis osteodystrophy" should be used in these patients for the progressive metabolic bone disease to distinguish it, and some aspects of its pathogenesis, from renal osteodystrophy in the undialyzed patient. Osteomalacia is one of the major components of dialysis osteodystrophy and its occurrence is associated with a high incidence of fractures. One form of osteomalacia in patients with chronic renal failure on treatment is unresponsive to treatment with either vitamin D or its biologically active metabolites. The incidence and rate of progression of dialysis osteodystrophy appears to vary not only between countries but also between dialysis centers within a country despite apparently similar dialysis techniques. One of the characteristics of dialysis osteodystrophy is that it progresses or develops despite the maintenance of serum concentrations of calcium and magnesium, and their ionized fractions, at values which in a healthy person would not interfere with the normal process of bone mineralization (Varghese et al., 1973). The availability of adequate amounts of phosphate, for the formation of hydroxyapatite, is of critical importance in the normal process of bone mineralization. The administration of aluminium-containing medications can induce a state of phosphate deficiency in patients with either normal renal function or in those with chronic renal failure. A state of phosphate deficiency, manifested by hypophosphatemia, is one of the recognized causes of osteomalacia.

(d) Hypophosphatemic Osteomalacia

Hypophosphatemic osteomalacia is a recognized iatrogenic complication of the oral administration of aluminium hydroxide in patients with end-stage chronic renal failure on hemodialysis treatment (Baker et al., 1974). Hypophosphatemic osteomalacia has also been reported in patients on hemodialysis who were not taking oral aluminium hydroxide; in these patients, there was evidence of impairment in the intestinal absorption of phosphate (Ahmed et al., 1976). Ahmed and his colleagues proposed that in this group of patients, the osteomalacia was the result of the combination of intestinal phosphate malabsorption, phosphate loss during dialysis, and a low phosphate dietary intake which was a part of their therapeutic regimen. The majority of patients with chronic renal failure have phosphate retention with hyperphosphatemia, and aluminium-induced hypophosphatemic osteomalacia may be considered to be a rare complication of end-stage chronic renal failure.

(e) Osteomalacic Dialysis Osteodystrophy

Attention was first drawn to potential toxic role of aluminium in the etiology of metabolic bone disease in patients with chronic renal failure on treatment in 1971 (Parsons et al., 1971). Parsons and his colleagues reported that the bone aluminium content was increased and that the values were higher in those patients who had been on hemodialysis for the longest period of time. They also reported that they could find no correlation between the bone aluminium content and the amount of oral aluminium hydroxide that had been administered. Parsons and his colleagues did not speculate on the potential sources of the aluminium in their patients; it could have been implied from their findings, however, that the hemodialysis procedure itself was potentially involved. Varghese and his colleagues in 1973 proposed that treatment by hemodialysis itself may play an important role in the etiology of the progressive metabolic bone disease and extraosseous calcification in patients on that form of treatment. Their proposal was based on the fact that could find no differences between a number of serum biochemical variables in a large group of patients with end-stage chronic renal failure on hemodialysis treatment when sub-divided into those with and those without radiological evidence of metabolic bone disease. However, they did find that the patients with evidence of metabolic bone disease had been receiving intermittent hemodialysis treatment for a significantly longer period than those without radiological evidence of metabolic bone disease. Varghese and his colleagues did not speculate on the potential factor or factors involved, but they did note that all of the patients were on regular treatment with aluminium hydroxide to reduce intestinal phosphate absorption.

Prevalence. The prevalence of dialysis encephalopathy and spontaneous fractures among 202 patients with chronic renal failure who were receiving hemodialysis treatment in their homes was reported on by Platts et al., 1977. They had found that there was an uneven geographical distribution of these complications which led them to investigate the home tap water supplies that were being used to prepare the dialysate for the hemodialysis. The home tap water used by the patients who developed either fractures or dialysis encephalopathy had a lower concentration of calcium and fluorine and a higher concentration of aluminium and manganese than the home tap water used by those patients without these complications. They also found that the patients with multiple fractures had been using tap water that contained higher aluminium and manganese contents to prepare their dialysate than those with a single fracture. Aluminium hydroxide ingestion was not considered by Platts and his colleagues to be a contributing factor in the development of these complications of hemodialysis because they had found a low compliance rate for this medication; only some of the patients took the gel and even they did not do so consistently. Platts and his colleagues concluded that "some contaminant in the water used for dialysis is very probably responsible for the development of dialysis encephalopathy and pathological fractures". Although they did not definitively incriminate aluminium as the specific toxic contaminant they proposed that patients who were being dialyzed in areas with a high aluminium content in the tap water should be supplied with water deionizers. It had been reported previously that the trace metals copper and zinc were actively taken up by blood during hemodialysis from the dialysis fluid even against a concentration gradient (Blomfield et al., 1969). Alterations in copper and zinc status, in patients with chronic renal failure, have been implicated in the anemia that is a feature of the syndrome of uremia.

Prevention. The use of deionized water to prepare the dialysate has been reported to reduce the incidence of the osteomalacic component of dialysis osteodystrophy. After one to four years of hemodialysis, osteomalacia was evident in only 15% of a group of patients using deionized water to prepare the dialysate, compared with 70% in a group using water from the same source which had not been deionized (Ward et al., 1978). Ward and his colleagues also reported that osteomalacia and dialysis encephalopathy occurred in their dialysis patients who used tap water with a high aluminium content; they proposed that the close association between these two complications of hemodialysis was consistent with a common etiology. In Great Britain, an epidemiological survey of 1,293 patients in 18 hemodialysis centers found that there was a highly significant correlation between the incidence of osteomalacic dialysis osteodystrophy and dialysis encephalopathy with the aluminium content of the water used to prepare the dialysate (Parkinson et al., 1979). It would now appear to be generally agreed that aluminium is one of the principal factors in the etiological mechanisms of osteomalacic dialysis osteodystrophy. However, it must be recognized that aluminium toxicity does not account entirely for all of the osteomalacia found in patients on hemodialysis but it is responsible for one particular type of osteomalacia. The particular type of osteomalacia which is aluminium-induced is progressive, associated with a myopathy, and is relatively resistant to treatment with either vitamin D or its biologically active metabolites.

Source of Aluminium. The source of the aluminium in patients with osteomalacic dialysis osteodystrophy is either from the water used to prepare the dialysate (and subsequent absorption across the dialysis membrane) or from the intestinal absorption of aluminium that has been given orally in the form of aluminium-containing phosphate-binding gels. The oral absorption of aluminium from the gastrointestinal tract has been the sole-source in some patients; usually children (Griswold et al., 1983). During hemodialysis, the all important factor or driving force for aluminium transfer across the dialysis membrane, appears to be the effective concentration gradient between the dialysate aluminium and the free diffusible serum aluminium concentration. The mechanisms involved in the intestinal absorption are not clear. It is, however, possible that individual absorptions vary widely, especially in disease states such as chronic renal failure, and that some patients may absorb excessive amounts from the intestinal tract after oral administration. In these patients, the intestinal absorption of aluminium from phosphate-binding gels could play the dominant role.

Mechanism of Toxicity. The mechanism whereby aluminium induces disorderly calcification with consequent osteomalacia has not been clearly defined. It has been established that aluminium ions in an in vitro buffer system can destabilize the buffer and form an insoluble phosphate (Bachra and Van Harskamp, 1970). The formation of insoluble aluminium phosphate could initiate the precipitation of calcium apatite and this effect in vivo would interfere with the normal orderly deposition of the mineral phase of bone. After the intraperitoneal injection of aluminium chloride it has been reported that there is a marked increase in the bone content of aluminium (Ellis et al., 1979). Ellis and his colleagues reported that the increase in the bone content of aluminium was associated with a disturbance in the normal mineralization process and that this was associated with the development of osteomalacia. They also reported that there was a relatively long lag period between the marked increase in the bone content of aluminium and the subsequent development of a defect in the bone mineralization process. Using a variety of different techniques, several different groups of workers have reported that in patients with osteomalacic dialysis osteodystrophy, the aluminium is mainly localized at the interface between the uncalcified osteoid and the calcified matrix. The interface, or calcification front, between the mineralized and non-mineralized matrix is normally the first site of bone mineral deposition and any accumulation of a "foreign" ion at that site would undoubtedly interfere with the normal formation of the calcified matrix. The mechanism by which aluminium interferes with the normal calcification process may involve the action of aluminium on the activities of bone acid and alkaline phosphatases and the response of these enzymes to parathyroid hormone and 1,25-dihydroxy-cholecalciferol (Lieberherr et al., 1982). It is of interest that although the bone content of aluminium in patients with dialysis osteodystrophy of predominantly the osteitis fibrosa type is increased, it is not localized in one specific area but appears to be diffusely distributed throughout the bone tissue (Cournot-Witmer et al., 1981).

(f) Anemia

In patients with chronic renal failure, anemia is a common finding. The anemia is of the normochromic normocytic type, it has a multifactorial etiology and it is not simply the end-result of a decrease in the renal production of erythropoietin (Wills, 1978). In the etiology of anemia in patients on long-term treatment with hemodialysis, a variety of toxic agents has been incriminated in the past and the list includes copper, zinc, nitrates, formaldehyde and chloramines. In recent years aluminium has been added to the list. Aluminium-induced anemia has been reported with a high prevalence in patients with osteomalacic dialysis osteodystrophy or dialysis encephalopathy. In patients with the latter complication of hemodialysis, it has been reported to precede the development of neurological symptoms (Elliott and MacDougall, 1978). The characteristics of aluminium-induced anemia are that it is non-iron-deficient, microcytic, hypochromic and responds to the use of a low-aluminium dialysate (O'Hare and Murnahagan, 1982). The various techniques, such as deionization and reverse osmosis, which are used to reduce or remove the aluminium content of domestic tap water supplies in the preparation of a low-aluminium dialysate could also remove some, if not all, of the other toxic agents that have been incriminated in the past. The nature of the anemia induced by aluminium has led O'Hare and Murnahagan (1982) to propose that it may be due to a disturbance in heme synthesis, specifically involving δ-aminolevulinic acid dehydrogenase.

(g) Other Toxic Effects

There is a high prevalence of soft-tissue or metastatic extra-skeletal calcification in patients with chronic renal failure. In the past the explanation for the soft-tissue or metastatic calcification has been that it is the direct consequence of super-saturation of body fluids with calcium and phosphorus and the consequent precipitation of calcium phosphate salts, especially in tissues where there is a localized alkalosis. In this concept of the mechanism for metastatic calcification, the most important factor has always been considered to be phosphate retention with consequent hyperphosphatemia. The chemical and crystalline composition of the calcium phosphate deposits in metastatic calcification in patients with chronic renal failure is of two distinct types (Contiguglia et al., 1973). One type is in the form of hydroxyapatite while the other is seen as an amorphous or microcrystalline compound, composed of calcium, magnesium and phosphorus in a constant molar ratio. In view of the observations that aluminium can initiate the precipitation of calcium apatite (Bachra and Van Harskamp, 1970) it could be proposed that aluminium may play a role in the pathogenesis of the metatastic calcification of chronic renal failure.

Aluminium toxicity has been incriminated in the sudden cardiac arrest that occurs in some patients with dialysis encephalopathy (Elliott et al., 1978). The patients reported by Elliott and his colleagues were in good fluid and electrolyte balance, and had no recent history of problems from either hypertension or ischemic heart disease; the cardiac arrest usually presented as acute pulmonary edema. Elliott and his colleagues (1978) proposed that, as aluminium is known to inhibit enzyme systems involving ATP and magnesium, high serum and tissue aluminium concentrations could exert a cardiotoxic effect.

4. CONCLUSION

In the past decade, aluminium has been recognized as a toxic metal and the clinical phenomena associated with a tissue excess of this metal may be truly classified as of iatrogenic origin in patients with chronic renal failure on long-term treatment. The increased tissue load of aluminium may be derived either from the concentrate itself or the tap water used to prepare the dialysate, or from intestinal absorption following the oral administration of aluminium containing gels. The latter are the main therapeutic agents for the control of the hyperphosphatemia of chronic renal failure. The increased tissue content of aluminium appears to be a major, if not the major, factor in the etiology of dialysis encephalopathy or dementia and a specific type of osteomalacia; these are long-term treatment complications of hemodialysis. Aluminium toxicity also appears to be an etiological factor in a non-iron deficient, microcytic, hypochromic anemia that occurs in the same group of patients. It is possible that an excess of aluminium also plays a role in some of the other complications of the long-term treatment of end-stage chronic renal failure. The prevention of iatrogenic aluminium poisoning involves the cautious use of aluminium-containing oral phosphate binders, together with regular monitoring of (i) the aluminium content of the dialysate (ii) the domestic tap water used to prepare the dialysate and (iii) serum aluminium concentration. Aluminium poisoning, with its multiple toxic sequelae, is a topic of current major toxicological importance which requires considerable further evaluation not only with regard to its prevalence but also with regard to the mechanisms by which aluminium exerts its toxic effects.

REFERENCES

K.Y. Ahmed, Z. Varghese, M.R. Wills, E. Meinhard, R.K. Skinner, R.A. Baillod and J.F. Moorhead (1976) Lancet, 2, 439-4432.

A.C. Alfrey, G.R. LeGendre and W.D. Kaehny (1976) N. Engl. J. Med., 294, 184-188.

A.C. Alfrey, J.M. Mishell, J. Burks, S.R. Contiguglia, H. Rudolph, E. Lewin and J.H. Holmes (1972) Trans. Am. Soc. Artif. Int. Organs, 18, 257-261.

B.N. Bachra and G.A. Van Harskamp (1970) Calcified Tissue Res., 4, 359-365.

L.R.I. Baker, P. Ackrill, W.R. Cattell, T.C.B. Stamp and L. Watson (1974) Br. Med. J., 3, 150-152.

G.M. Berlyne, J. Ben-Ari, D. Pest, J. Weinberger, M. Stern, G.R. Gilmore and R. Levine (1970) Lancet, 2, 494-496.

G.M. Berlyne, R. Yagil, J. Ben-Ari, G. Weinberger, E. Knopf and G.M. Danovitch (1972) Lancet, 1, 564-568.

J. Blomfield, J. McPherson and C.R.P. George (1969) Br. Med. J., 2, 141-145.

E.M. Clarkson, V.A. Luck, W.V. Hynson, R.R. Bailey, J.B. Eastwood, J.S. Woodhead, V.R. Clements, J.L.H. O'Riordan and H.E. DeWardener (1972) Clin. Sci., 43, 519-531.

S.R. Contiguglia, A.C. Alfrey, N.L. Miller, D.E. Runnells and R.Z. LeGros (1973) Kidney Intl., 4, 229-235.

G. Cournot-Witmer, J. Zingraff, J.J. Piacnot, F. Escaig, R. Letevre, P. Boumati, A. Bourdeau, M. Garabedian, P. Galle, R. Bourdon, T. Drueke and S. Balsan (1981) Kidney Intl., 20, 375-385.

H.L. Elliott and A.I. MacDougall (1978) Proc. Eur. Dial. Transpl. Assoc., 15, 157-163.

H.L. Elliott, A.I. MacDougall and G.S. Fell (1978) Lancet, 1, 1203.

H.A. Ellis, J.H. McCarthy and J. Herrington (1979) J. Clin. Pathol., 32, 832-844.

W.R. Griswold, V. Reznik, A. Mendoza, D. Trauner and A.C. Alfrey (1983) Pediatrics, 71, 56-58.

M. Lieberherr, B. Grosse, G. Cournot-Witmer, C.L. Thil and S. Balsan (1982) Calcif. Tissue Intl., 34, 280-284.

Y. Mauras, P. Allain and P. Riberi (1982) Therapie, 37, 593-605.

G.H. Mayor, S.M. Sprague and T.V. Sanchez (1981) Am. J. Kidney Dis., 1, 141-145.

J.A. O'Hare and D.J. Murnahagan (1982) N. Engl. J. Med., 306, 654-656.

I.S. Parkinson, M.K. Ward, T.G. Feest, R.W.P. Fawcett and D.N.S. Kerr (1979) Lancet, 1, 406-409.

I.S. Parkinson, M.K. Ward and D.N.S. Kerr (1981) J. Clin. Pathol., 34, 1285-1294.

V. Parsons, C. Davies, C. Goode, C. Ogg and J. Siddiqui (1971) Br. Med. J., 4, 273-275.

M.M. Platts, G.C. Goode and J.S. Hislop (1977) Br. Med. J., 2, 657-660.

J.C. Prior, E.C. Cameron, W.J. Knickerbocker, V.P. Sweeney and O. Suchowersky (1982) Am. J. Med., 72, 33-42.

S. Sideman and D. Manor (1982) Nephron, 31, 1-10.

J.R.J. Sorenson, I.R. Campbell, L.B. Tepper and R.D. Lingg (1974) Environ. Health Perspectives, 8, 3-95.

Z. Varghese, J.F. Moorhead and M.R. Wills (1973) Lancet, 2, 985-988.

M.K. Ward, I.G. Feest, H.A. Ellis, I.S. Parkinson, D.N.S. Kerr, J. Herrington and G.L. Goode (1979) Lancet, I, 841-845.

M.R. Wills (1978) Metabolic Consequences of Chronic Renal Failure, Harvey Miller and Medcalf, Aylesbury, England, 2nd edn.

CARCINOGENICITY OF NICKEL COMPOUNDS IN ANIMALS*

F. William Sunderman Jr.

Department of Laboratory Medicine and Pharmacology
University of Connecticut School of Medicine
263 Farmington Ave., Farmington, CT 06032, U.S.A.

1. INTRODUCTION

Campbell (1943) reported that chronic inhalation of nickel dust caused two-fold increase of lung tumor incidence in mice. During the forty years since Campbell's pioneering study, the carcinogenicity of nickel compounds in experimental animals has been studied intensively in sixteen laboratories, directed by Bruni, Furst, Gilman, Hildebrand, Hueper, Jasmin, Kasprzak, Mason, Nettesheim, Payne, Shimkin, Sunderman Sr., Sunderman Jr., Tjalve, Webb, and Wehner. The voluminous literature on nickel carcinogenesis has been compiled in monographs (NAS, 1975; IARC, 1976; NIOSH, 1977) and updated in recent reviews (Furst and Radding, 1980; Leonard et al., 1981; Raithel and Schaller, 1981; Sunderman, 1981).

The carcinogenicity of nickel subsulfide, αNi_3S_2, has been evaluated more thoroughly than any other nickel compound (Sunderman, 1983), and its metabolism has been elucidated by radiotracer techniques and X-ray diffractometry (Kasprzak, 1974; Sunderman et al., 1976; Kasprzak and Sunderman, 1977; Oskarsson et al., 1979; DeWally and Hildebrand, 1980; Hui and Sunderman, 1980; Onkelinx and Sunderman, 1980). A dose-effect relationship has been clearly established for induction of sarcomas in rats by im injection of αNi_3S_2 (Gilman, 1965; Sunderman, 1979a). At the site of im injection, αNi_3S_2 particles enter mononuclear cells and fibroblasts by phagocytosis or pinocytosis, and αNi_3S_2 is progressively transformed to αNi_7S_6, $Ni(OH)_2$, βNiS, and $NiSO_4$, which slowly dissolves to liberate Ni[II] ions (Sunderman et al., 1976; Kasprzak and Sunderman, 1977; Oskarsson et al., (1979). In rhabdomyosarcomas induced in rats by im injection of metallic Ni dust, nickel is localized in tumor cell nuclei, and especially in nucleoli (Webb et al., 1972, suggesting that Ni[II]-binding to chromatin and nucleo-

* This paper is a condensation and adaptation of an article to be published in "Nickel in the Human Environment" (Ed., W. Davis), International Agency for Cancer Research, Lyon, 1983.

Chemical Toxicology and Clinical Chemistry of Metals
ISBN 0.12.137520.X
Copyright © 1983 by IUPAC
All rights of reproduction in any form reserved

proteins may be involved in the molecular mechanism of nickel carcinogenesis (Sunderman, 1979b; Hui and Sunderman, 1980). The ligand interactions of Ni[II] in rat tissues, including Ni[II]-binding to nucleic acids and proteins, have been partially clarified by recent investigations in several laboratories (Sarkar, 1980; Sunderman et al., 1981; Ciccarelli et al., 1981; Ciccarelli and Wetterhahn, 1982; Herlant Peers et al., 1982; Sunderman et al., 1983).

Research on the relationship between chemical attributes and carcinogenic activities of nickel compounds has been stimulated by observations that certain nickel compounds, which are closely related in elemental composition and similar in physical-chemical properties, differ substantially in their carcinogenic activities in experimental animals (Gilman, 1962, 1965; Payne, 1964; Sunderman and Maenza, 1976; Sunderman et al., 1979b; Sunderman and McCully, 1983a) and in their capacities to induce morphological transformation of cell cultures (DiPaolo and Casto, 1979; Costa et al., 1979; Costa and Mollenhauer, 1980). For example, αNi_3S_2 and crystalline βNiS are potent inducers of neoplastic transformation in vivo and in vitro, whereas amorphous NiS is relatively inactive under the same experimental conditions (DiPaolo and Casto, 1979; Costa et al., 1979, 1981a). Some workers have proposed that differences in dissolution rates of nickel compounds in body fluids may account for the differences in their carcinogenic activities (Payne, 1964; Weinzierl and Webb, 1972; Kasprzak, 1974; Kasprzak et al., 1983); others have suggested that differences in susceptibility of nickel compounds to phagocytosis may be responsible (Costa and Mollenhauer, 1980; Abbracchio et al., 1982).

Previous attempts to relate physical, chemical, and biological properties of nickel compounds to their carcinogenic activities have been hampered by lack of carcinogenesis data that were obtained under standardized experimental conditions. To remedy this situation, 18 nickel compounds have been tested in the author's laboratory by im administration to male Fischer rats with equivalent dosages and uniform protocols. Some of these experiments have been published previously (Sunderman and Maenza, 1976; Sunderman, 1979a; Sunderman et al., 1979b; Sunderman and McCully, 1983a).

Dissolution rates of 17 of the nickel compounds have been measured in rat serum and renal cytosol (Kuehn and Sunderman, 1982); phagocytosis of the compounds has been quantified in monolayer cultures of rat peritoneal macrophages (Kuehn et al., 1982), and biological activity of the compounds has been assessed by intrarenal (ir) administration to rats in order to measure Ni-stimulation of erythropoiesis (Sunderman and Hopfer 1983). The present paper presents the outcome of our search for rank-correlations between (a) incidences of sarcomas induced by the compounds in rats, (b) mass-fractions of nickel in the compounds, (c) dissolution half-times of the compounds in rat serum and renal cytosol, (d) phagocytic indices of the compounds in rat macrophages, and (e) blood hematocrits of rats at two months after ir injection of the compounds.

2. MATERIAL AND METHODS

Physical, chemical and biological characteristics and sources of the test compounds have been described previously (Kuehn and Sunderman, 1982; Kuehn et al., 1982; Sunderman and Hopfer, 1983). The elemental purity of each compound was > 99.5%, based upon the producer's specifications. The identities of the compounds were checked by X-ray powder diffraction, as previously described (Sunderman and Maenza, 1976; Kasprzak and Sunderman, 1977). Amorphous NiS was devoid of crystal structure, based upon absence of X-ray diffraction spectrum. The compounds were ground in an agate mortar to median particle diameters < 2 µm, as verified by scanning electron microscopy.

Carcinogenesis tests were performed in 414 male Fischer-344 rats (Charles River Breeding Laboratories, Inc., Wilmington, MA), housed in stainless-steel or polypropylene cages, and fed laboratory rat chow (Ralston-Purina Co., St. Louis, MO) and water ad libitum. The rats were 2-3 months old when the test compounds were administered by injection deep into the extensor musculature at the mid-length of the right thigh (14 mg Ni/rat). The injection vehicles were 0.3-0.5 ml of glycerol:water (1:1 mixture, v/v) or procaine penicillin G suspension (3 x 10^7 units/ml, Wyeth Laboratories, Inc., Philadelphia, PA). The rats were weighed and examined at weekly or biweekly intervals. Most rats died spontaneously; some were killed when they became so cachectic that they could not move around in their cages and hence could not obtain food or water. Rats that died within two months were dropped from the study. The carcinogenesis tests were terminated two years after im injection of the test compounds. Necropsies and histologic examinations were performed as previously described (Sunderman and Maenza, 1976; Sunderman et al., 1979a; Sunderman and McCully, 1983b). Classification of sarcomas was based on the histological criteria of Stout and Lattes (1967).

Statistical comparisons of tumor incidences and proportions of 2-year survivors were performed as described by Peto et al. (1980); comparisons of median survival periods, based upon biweekly tabulations of mortality, were performed by the Mann-Whitney test (Conover, 1980). Correlations between sarcomas incidences and Ni mass-fractions, dissolution half-times, phagocytic indices, and blood hematocrits were evaluated by Kendall's rank-correlation test (Siegel, 1956).

3. RESULTS

Carcinogenesis tests of the nickel compounds in male Fischer rats are summarized in Table 1. The nickel compounds fell into

TABLE 1

Survival Data and Sarcoma Incidences in Carcinogenesis Tests of Eighteen Nickel Compounds

Category	Test substance	Survivors at 2 yr/ total no. of rats	Rats with local sarcomas/ total no. of rats	Median tumor latency (weeks)	Median survival period (weeks)	Rats with metastases/ rats with sarcomas
Controls	Vehicle	49/84 [58%]	0/84 [0%]		>100	
Class A	αNi_3S_2	0/9[c] [0%]	9/9[c] [100%]	30	39[b]	5/9 [56%]
	βNiS	0/14[c] [0%]	14/14[c] [100%]	40	48[b]	10/14 [71%]
	Ni_4FeS_4	0/15[c] [0%]	15/15[c] [100%]	16	32[b]	10/15 [67%]
Class B	NiO	0/15[c] [0%]	14/15[c] [93%]	49	58[b]	4/14 [29%]
	Ni_3Se_2	0/23[c] [0%]	21/23[c] [91%]	28	38[b]	18/21 [86%]
	NiAsS	0/16[c] [0%]	14/16[c] [88%]	40	57[b]	10/14 [71%]
	NiS_2	0/14[c] [0%]	12/14[c] [86%]	36	47[b]	6/12 [50%]
	Ni_5As_2	0/20[c] [0%]	17/20[c] [85%]	22	44[b]	9/17 [53%]
Class C	Ni dust	4/20[b] [20%]	13/20[c] [65%]	34	42[b]	6/13 [40%]
	NiSb	9/29[b] [31%]	17/29[c] [59%]	20	66[b]	10/17 [59%]
	NiTe	12/26 [46%]	14/26[c] [54%]	17	80[b]	8/14 [57%]
	NiSe	7/16 [44%]	8/16[c] [50%]	56	72[b]	3/8 [38%]
	$Ni_{11}As_8$	5/16[a] [31%]	8/16[c] [50%]	33	88[b]	6/8 [75%]
Class D	amorphous NiS	5/25[b] [20%]	3/25[b] [12%]	41	71[b]	3/3 [100%]
	$NiCrO_4$	10/16 [63%]	1/16 [6%]	72	>100	1/1 [100%]
Class E	NiAs	13/20 [65%]	0/20 [0%]		>100	
	$NiTiO_3$	11/20 [55%]	0/20 [0%]		>100	
	$NiFe_{1.6}$ alloy	11/16 [75%]	0/20 [0%]		>100	

[a] $P < 0.05$, [b] $P < 0.01$, [c] $P < 0.001$ versus corresponding vehicle controls.

five clear-cut categories: Compounds in Class A induced sarcomas at the injection site in 100% of the rats; compounds in Class B induced sarcomas in 85 to 93% of the rats; compounds in Class C induced sarcomas in 50 to 65% of the rats; compounds in Class D induced local sarcomas in 6 to 12% of the rats; and compounds in Class E did not induce any sarcomas. No sarcomas occurred at the injection site in 84 control rats that received im injection of the vehicles.

A total of 180 sarcomas developed at the site of the im injection of the nickel compounds. These tumors included 94 rhabdomyosarcomas (52%), 32 fibrosarcomas (18%), 23 undifferentiated sarcomas (13%), 15 osteosarcomas (8%), 14 unclassified sarcomas (8%), a neurofibrosarcoma, and a fibrous histiocytic sarcoma. Distant metastases were found in 61% of the sarcoma-bearing rats. No significant differences were observed among the test groups in the proportions of histologic types of sarcomas or in the frequencies of metastases. Incidences of malignant tumors distant from the injection site did not differ significantly among the experimental groups. Benign interstitial cell adenomas of the testis, which occur spontaneously in most senescent male Fischer rats (Jacobs and Huseby, 1967; Haseman, 1983) were found in practically all rats that lived more than 100 weeks. Foreign-body granulomas with multinucleated giant cells and mononuclear cells that contained metallic particles were evident at the injection site in most rats that received nickel compounds, and were not seen in the controls. Mild pyelonephritis with hyaline casts was common in rats of all experimental groups. Peribronchial foci of mononuclear cells were also frequently observed.

As shown in Fig. 1, the incidences of sarcomas in the 18 groups of rats that received im injections of nickel compounds were correlated with the mass-fractions (MF) of nickel in the respective compounds ($p = 0.02$). Sarcoma incidences were not significantly related to the dissolution rates of the nickel compounds in rat serum (Fig. 2) or renal cytosol, or to the susceptibilities of the compounds to phagocytosis by rat macrophages in vitro (Fig. 3). On the other hand, striking rank-correlation was evident between the sarcoma incidences and the capacities of the compounds to induce erythrocytosis after ir administration to rats ($p<0.001$, Fig. 4). Nickel monoselenide (NiSe) was a notable exception to the over-all concordance between the rankings of blood hematocrits and sarcoma incidences.

The entire set of rank-correlations between chemical and biological characteristics of the nickel compounds is listed in Table 2. Phagocytic indices of the nickel compounds were correlated with their dissolution half-times in rat serum ($p = 0.03$) and their capacities to induce erythrocytosis after ir administration to rats ($p = 0.04$). Strong rank-correlation was observed between the mass-fractions of nickel in the compounds and their potencies for induction of erythrocytosis ($p < 0.01$).

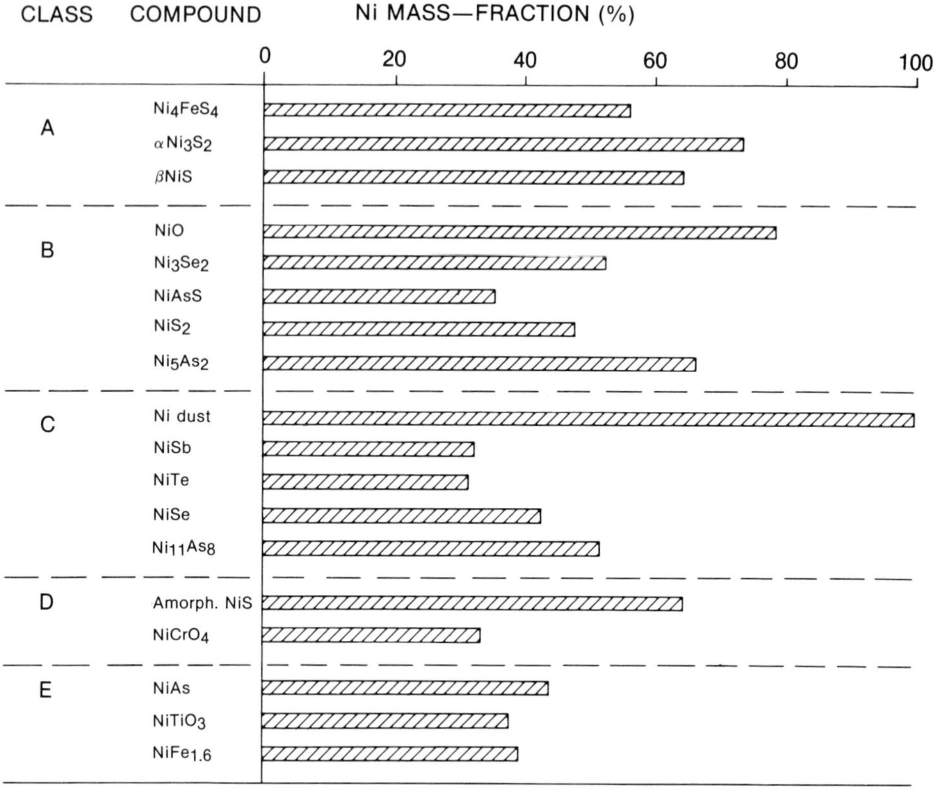

FIG. 1. Relationship between carcinogenic activities of the test substances and their nickel mass-fractions (i.e., the proportional weight of nickel per unit weight of substance). Rank comparisons of sarcoma incidences at 2 years after im injection of the test substances versus their nickel mass-fractions yielded a rank correlation coefficient of 0.35 ($P < 0.02$).

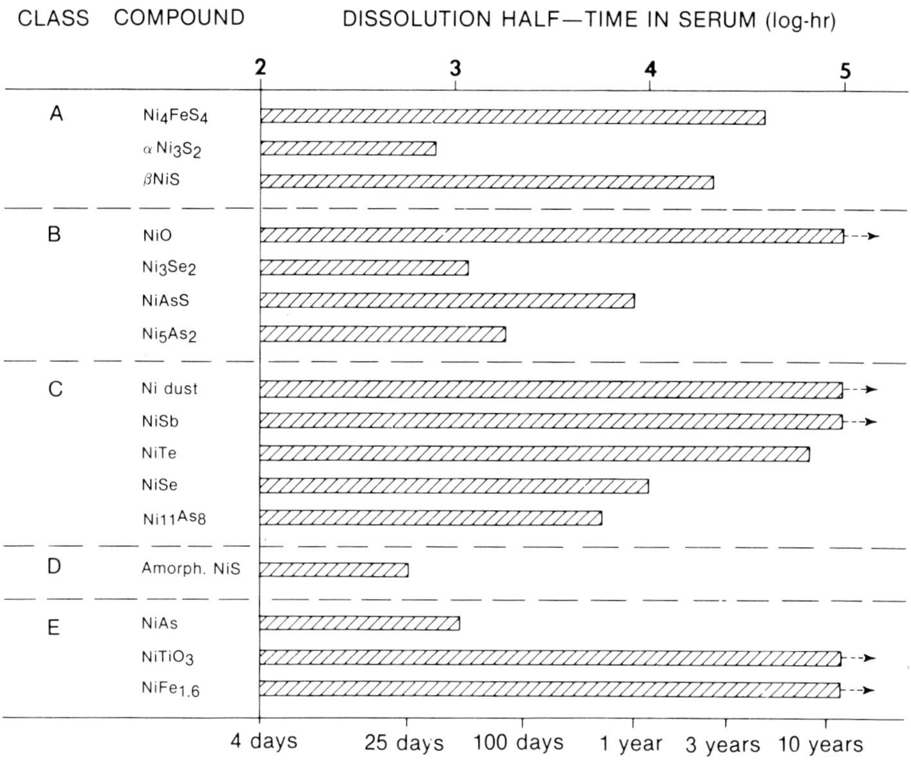

FIG. 2. Relationship between carcinogenic activities of the test substances and their dissolution half-times in rat serum. The dissolution half-time represents the estimated time for dissolution of 50% of nickel-containing particles during in vitro incubation in rat serum (37°C, 2 mg Ni/ml (Kuehn and Sunderman, 1982). Formation of flocculent precipitates during incubation of NiS_2 in rat serum and renal cytosol precluded measurements of its dissolution half-times.

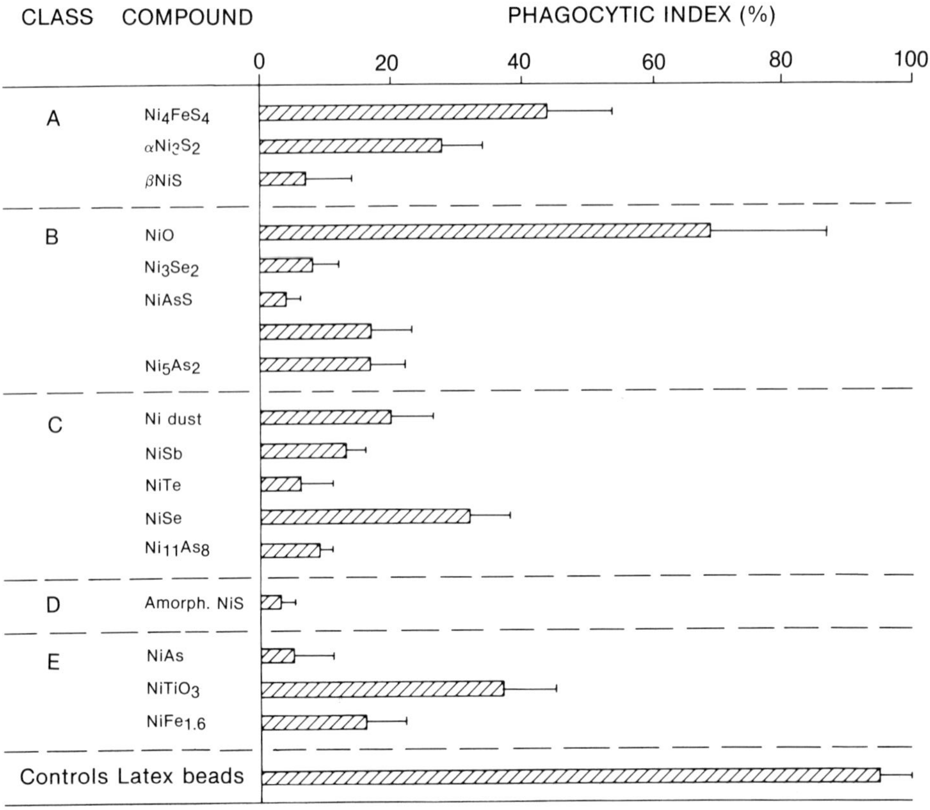

FIG. 3. Relationship between carcinogenic activities of the test substances and their susceptibilities to phagocytosis by rat peritoneal macrophages *in vitro*. The phagocytic index represents the percent (mean ± SD) of rat peritoneal macrophages that phagocytized one or more particles during incubation for 1 h at 37°C in medium that contained nickel compounds (10 μg/ml) (Kuehn *et al.*, 1982).

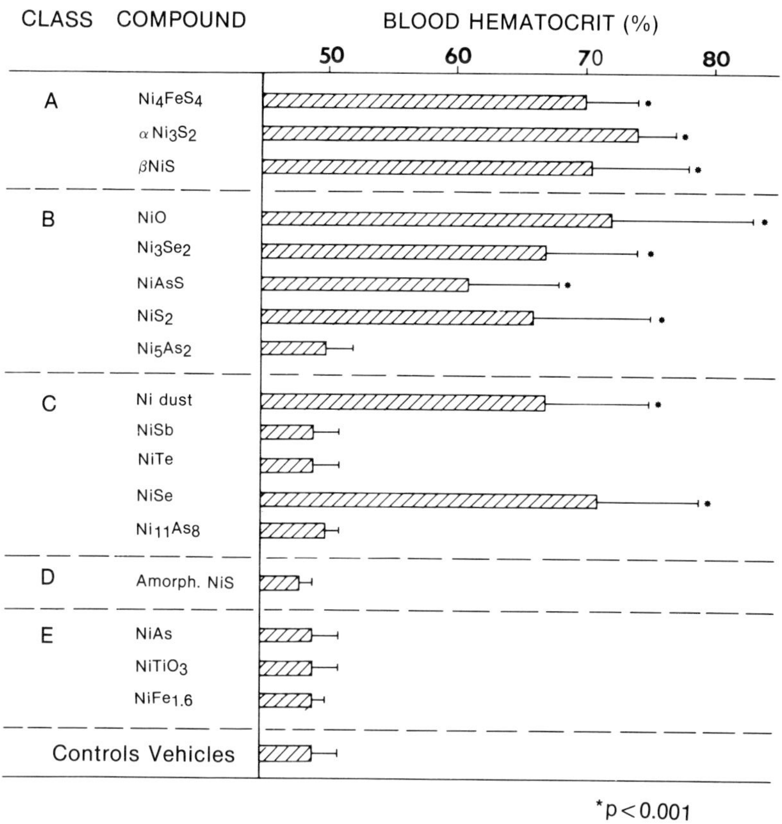

FIG. 4. Relationship between carcinogenic activities of the test substances and their capacities to induce erythrocytosis, as measured by blood hematocrits (%, mean ± SD) in groups of 11 to 57 rats at 2 months after intrarenal injection of nickel compounds (7 mg Ni/rat). The corresponding mean hematocrit in 79 control rats at 2 months after intrarenal injection of vehicle was 49 ± 3% (Sunderman and Hopfer, 1983).

*P < 0.01 versus vehicle controls.

TABLE 2

Rank Correlations between Chemical and Biological Parameters of Nickel Compounds

Parameters compared by rank	No. of compounds compared	Kendall corr. coef.	p
Sarcoma incidence[a] *vs* Ni mass-fraction[b]	18	0.35	0.02
vs serum T50[c]	16	0.07	
vs cytosol T50[c]	16	0.11	
vs phagocytic index[d]	17	0.17	
vs hematocrit[e]	17	0.72	< 0.0001
Hematocrit *vs* Ni mass-fraction	17	0.51	< 0.01
vs serum T50	16	0.06	
vs cytosol T50	16	0.08	
vs phagocytic index	17	0.32	0.04
Phagocytic index *vs* Ni mass-fraction	17	0.15	
vs serum T50	16	0.35	0.03
vs cytosol T50	16	0.28	
Ni mass-fraction *vs* serum T50	16	-0.17	
vs cytosol T50	16	-0.09	
Serum T50 *vs* cytosol T50	16	0.79	< 0.0001

[a] Sarcoma incidence in rats at 2 years after im injection (14 mg Ni/rat).
[b] Proportional weight of Ni per unit weight of substance.
[c] Dissolution half-time during *in vitro* incubation at 37° (2 mg Ni/ml).
[d] Phagocytosis by rat peritoneal macrophages *in vitro* (10 μg/ml).
[e] Mean blood hematocrit of rats at 2 months after ir injection (7 mg Ni/rat).

4. DISCUSSION

This study indicates that carcinogenic activities of particulate nickel compounds are systematically related to a physical property - the mass-fraction of nickel, suggesting that physical-chemical attributes of nickel compounds modulate the formation of the electrophilic nickel species that supposedly initiates neoplastic transformation (Sunderman, 1979b). The outcome of this study dashed our hopes that carcinogenic activities of particulate nickel compounds could be predicted by *in vitro* measurements of their rates of dissolution in body fluids or their susceptibilities to phagocytosis by macrophages. On the other hand, the study revealed notable rank-correlations between certain attributes of the nickel compounds. Those compounds that dissolved most readily in rat serum tended to have the lowest phagocytic indices in rat macrophages, suggesting that slow release of Ni[II] from the surface of the more soluble nickel compounds inhibits phagocytic activity. Furthermore, phagocytic indices of the nickel compound were correlated with their capacities to induce erythrocytosis in rats, suggesting that susceptibility to phagocytosis influences this biological effect. In this respect, the present study indirectly supports the conclusions of Costa and Mollenhauer (1980), Costa *et al.* (1981b), and Evans *et al.* (1982), based upon tissue culture experiments, that phagocytosis plays an important role in cellular uptake and metabolism of particulate nickel compounds.

This study demonstrates that the capacity to stimulate erythropoiesis in rats is predictive of the carcinogenic activity of nickel compounds. The occurrence of rank-correlation between the erythropoietic and carcinogenic effects of nickel compounds in rats does not necessarily imply that these two phenomena are related in their pathogenesis. However, our finding that αNi_3S_2-induction of erythrocytosis and carcinogenesis in rats are both antagonized by concurrent administration of Mn dust (Sunderman *et al.*, 1976; Hopfer and Sunderman, 1978; Sunderman *et al.*, 1979a; Sunderman and McCully, 1983b) suggests that the two effects may have a common initiating mechanism. Enhanced renal production of erythropoietin, which occurs following ir injection of nickel compounds (Jasmin and Riopelle, 1976; Solymoss and Jasmin 1978, Hopfer *et al.*, 1979; Sunderman *et al.*, 1982), may be a marker of renal dedifferentiation, similar to the increased plasma concentration of α-fetoprotein that occurs in rats during the early stage of hepatocarcinogenesis by N-2-fluorenylacetamide (Sell, 1980). We are investigating whether the delay of two months to quantify the hematocrit response of rats can be obviated by radioimmunoassay of erythropoietin in plasma obtained from rats a few days following ir injection of nickel compounds. Such a rapid assay would facilitate the search for correlations between *in vivo* effects and physical-chemical attributes of crystalline nickel compounds.

5. SUMMARY

Eighteen nickel compounds were tested for carcinogenicity in male Fischer rats by a single im injection at equivalent dosages (14 mg Ni/rat). Within two years, the following incidences of sarcomas occurred at the injection site: αNi_3S_2, 100%; crystalline βNiS, 100%; Ni_4FeS_4, 100%; NiO, 93%; Ni_3Se_2, 91%; NiAsS, 88%; NiS_2, 86%; Ni_5As_2, 85%; Ni dust, 65%; NiSb, 59%; NiTe, 54%; NiSe, 50%; $Ni_{11}As_8$, 50%; amorphous NiS, 12%; $NiCrO_4$, 6%; NiAs, 0%; $NiTiO_3$, 0%, $NiFe_{1.6}$ alloy, 0%; vehicle controls (N = 84), 0%. Kendall's rank-correlation test showed that carcinogenic activities of the compounds were correlated ($p = 0.02$) with their Ni mass-fractions, but not with dissolution half-times in rat serum or renal cytosol, or with phagocytic indices by rat peritoneal macrophages *in vitro*. Rank-correlation ($p < 0.0001$) was found between carcinogenic activities and potencies of the compounds to induce erythrocytosis in rats. Discovery that carcinogenic activities of particulate nickel compounds are correlated with a physical property, *i.e.*, the Ni mass-fraction, may help to elucidate the mechanisms of nickel carcinogenesis; observation that Ni-stimulation of erythropoiesis is correlated with carcinogenic activity provides a new *in vivo* screening test for carcinogenic risk of nickel compounds.

6. REFERENCES

M.P. Abbracchio, J. Simmons-Hansen and M. Costa (1982) *J. Toxicol. Environ. Health* *9*, 663-676.

J. A. Campbell (1943) *Brit. Med. J.* *1*, 179-183.

R. B. Ciccarelli and K. E. Wetterhahn (1982) *Cancer Res.* *42*, 3544-3549.

R. B. Ciccarelli, T. H. Hampton and K. W. Jennette (1981) *Cancer Lett.* *12*, 349-354.

M. Costa and H. H. Mollenhauer (1980) *Cancer Res.* *40*, 2688-2694.

W. J. Conover (1980) in *Practical Nonparametric Statistics*. John Wiley and Sons, New York. pp. 216-223.

M. Costa, J. W. Nye, F. W. Sunderman Jr., P. R. Allpass and B. Gondos (1979) *Cancer Res.*, *39*, 3591-3597.

M. Costa, M.P. Abbracchio and J. Simmons-Hansen (1981a) *Toxicol. Appl. Pharmacol.*, *60*, 313-323.

M. Costa, J. Simmons-Hansen, C.W.M. Bedrossian, J. Bonura and R. M. Caprioli (1981b) *Cancer Res.*, *41*, 2868-2876.

D. DeWally and H. F. Hildebrand (1980) in *Nickel Toxicology*, (Eds, S.S. Brown and F.W. Sunderman, Jr.), Academic Press, London, pp. 51-54.

J. A. DiPaolo and B. C. Casto (1979) *Cancer Res.* *39*, 1008-1013

R. M. Evans, P.J.A. Davies, and M. Costa (1982) *Cancer Res.* *42*, 2729-2735.

A. Furst and S. B. Radding (1980) in *Nickel in the Environment*, (Ed, J. Nriagu), John Wiley and Sons, New York, pp. 585-600.

J.P.W. Gilman (1962) Cancer Res. 22, 158-165.
J.P.W. Gilman (1965) in Proceedings of the Sixth Canadian Cancer Research Conference (Honey Harbor, Ontario, 1964), Pergamon Press, Oxford, pp. 209-233.
J.K. Haseman (1983) Fundam. Appl. Toxicol., 3, 1-9.
M.C. Herlant-Peers, H.F. Hildebrand and G. Biserte (1982) Zbl. Bakt. Hyg., I. Abt. Orig. B. 176, 368-382.
S.M. Hopfer and F.W. Sunderman Jr. (1978) Res. Commun. Chem. Pathol. Pharmacol., 19, 337-345.
S.M. Hopfer, F.W. Sunderman Jr., T.N. Fredrickson and E.E. Morse (1979) Res. Commun. Chem. Pathol. Pharamacol., 23, 155-170.
G. Hui and F.W. Sunderman Jr. (1980) Carcinogenesis 1, 297-304.
IARC (International Agency for Research on Cancer) (1976) in IARC Monographs on the Evaluation of Carcinogenic Risk of Chemicals to Man, Lyon, Vol. 11, pp. 75-112.
B.B. Jacobs and R. A. Huseby (1967) J. Natl. Cancer Inst., 39, 303-309.
G. Jasmin and J.L. Riopelle (1976) Lab. Invest., 35, 71-78.
K.S. Kasprzak (1974) Res. Commun. Chem. Pathol. Pharmacol., 8, 141-150.
K.S. Kasprzak and F.W. Sunderman Jr. (1977) Res. Comm. Chem. Pathol. Pharmacol. 16, 95-108.
K.S. Kasprzak, P. Gabryel and K. Jarczewska (1983) Carcinogenesis, 4, 275-279.
A. Kotowski (1965) in Gmelins Handbuch der Anorganischen Chemie, Weinheim, Verlag Chemie GmbH, Teil B, Lieferung 1, pp. 1-1181.
K. Kuehn and F.W. Sunderman Jr. (1982) J. Inorg. Biochem. 17, 29-39.
K. Kuehn, C.B. Fraser and F.W. Sunderman Jr. (1982) Carcinogenesis, 3, 321-326.
A. Leonard, G.B. Gerber and P. Jacquet (1981) Mutat. Res. 87, 1-15.
NAS (National Academy of Sciences) (1975) Nickel, NAS Publishing Office, Washington, DC, pp. 144-190.
NIOSH (National Institute for Occupational Safety and Health) (1977) Criteria for a Recommended Standard: Occupational Exposure to Inorganic Nickel. U.S. Government Printing Office, Washington, D.C., pp. 1-282.
C. Onkelinx and F.W. Sunderman Jr. (1980) in Nickel in the Environment, (Ed. J.O. Nriagu) Wiley/Interscience Co., New York, pp. 525-545.
A. Oskarsson, Y. Andersson, and H. Tjalve (1979) Cancer Res. 39, 4175-4182.
W.W. Payne (1964) Proc. Am. Ass. Cancer Res. 5, 50.
R. Peto, M.C. Pike, N.E. Day, R.G. Gray, P.N. Lee, S. Parish, J. Peto, S. Richards and J. Wahrendorf (1980) in IARC Monographs on the Evaluation of the Carcinogenic Risks of Chemicals to Humans, Lyon, Supplement 2, pp 311-426.
H.-J. Raithel and K. H. Schaller (1981) Zbl. Bakt. Hyg. I. Abt.

Orig. B. 173:63-91.
B. Sarkar (1980) in Nickel Toxicology (Eds. S.S. Brown and F.W. Sunderman Jr.) Academic Press, London, pp. 81-84.
S. Sell (1980) in Cancer Markers: Diagnostic and Developmental Significance, (Ed, S. Sell), Humana Press, Clifton, N.J., pp. 249-293.
S. Siegel (1956) Nonparametric Statistics for the Behavioral Sciences. McGraw-Hill Book Co., New York, pp. 116-127 & 213-223.
B. Solymoss and G. Jasmin (1978) Exp. Hemat. 6, 43-47.
A.P. Stout and R. Lattes (1967). in Atlas of Tumor Pathology, Washington, D.C., American Registry of Pathology, Series 2, Fascicle 1, pp. 1-197.
F.W. Sunderman Jr. (1979a) in Environmental Carcinogenesis (Eds. P. Emmelot and E. Kriek) Elsevier/North Holland Biomedical Press, Amsterdam, pp. 165-192.
F.W. Sunderman Jr. (1979b) Biol. Trace Element Res. 1, 63-86.
F.W. Sunderman Jr. (1981) Envir. Health Persp. 40, 131-141.
F.W. Sunderman Jr. (1983) in Organ and Species Specificity in Chemical Carcinogenesis (Eds. R. Langenbach, S. Nesnow and J.M. Rice) Plenum Press, New York, pp. 107-126.
F.W. Sunderman Jr. and S.M. Hopfer (1983) in Biological Aspects of Metals and Metal-Related Diseases (Ed. B. Sarkar), Raven Press, New York, pp. 171-181.
F.W. Sunderman Jr. and R.M. Maenza (1976) Res. Commun. Chem. Pathol. Pharmacol. 14, 319-330.
F.W. Sunderman Jr. and K.S. McCully (1983a) Cancer Invest. (in press).
F.W. Sunderman Jr. and K.S. McCully (1983b) Carcinogenesis 4, 461-465.
F.W. Sunderman Jr., K.S. Kasprzak, T.J. Lau, P.P. Minghetti, R. M. Maenza, N. Becker, C. Onkelinx and P.J. Goldblatt (1976) Cancer Res. 36, 1790-1800.
F.W. Sunderman Jr., R.M. Maenza, S.M. Hopfer, J.M. Mitchell, P.R. Allpass, and I. Damjanov (1979a) J. Envir. Pathol. Toxicol. 2, 1511-1527.
F.W. Sunderman Jr., S.B. Taubman, and P.R. Allpass (1979b) Ann. Clin. Lab. Sci. 9, 441.
F.W. Sunderman Jr., E.R. Costa, C. Fraser, G. Hui, J.L. Levine, and T.P.H. Tse (1981) Ann. Clin. Lab. Sci. 11, 488-496.
F.W. Sunderman Jr., S.M. Hopfer, M.C. Reid, S.K. Shen and C.B. Kevorkian (1982) Yale J. Biol. Med. 55, 123-136.
F.W. Sunderman Jr., B.L.K. Mangold, S.H.Y. Wong, S.K. Shen, M.C. Reid and I. Jansson (1983) Res. Commun. Chem. Pathol. Pharmacol. 39, 477-492.
M. Webb, J.C. Heath and T. Hopkins (1972) Brit. J. Cancer 26, 274-278.
S. Weinzierl and M. Webb (1972) Brit. J. Cancer 26, 279-291.

NICKEL-INDUCED ERYTHROCYTOSIS IN RATS

Gaëtan Jasmin, Béla Solymoss and Libuse Proschek

Départment de pathologie, Faculté de médecine, Université de Montréal, Montreal, Canada, H3C 3J7

1. INTRODUCTION

Some 10 years ago, we demonstrated that nickel subsulfide (Ni_3S_2) injected into the rat kidney parenchyma causes within one month a severe and sustained polycythemia (Fig. 1) which usually subsides after 8 months with emergence of a neoplastic growth (Jasmin 1973; Jasmin and Solymoss 1975; Jasmin and Riopelle 1976). The mitochondrial changes in kidney tubular cells have led us to believe that this polycythemia derives from a cellular hypoxia with overproduction of a renal factor that stimulates erythropoiesis in bone marrow (Jasmin 1973; Jasmin and Riopelle 1976). To acquire a better insight of the mechanism of this experimental erythrocytosis, we undertook evaluating the specificity and the cytopathic effects of nickel preparations on kidney parenchyma.

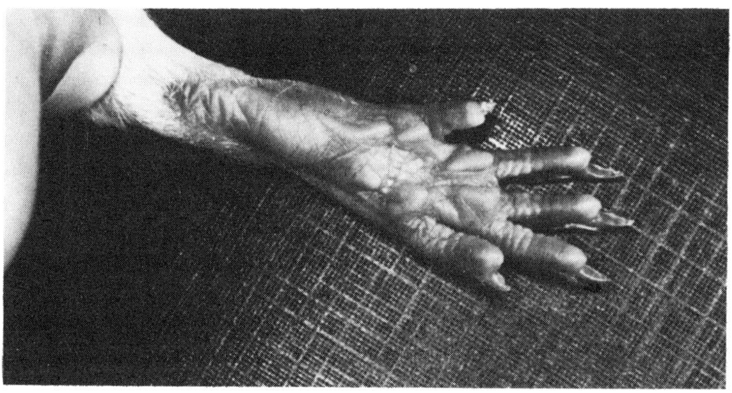

FIG. 1. Reddish extremities as seen during the course of nickel-induced erythrocytosis in the rat.

2. EXPERIMENTAL PROCEDURE

Doses and routes of administration. We first gave our attention to the induction procedure in rats. A maximal response was obtained by injecting each pole of the same kidney, 3 mm below the surface, with a 5 mg dose of Ni_3S_2. The erythrocytosis also develops with doses as low as 0.25 mg (Table 1) but for reasons of consistency, we adopted the double 5 mg treatment considering the possible differences in the size of particles (0.8-2.5 μ) and the relative insolubility of the nickel salt (Jasmin and Solymoss 1978).

Table 1

DOSE-RESPONSE RELATIONSHIP CONCERNING HEMATOLOGIC EFFECTS AND TUMOR INCIDENCE IN NICKEL-TREATED RATS*

Ni_3S_2 treatment	Hemogram at intervals after injection of nickel[1]								Tumor incidence in 12 months
	1 month		3 months		8 months		12 months		
	Hgb[2]	RBC[3]	Hgb	RBC	Hgb	RBC	Hgb	RBC	
0.25 mg/one pole	17.6±1.6	9.02±.75	19.3±2.6	10.75±.97	14.4±0.5	7.27±.35	13.1±0.3	6.98±.14	0/6
0.25 mg/each pole	16.9±1.4	8.91±.69	19.8±1.8	10.07±.81	14.7±0.3	7.35±.19	14.0±0.2	7.47±.15	0/6
0.5 mg/one pole	15.3±0.2	8.12±.06	19.4±1.7	10.79±.89	13.8±0.4	6.94±.27	12.4±0.1	6.85±.12	0/6
0.5 mg/each pole	18.2±1.7	8.54±.79	22.8±1.7	11.45±.43	14.2±1.4	6.84±0.8	13.9±0.2	7.18±.18	0/6
1 mg/one pole	15.7±0.6	8.44±.21	18.9±1.4	9.50±.61	14.4±0.1	7.27±0.1	14.2±0.3	7.39±.16	0/5
1 mg/each pole	18.2±1.0	8.66±.36	20.9±1.5	10.87±.24	15.5±0.9	7.72±.43	13.0±1.0	6.83±.43	0/5
3 mg/one pole	19.9±1.5	10.09±.87	22.9±2.4	11.52±.48	18.2±0.2	9.73±.97	13.7±0.4	7.34±.27	1/5
3 mg/each pole	18.4±1.4	8.77±.56	17.6±.46	9.18±.23	14.9±0.4	7.39±1.0	13.6±0.2	7.11±.25	2/5
5 mg/one pole	16.1±0.2	8.83±.51	19.2±.05	9.30±.29	14.4±0.1	7.12±.16	11.3±2.0	6.89±.77	5/8
5 mg/each pole	17.1±1.4	9.53±.63	20.4±1.8	10.63±.46	13.6±1.2	6.88±.61	13.7±0.2	7.20±.32	6/8
5 mg/one pole of both kidneys	18.9±1.3	9.63±.99	21.0±2.6	10.84±1.04	15.3±1.5	9.13±1.01	14.6±.26	7.93±.18	4/6[4]

* Group of 5-8 Sprague-Dawley female rats averaging 140 ± 6 g body weight

1 Means ± SE

2 Hgb (g/dl)

3 RBC ($x10^6/mm^3$)

4 Two with tumors in each kidney

Glycerol has proven to be the most appropriate vehicle to inject (23 gauge needle) the metallic particles as an homogenous suspension. The topical treatment in other tissues or even the intravenous or intraarterial injection of an aqueous Ni_3S_2 suspension had dissimilar effects (Jasmin and Solymoss 1978) (Table 2).

Other prospective nickel compounds. Other nickel compounds or divalent metal salts failed to produce the erythrocytosis (Jasmin and Riopelle 1976); nickel monosulfide, for instance, was inactive. The lack of erythrogenic effect of these metal salts however did not exclude their oncogenic potency. In this connection, Sunderman and Hopfer (1982) have identified 9 other nickel compounds which stimulate erythropoiesis and exert some carcinogenic action when injected intramuscularly in rats. According to these workers, the erythrogenic response to various nickel preparations could be used as an oncodevelopment marker.

Table 2

Blood hemogram of rats one month after injection of Ni_3S_2 by different routes

	Group	Hematocrit (%)	Hemoglobin (g/dl)	RBC ($x10^6/mm^3$)
I	Controls[1]	45.5 ± 0.5	15.2 ± 0.4	7.9 ± 0.4
II	Ni_3S_2 i.m.[2]	42.5 ± 0.8	15.2 ± 0.3	7.8 ± 0.2
III	Ni_3S_2 i.v.[3]	44.9 ± 0.4	14.7 ± 0.3	7.9 ± 0.1
IV	Ni_3S_2 into the liver[4]	46.0 ± 0.6	15.1 ± 0.3	7.9 ± 0.2
V	Ni_3S_2 into the kidney[5]	68.0 ± 2.0[6]	21.6 ± 0.7[6]	10.6 ± 0.4[6]

Each group consisted of at least six female Sprague-Dawley rats with an initial body weight of 110 ± 10 g. Values are expressed as the Mean ± S.E.

1 Controls -- untreated.

2 10 mg in 0.2 ml 0.85% NaCl injected into m. gluteus max.

3 10 mg in 1 ml 0.85% NaCl injected into the left jugular vein.

4 2x5 mg in 2x0.05 ml glycerin injected into the right and left lobes of the liver.

5 2x5 mg in 2x5.05 ml glycerin injected into each pole of the left kidney.

6 Significantly different ($P < 0.05$) as compared to the controls and all other groups.

The question however remains debatable since subcarcinogenic doses of Ni_3S_2 were efficient in inducing erythrocytosis in rats (Table 1).

Susceptibility of animals. In examining the susceptibility of animals, we found that both Fisher inbred and closely related Sprague-Dawley rats were equally responsive to Ni_3S_2. For yet undetermined reasons, the Wistar or Wistar-Lewis inbred strains were less sensitive. Interestingly, the reaction developed earlier and was more severe in 28-day-old Fisher rats than in slightly older animals. Trials in other species were at first unsuccessful but recently, the guinea pig proved to be a good reactor (Sunderman *et al.* 1982). It is not unlikely that further investigations on species susceptibility could lead to the demonstration of an immune factor underlying this experimental erythrocytosis.

3. EVIDENCE FOR KIDNEY PRODUCTION OF ERYTHROPOIETIN

We then extended our studies to the pathophysiology of the erythrocytosis. We soon found that the removal of the treated kidney was followed by a progressive drop in blood hematocrit, suggesting that some humoral factor is released by the injected kidney (Jasmin and Solymoss 1973, 1975). Of relevance, is also the fact that a parabiotic non-injected partner rapidly develops an erythemia when coupled with an Ni_3S_2-injected littermate. This is further supported by hyperplasia of the tibial bone marrow (Fig. 2).

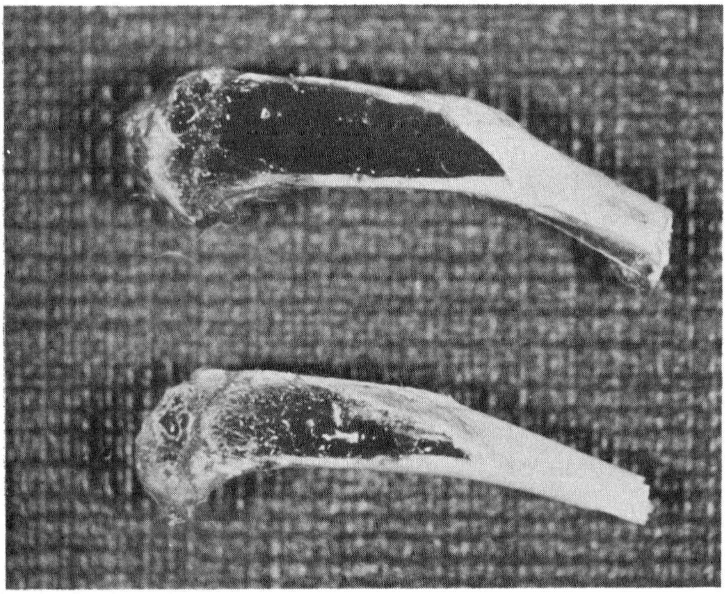

FIG. 2. Bone marrow hyperplasia in a tibia of a parabiotic rat which has been coupled with a littermate previously injected with Ni_3S_2.

A biological erythropoietin assay in polycythemic mice as determined by incorporation of radioactive iron into circulating erythrocytes has revealed a significant rise of an erythrogenic hormone between the 2nd and the 8th week following injection of Ni_3S_2 in renal tissue (Solymoss and Jasmin,1978)(Table 3). Removal of the injected kidney, on the other hand, entails a rapid drop of erythrogenic activity in the blood. Enzyme immunoassay recently carried out in our laboratory, using antibodies prepared against human erythropoietin, indicates a 5-fold increased erythropoietin production 2 weeks after injecting Ni_3S_2 into the kidney (unpublished observation).

4. INDUCTION AND MORPHOLOGY OF RENAL TUMORS

Let us at this moment briefly discuss the neoplastic growth in these animals. The kidney tumors are rarely detected before 8 months. They are easily palpable and occur with a 40% incidence after 12 months. Over 24 months, we gained the impression that the frequency would not increase further, presumably because the nickel deposits are sequestered within scar tissue.

Table 3

Assessment of rat plasma erythropoietin activity in polycythemic mice[1] (Means ± S.E.)

	Group	Time interval after intra-renal Ni treatment[3]	Hematocrit of rats at the time of plasma collection (%)	Fe^{59} incorporation into erythrocytes of mice in 48 h (% of dose injected)[4]
I	Polycythemic mice injected twice with 0.5 ml control rat plasma[2]		42.0 ± 1.4	0.58 ± 0.09
II	Polycythemic mice injected twice with 0.5 ml of Ni-treated rats	1 week	41.3 ± 0.4	0.64 ± 0.11
III	" " "	2 weeks	51.3 ± 3.1[5]	1.11 ± 0.23[5]
IV	" " "	4 weeks	61.7 ± 3.1[5]	1.44 ± 0.20[5]
V	" " "	8 weeks	67.8 ± 3.3[5]	1.14 ± 0.11[5]

1 Polycythemia was induced by blood transfusions.
2 Glycerin into each pole of the left kidney.
3 Ni_3S_2 (5 mg) in glycerin injected into each pole of the left kidney.
4 0.5 μCi in 0.2 ml of 0.85% NaCl, i.v.
5 Significantly different ($P < 0.05$) as compared to groups IV and V.

The neoplastic cells expand contiguously in the renal parechyma, ma, invading the hilar region or bulging on the surface of the kidney. Some tumors reach 6 cm^3 and weigh over 100 g. After routine staining, the more compact neoplastic growths show a spindle-cell arrangement in a sarcoma pattern. The cytologic aspect of these tumors, as is the case with anaplastic variance of human renal carcinomas, may be a source of confusion. Staining paraffin sections with periodic acid silver methenamine or thin Epon sections with toluidine blue better delineates the anaplastic carcinoma cells, showing a trabecular or tubular arrangement. In less indurated growths, the epithelial pattern seems more evident just as in lung or liver metastases. Ultrastructurally, the tumor cells show surface specialization with numerous villi and tubular arrangement reminiscent of immature nephrons. The anaplastic cells contain polysomes and microfilaments while the irregularly shaped nuclei show a marginal chromatin with a prominent nucleolus (Jasmin and Riopelle 1976; Jasmin and Solymoss 1978).

5. CYTOPATHIC EFFECTS OF Ni_3S_2.

Arteriosclerosis. The second part of this paper deals with the cellular substrate of the adverse effects of Ni_3S_2 on the kidney tissue. Before doing so, it is worth mentioning that McCully *et al.* (1982) have noted arteriosclerotic changes and mesengial cell hyperplasia. These would result from endothelial cell injury in arteries and arterioles due to early dispersion of nickel particles in the blood stream. This finding may correlate with the increased vascular permeability that follows i.v. injection of soluble nickel or cobalt salts (Jasmin 1974). On the other hand, we noticed an abnormal enlargement of the aorta in erythremic rats (Fig. 3); at the time this was interpreted as being secondary to blood volume expansion and we failed to demonstrate any arteriosclerotic changes.

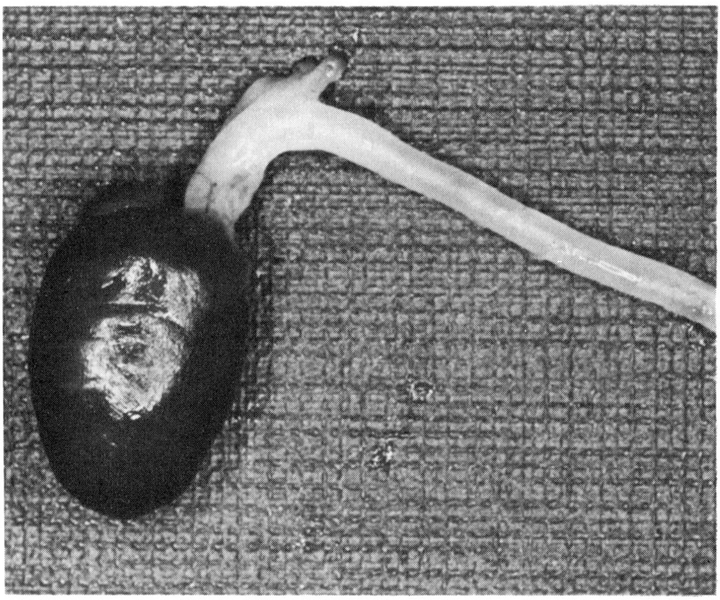

FIG. 3. Dilatation of aorta more prominent 6 months after intrarenal injection of Ni_3S_2. Note the dilatation and stiffness of the descending aorta.

Distribution of nickel in renal cell fractions. Regarding early lesions in the renal tubular cells, we were struck by the presence of abnormal chromatin aggregates in nuclei and crystalline inclusions in mitochondria (Jasmin 1973) (Fig. 4). Quantitative estimates of nickel contents in cell fractions have revealed that the metal concentrates quite early in the nuclear fractions and in mitochondria by order of importance (Jasmin and Solymoss 1978). We thus postulated that the altered chondriome in kidney tubular cells elicit an hypoxydosis which triggers the production of erythropoietin.

We then considered the possibility of a preferential uptake of nickel ions by these organelles. It would appear as suggested by other workers that the penetration of nickel ions is favored by its molecular association with sulfur (Dewally and

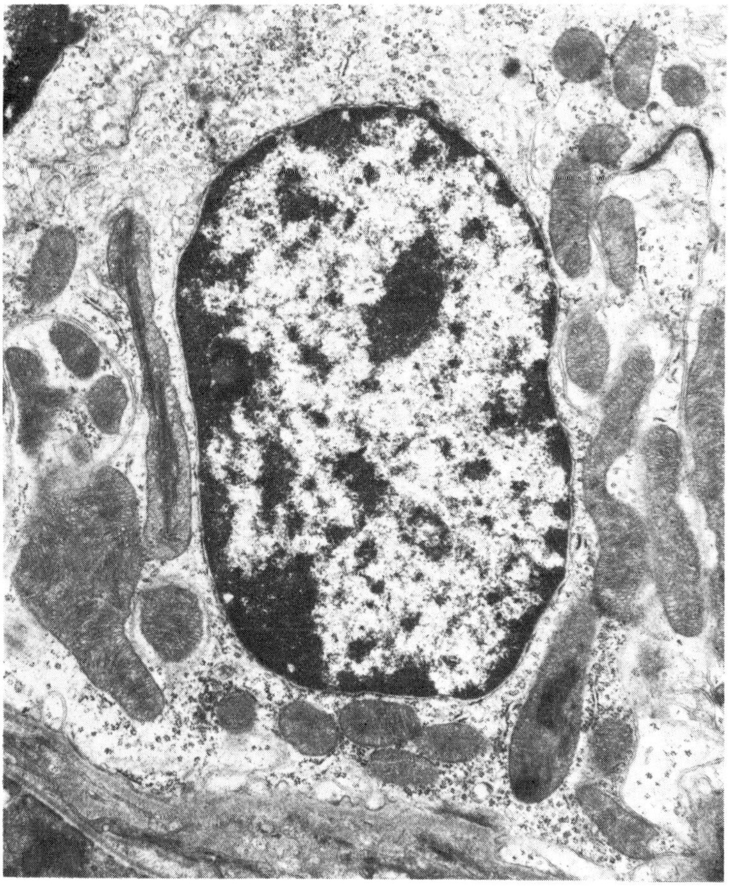

FIG. 4. Nuclear and mitochondrial ultrastructural changes in kidney tubular cells 4 months after injecting Ni_3S_2. Note the chromatin aggregates in nucleus and numerous crystalline inclusions in mitochondria. X5, 685 before reduction.

and Hildebrand 1980). Recent studies carried out by Sunderman and Hopfer (1982) on the metabolic fate of Ni_3S_2 would indicate that the carcinogenic properties of particulate nickel compounds seemingly are not related to their susceptibility to phagocytosis (Costa 1980; Costa *et al.* 1981; Kuehn *et al.* 1982) or else to their rate of dissolution in body fluids (Hopfer *et al.* 1978).

X-ray microanalysis. To further evaluate the cytopathic effects of Ni_3S_2, we investigated the elementary composition of this metal compound by x-ray dispersive microanalysis. Surprisingly, we found beside the expected proportion of elemental nickel and sulfur, some silicon, niobium, molybdenum, titanium and chlorine, the latter as part of the supporting membrane. In searching for the presence of these compounds in tubular cells, we found that the relative amount of sulfur within the whole cell was extremely low when compared to niobium and molybdenum which were particularly high (Table 4). This also applies to the nucleus, which retains a substantial amount of nickel but little sulfur or silicon. Nickel as well as other metals were also detected in lysosomes and mitochondria.

Table 4

RELATIVE CONCENTRATION AND CELLULAR DISTRIBUTION OF ELEMENTS IN Ni_3S_2 PARTICLES AS DETERMINED BY X-RAY MICROANALYSIS (NORMALIZED VALUES IN %)*

Element	Relative content in particles	Cell analysis	Subcellular analysis		
			Nucleus	Lyso	Mito
Ni	43.8	8.7	30.3	10.8	4.6
S	18.0	-6.3	-5.3	7.5	-0.6
Si	10.0	1.6	-15.8	15.0	3.3
Nb	10.4	13.4	15.8	43.0	39.9
Mo	4.1	44.9	28.9	11.8	45.0
Ti	4.3	0.0	0.0	-5.4	5.2
Cl	9.4	37.8	46.0	17.2	2.6

* Determination was made in a JEOL STEM equipped with an Ortec energy dispersive x-ray microanalyser. Ni_3S_2 particles were dispersed on a formvar-coated nylon grid. Epon thin sections of renal cortex were stained with uranyl only; 4-month-injected kidneys were utilized.

6. MITOCHONDRIAL PROTEIN INVOLVEMENT.

Finally, we shall offer and discuss some data concerning the chemical composition of the mitochondrial inclusions. Electron cytochemical studies by subjecting tissue sections to protease digestion have shown that these inclusions were very sensitive to the lytic effect of pepsin (Jasmin 1978); these structures were also devoided of cytochrome oxidase activity (Jasmin 1978). Electrophoresis of the mitochondrial matrix proteins has disclosed two separate bands (Fig. 5), one of which seems related to cytochrome C as further evidenced by electro-focusing (Proschek and Jasmin 1979). The newly-formed protein would develop during remodeling of mitochondrial membranes (Jasmin 1978; Jasmin *et al.* 1979). It was reported several years ago that mitochondria in kidney tubules are very sensitive to changes in ionic environment with progressive loss of

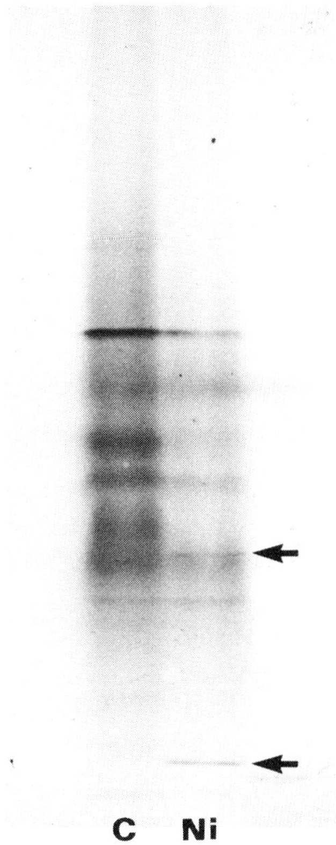

FIG. 5 Gradient electrophoretic patients of kidney mitochondrial matrix proteins in control rats (C) and in Ni_3S_2-treated rats (Ni); newly-formed bands are indicated by arrows (polyacrylamide gel gradient 4%-30%, Tris-EDTA-boric acid buffer pH 9.6).

oxidative enzyme activity (Karnovsky 1963). This may be interpreted as an expression of cellular hypoxydosis. The question as to whether erythropoietin or some of its precursors derive from a light mitochondria fraction (Contrera and Gordon 1966) or from the glomerular (Fisher and Busuttil 1977) or mesengial cells (McCully *et al.* 1982) is still debated. Our findings provide some evidence that an abnormal protein fraction exists in mitochondria with a depressed respiratory rate. It remains to be shown however if this fraction activates erythropoietin formation.

7. SUMMARY

In summary, we have shown that Ni_3S_2 injected in mg doses into renal parenchyma stimulated erythropoiesis in rats. This experimental model of polycythemia has opened new avenues for investigating the mode of action of nickel at cellular and subcellular levels. We originally found that the erythrogenic response is selective to the rat and to renal tissue. An early and optimal reaction is better obtained in 28-day-old closely related Sprague-Dawley or inbred Fisher-344 rats, using doses as low as 0.25 mg Ni_3S_2 and glycerin as a vehicle. More recently Sunderman et al (1982) have found that 9 other nickel compounds exerted similar effects and that erythrocytosis could be also induced in guinea pigs. The nickel-induced erythrocytosis is transmissible to a parabiotic partner and retrocedes after extirpation of the injected kidney. We further observed a small but significant rise in erythropoietin plasma level 2 to 4 weeks after injecting Ni_3S_2. Quantitative determinations of nickel ion in renal cell fractions have revealed that the metal concentrates quite early in the nuclear fraction. The occurrence of intramitochondrial crystals in tubular epithelial lining cells is consistent with the hypothesis of cellular hypoxia that may contribute to activating the production of an erythrogenic factor apparently generated by the mesengial cells. It is believed that the carcinogenic properties of nickel particulates are unrelated to their propensity of being phagocytozed or to their rates of dissolution. It has also been suggested that the erythrocytic reaction could serve as an oncodevelopment marker. The latter is at variance with previous findings whereby subcarcinogenic doses of Ni_3S_2 were efficient in inducing erythrocytosis in rats.

8. ACKNOWLEDGEMENTS

This work was supported by The Medical Research Council of Canada, The Muscular Dystrophy Association of Canada and The MacDonald-Stewart Foundation.

9. REFERENCES

J.F. Contrera and A.S. Gordon (1966) *Science* 152, 653-654.

M. Costa (1980) *Science* 209, 515-517.

M. Costa, J. Simmons-Hansen, C.W.M. Bedrossian, J. Bonura and R.M. Caprioli (1981) *Cancer Res.* 41, 2868-2876.

D. Dewally and H.F. Hildebrand (1980) in *Nickel Toxicology*. (Eds, S.S. Brown and F.W. Sunderman, Jr), Academic Press, London, pp. 51-54.

J.W. Fisher and R.W. Busuttil (1977) in *Kidney Hormones (Ed, J.W. Fisher)*, Vol. II. Academic Press, London, N.Y., San Francisco, pp. 165-185.

S.M. Hopfer, F.W. Sunderman, Jr., T.N. Fredrickson and E.E. Morse (1978) *Ann. Clin. Lab. Sci.* 8, 396-402.

G. Jasmin (1973) *J. Clin. Invest.* 21, 1068.

G. Jasmin (1973) *J. Microscopie* 17, 68a.

G. Jasmin (1974) *Proc. Soc. Exp. Biol. Med.* 147, 289-292.

G. Jasmin (1975) *J. Microscopie et Biol. Cell.* 23, 54a.

G. Jasmin (1978) *Exp. Molec. Path.* 29, 199-210.

G. Jasmin, R. Bonneau and J. André (1979) *Biol. Cell.* 35, 81-86.

G. Jasmin and J.L. Riopelle (1976) *Lab. Invest.* 35, 71-78.

G. Jasmin and B. Solymoss (1973) *Proc. Can. Fed. Biol. Soc.* 16, 125.

G. Jasmin and B. Solymoss (1975) *Proc. Soc. Exp. Biol. Med.* 148, 774-776.

G. Jasmin and B. Solymoss (1978) in *Advances in Experimental Medicine and Biology*. (Ed, N.G. Schrauzer), Vol. 91. Plenum Press, N.Y., London, pp. 69-83.

M.J. Karnovsky (1963) *Exp. Molec. Path.* 2, 347-366.

K. Kuehn, C.B. Fraser and F.W. Sunderman, Jr. (1982) *Carcinogenesis* 3, 321-326.

K.S. McCully, L.A. Rinehimer, C.G. Gillies, S.M. Hopfer and F.W. Sunderman, Jr. (1982) *Virchows Arch* [*Pathol Anat*] 394, 207-220.

L. Proschek and G. Jasmin (1979) in *Development in Biochemistry*. (Eds, H. Haglund, J.G. Westerfeld and J.T. Ball, Jr.), Vol. 7. Elsevier North-Holland, pp. 93-98.

B. Solymoss and G. Jasmin (1978) *Exp. Hemat.* 6, 43-47.

F.W. Sunderman, Jr. and S.M. Hopfer (1982) in *Biological Aspects of Metals and Metal Related Diseases* (Ed, B. Sarkar), Raven Press, N.Y., pp. 1-23.

F.W. Sunderman, Jr., S.M. Hopfer, M.C. Reid, S.K. Shen and C.B. Kevorkian (1982) *The Yale J. Biol. Med.* 55, 123-136.

MONITORING TRACE ELEMENT CONCENTRATIONS IN BLOOD FROM PREMATURE INFANTS IN A NEONATAL INTENSIVE CARE UNIT

P.A. Pleban, B. Numerof*, B.A. Chrenka*, and F. Wirth*

Department of Chemical Sciences, Old Dominion University, Norfolk, VA 23508 U.S.A.
**Children's Hospital of the King's Daughters Norfolk, VA 23507 U.S.A.*

1. INTRODUCTION

It has been reported that premature infants have low body stores of essential trace elements and are at risk of developing deficiencies of these metals if maintained on a diet not supplemented with them (Easton et al., 1982). However, because of variable renal and liver function and small size, supplemented infants should be monitored closely to insure that at least plasma and erythrocyte concentrations remain appropriate. We have determined plasma copper, selenium, and zinc and erythrocyte manganese and selenium concentrations in 88 infants of varying gestational age, weight, and clinical condition. Twenty one of these infants were monitored for periods of up to eight months. The efficacy of monitoring plasma and erythrocyte zinc concentrations in premature neonates on total parenteral nutrition was investigated. In addition, possible correlations between trace metal concentratrations and postconceptual age, vitamin E supplementation, sex, and other trace metal concentrations were examined.

2. EXPERIMENTAL

Trace element concentrations were usually determined within two weeks of admission to the neonatal intensive care unit, depending on the patient's clinical appearance. Specimens were collected between 0600 and 0800 a.m., usually by heel puncture. Microtainer capillary collection tubes without additives (Becton Dickinson) were found to be adequate. For venous or arterial punctures, butterfly needles (Abbott Laboratories) and plastic syringes (W.G. Sarstedt, Inc.) were used for specimen collection. Thirty µL of 30 mM Na_2EDTA (recrystallized from hot water) per mL of blood was used as the anticoagulant. Except for plasma zinc concentrations in specimens collected by heel puncture agreed well with those collected by arterial puncture or from an umbilical artery line. Mean plasma zinc concentrations tended to be

Chemical Toxicology and Clinical Chemistry of Metals
ISBN 0.12.137520.X
Copyright © 1983 by IUPAC
All rights of reproduction in any form reserved

about 10% higher in specimens collected using heel puncture. The total blood volume required to perform all of the metals analyses was 0.2 to 0.7 mL, depending on the patient's hematocrit.

Specimens were separated within one hour of collection to prevent leakage of trace elements from the cells into the plasma. Erythrocytes were diluted fourfold with deionized water and hemolyzed by freezing at -20 °C.

(a) Analytical Methods

All metals were quantitated directly in the diluted specimen, using Zeeman-effect atomic absorption spectroscopy (Model 180-70, N.S.A. Hitachi, Ltd.). Details of the methodologies have been published elsewhere (Pleban and Pearson, 1979); Pleban and Munyani, 1981; Pleban et al., 1982). Plasma copper and zinc concentrations and erythrocyte zinc concentrations were determined using flame techniques. The absorbance obtained for the diluted specimen (tenfold dilution for plasma copper and zinc analyses, hundredfold dilution for erythrocyte zinc analyses) was compared to a standard curve prepared from aqueous standards. Plasma selenium, erythrocyte manganese, and erythrocyte selenium concentrations were determined using graphite furnace techniques. The peak areas obtained for the atomic absorption signals from the diluted specimens (twofold dilution for plasma selenium analyses, tenfold for erythrocyte manganese analyses, and fourfold for erythrocyte selenium analyses) were compared to a matrix-based standard curve. Between-run coefficients of variation ranged between 4 and 8 % for the assays.

Hemoglobin concentrations were determined in the diluted hemolysates using a standard cyanmethemoglobin method. Erythrocyte trace element concentrations were reported as micrograms per gram of hemoglobin (μg/g Hb). Reporting concentrations in this manner should correct for transfer errors when pipetting erythrocytes and also minimizes the effects of irregular cell volumes.

(b) Statistical Methods

Statistical analyses were performed on a DECsystem 10 computer utilizing Statistical Packages for the Social Sciences (SPSS). A two-tailed t test was used to examine possible correlation between metal concentrations and postconceptual age and vitamin E supplementation. The Spearman Rank Sum Correlation Test was used to examine correlations between metals concentrations and the other parameters.

3. RESULTS AND DISCUSSION

Mean plasma copper, selenium, and zinc concentrations, and erythrocyte manganese, selenium, and zinc concentrations were determined for 88 infants in the intensive care unit. Only the metal concentrations initially determined in each infant were used for the statistical calculations. Plasma copper and erythrocyte zinc concentrations appeared to depend significantly on postconceptual age. Results for plasma copper and zinc and erythrocyte zinc concentrations are shown in Table 1.

TABLE 1

Effect of Gestational Age on Concentrations of Zinc in Plasma and Erythrocytes

Age (weeks)	N	Plasma copper (μg/dl)	Plasma zinc (μg/dl)	N	Erythrocyte zinc (μg/g Hb)
		Mean (SD)			
< 28	12	47.9(16.4)[a]	86.8(20.6)	11	43.3(10.3)[b]
28 - 32	26	45.4(13.1)	87.5(32.9)	21	38.3(9.7)
33 - 36	19	50.8(14.0)	67.0(26.3)	14	35.6(15.3)
> 36	19	66.6(27.6)[a]	71.7(20.7)	16	34.5(14.3)[b]
Normal Adult		101(15)	90(13)		40.7(4.0)

[a]$t = 7.1$, $p = .001$, [b]$t = 2.05$, $p = .05$

Mean (SD) plasma selenium concentrations for the neonates were 31.8 (16.3) μg/L, N = 80. For normal adults the range is 80 - 120 μg/L. Mean erythrocyte manganese and selenium concentrations were found to be .103 (.073) μg/g Hb, N = 65, and .50 (.17) μg/g Hb, N = 64, respectively. For normal adults the ranges are .030 - .095 and .30 - 1.20 μg/g Hb, respectively. Plasma copper concentrations are reported to correlate positively with postconceptual age (Hillman et al., 1981), as confirmed by our data. Conversely, plasma zinc concentrations are reported to correlate inversely with postconceptual age. While mean plasma zinc concentrations found in this study decrease with age, the trend is not statistically significant. Erythrocyte zinc concentrations, however, were significantly higher in infants less than 28 weeks postconceptual age when compared to those in infants greater than 36 weeks. Neonatal erythrocyte selenium concentrations were similar to those found in adults. Plasma selenium concentrations, however, were much lower in the infants. If erythrocyte levels adequately reflect tissue stores, the lower concentrations may reflect liver immaturity. Two infants with plasma selenium concentrations of 19 and 10 μg/L and erythrocyte selenium concentrations of .25 and .20 μg/g Hb, respectively, were supplemented with selenium in their total parenteral nutrition fluids. Plasma and erythrocyte levels rose to 81 and 88 μg/L and .53 and .81 μg/g Hb, respectively. Upon cessation of supplementation, plasma levels fell to their unsupplemented concentrations within two weeks. Erythrocyte selenium levels fell more slowly and remained within normal adult range for six weeks.

Erythrocyte manganese levels were elevated compared with concentrations found in normal adults. However, since the iron status of these infants is uncertain, it is possible that undetected iron deficiency led to the elevated levels.

The efficacy of monitoring plasma and erythrocyte zinc levels was determined by assaying zinc concentrations in these tissues for up to three weeks after supplementation. A total of 21 infants were monitored. Results are shown below.

TABLE 2

Effect of Lengths of Supplementation Time on Zinc Concentrations in Plasma and Erythrocytes

	Length of Time Between Start of Supplementation and assay — Weeks			
	<1	1	2	3
	Mean (SD)			
Plasma Zn, μg/dL				
Supplemented	68.8(31.9)	103.2(49.8)	87.5(45.4)	90.3(42.7)
N	11	12	11	6
Before Supplement	46.2(17.4)	48.8(14.1)	49.7(19.7)	46.3(16.1)
RBC Zn, μg/g Hb				
Supplemented	40.3(11.6)	43.4(10.1)	43.5(12.3)	36.3(10.2)
N	11	11	11	6
Before Supplement	37.8(16.2)	35.8(7.9)	36.7(14.0)	31.0(7.5)

Both tissues reflect supplementation, with plasma levels rising more rapidly.

Vitamin E supplementation and sex of infant did not significantly affect trace metal levels. In addition, plasma and erythrocyte zinc levels were significantly correlated ($p = .039$), while plasma copper and zinc levels were inversely related ($p = .002$).

4. REFERENCES

L.B. Easton, M.S. Halata, and H.S. Dweck (1982) Pediat. Clinics North Amer. 29, 1171 - 1190.

L.S. Hillman,(1981) J. Pediat. 98, 305 - 308.

P.A. Pleban and A. Munyani (1981) Va. J. Sci. 32, 110.

P.A. Pleban, A. Munyani, and J. Beachum (1982) Clin. Chem. 28, 311 - 316.

P.A. Pleban and K.H. Pearson (1979) Clin. Chem. 25, 1915 - 1918.

THE EFFECT OF VARYING ZINC SUPPLEMENTATION ON THE ACRODERMATITIS SYNDROME IN INFANTS ON PARENTERAL NUTRITION

Gillian Lockitch, Margaret M. Johnston,
Margaret R. Pendray and William J. Godolphin.

Departments of Pathology and Pediatrics, and Division of Dermatology, University of British Columbia, Vancouver General Hospital and Children's Hospital, 4480 Oak Street, Vancouver, B.C. Canada V6H 3V4

1. INTRODUCTION

A syndrome of acute dermatitis, hypozincemia and cutaneous candidiasis responsive to increased zinc supplementation was reported in several infants on zinc supplemented parenteral nutrition (PN) by Arlette and Johnston (1981). The lesions were like those seen in acrodermatitis enteropathica.

The infants were generally of low birth weight, required long term PN, and had complicated postnatal courses with respiratory distress syndrome, necrotising enterocolitis and sepsis. Concern that the zinc dose was inadequate prompted this study.

2. OBJECTIVES

To determine whether :

a) increased intravenous (IV) zinc supplement would prevent the Acute Dermatitis Syndrome;

b) weekly measurement of serum zinc concentration would aid prediction and prevention of the syndrome.

3. METHODS

a) Subjects

Baseline Group. This consisted of 28 infants needing PN during the three months preceding formal commencement of the study. They were examined by the dermatology team to establish the incidence and diagnostic criteria for diagnosis of acute dermatitis. They received 40 μg zinc/kg/day contained in the 24 hour infusion volume (Table 1).

Chemical Toxicology and Clinical Chemistry of Metals
ISBN 0.12.137520.X
Copyright © 1983 by IUPAC
All rights of reproduction in any form reserved

Study Group. This consisted of 219 infants needing PN over the 18 months of the study period. They were randomized within birth weight groups into seven zinc/copper dosage groups to receive either 20 or 40 μg/kg/day of copper and 40,100,200 or 400 μg/kg/day of zinc. This supplement was however added to the first 12 hour infusion volume to ensure that the infants received the prescribed dose (Table 1).

Table 1

Details of Study Subjects

Baseline group		Study Groups 1	2	3	4	5	6	7
n	28	39	36	36	35	38	18	17
BW*	1355	1455	1183	1276	1342	1332	1312	1293
	(618)	(727)	(368)	(487)	(601)	(441)	(460)	(506)
Zn#	40	40	100	100	200	200	400	400
Cu##	20	20	20	40	20	40	20	40

* Mean (S.D.) Birth Weight for Baseline/ Study Groups
# Zinc supplement (μg/kg/day)
Copper supplement (μg/kg/day)

b) Clinical and Biochemical Studies

Study Group infants had weekly serum zinc and copper levels analyzed by atomic absorption spectrophotometry. Details of study methodology have been described by Lockitch et al. (1983).

Each infant in both Baseline and Study Groups was examined weekly by the dermatologist whether or not they were still receiving PN. Acute dermatitis was noted and categorized as mild and severe. Mild dermatitis was defined as a marginated redness and scaling of the body folds. Severe acute dermatitis was defined as widespread erythematous, exudative and scalding lesions together with poor wound healing. The presence of Candidal infections poorly responsive to topical mycostatin was also noted.

4. RESULTS

Table 2
Incidence of Acute Dermatitis Syndrome

	Baseline Group	Study Group
Mild dermatitis	7 (25%)	43 (19%)
Severe dermatitis	9 (32%)	0
Total	16 (57%)	43 (19%)

The comparison between Baseline and Study Groups is seen in Table 2. No case of severe dermatitis was seen in the Study Group although the lowest zinc dose in the Study Group was identical to the standard dose for the Baseline Group (40 μg/kg/day).

There were 43 infants with mild acute dermatitis in the Study Group. Seven of 39 (18%) were in the 40 μg group, 12/72 (17%) were in the 100 μg group, 20/73 (28%) in the 200 μg group but only 4/38 (10%) in the 400 μg group.

However, when considering only the 22 infants actually still on PN at the time that dermatitis was diagnosed, the diagnosis was made in 8%, 14%, and 8% of the 100, 200, and 400 μg groups respectively, but only 5% of the 40 μg group.

Serum zinc concentration declined with time in all weight groups and in all zinc dose groups although the decline was minimal in the 400 μg group. Hypozincemia (serum zinc < 50 μg/dl) was recorded in 50 infants of the Study Group; in 10% of the 400 μg group but about 26% of each of the lower dosage groups. Twenty-four had no evidence of acrodermatitis despite zinc levels as low as 33 μg/dl. Four had a slight erythematous scaliness insufficient to fulfil the criteria for mild acrodermatitis.

Twenty-two infants with hypozincemia had at least 1 episode of acrodermatitis. In only 15 cases was the hypozincemia present concurrently with the acrodermatitis. Conversely, 22 infants with the diagnosis of acrodermatitis had no recorded hypozincemia with a mean serum zinc of 99 μg/dl.

Copper dose did not affect the prevalence at all.

5. DISCUSSION AND CONCLUSIONS

These findings suggest that the method of providing the zinc supplement may be as important as the dosage level. When zinc was added to the full calculated 24 hour fluid volume any reduction in infusion rate for fluid overload or intravenous problems resulted in a reduction in the amount of zinc actually received. Once the zinc supplement was added to the burette containing the first 12 hour infusion volume it was ensured that the prescribed level was infused. Under these circumstances it appeared that 40 μg/kg/day was adequate to prevent the severe acrodermatitis syndrome. Raising the dose from 40 to 200 μg/kg/day did not decrease the occurrence of mild acute dermatitis. Only in the 400 μg group was the decline in serum zinc minimized and a significant reduction in both dermatitis and hypozincemia episodes seen. This would tend to support the tentative recommendations made by the Expert Panel of the AMA (1979) of an IV dose of about 300 μg/kg/per day for small infants.

If zinc supplements are to be added to PN infusions, particularly where fluid overload may be a problem, adding the supplement to the first half of the day's infusion should be considered.

The occurrence of dermatitis did not correlate with the serum zinc concentration. There was, however, a positive correlation between the number of episodes of low zinc and the number of times that dermatitis was seen in a given infant. Thus, while a single low zinc level may be of marginal significance, several low values or a sharp decline may indicate need for additional zinc.

6. ACKNOWLEDGEMENT

Supported by a grant from the B.C. Health Care Research Foundation.

7. REFERENCES

J.P. Arlette and M.M. Johnston (1981) J. Am. Acad. Derm., 5, 37-42.

G. Lockitch, W.J. Godolphin, M.R. Pendray, D. Riddell, and G. Quigley (1983) J. Pediatr., 102, 303-308

Expert Panel for Nutrition Advisory Group, American Medical Association (1979) JAMA, 241, 2051-2054.

SERUM NICKEL CONCENTRATIONS IN WOMEN DURING PREGNANCY, PARTURITION, AND POST-PARTUM

S. Nomoto, T. Hirabayashi* and T. Fukuda*

Division of Clinical Chemistry, Shinshu University School of Medical Technology
Department of Obstet. and Gynecology, Shinshu University School of Medicine, Matsumoto, 390 Japan

1. INTRODUCTION

This investigation has been undertaken to confirm or refute the observation by Rubanyi and associates (1982) that concentrations of serum nickel are significantly increased immediately after delivery of the infant, but before delivery of the placenta, and that sixty minutes after parturition, the concentration of serum nickel is restored to the control value. Their report suggested a physiologic significance of transient hypernickelemia in the advanced stages of human labor, possibly related to support of placental separation and prevention of atonic bleeding.

2. MATERIALS AND METHOD

Blood specimens for measurement of nickel concentrations were obtained, without intake of food, from three groups of women: 23 healthy women, age 19 to 27 years, as normal controls (Group 1); 13 pregnant women, 32-38 weeks of gestation (Group 2); 16 parturient women (Group 3) (a) before labor (3-A), (b) during labor (3-B), (c) post-partum within 10 min. (3-C), (d) 60 min. after delivery (3-D), and (e) more than 2 hr after delivery (3-E). The pregnant women were patients at Shinshu University Hospital. Nickel analyses were performed by the IUPAC reference method (Brown et al., 1981) using a Perkin-Elmer electrothermal atomic absorption spectrometer (model 4000) equipped with HGA-500 atomizer.

3. RESULTS AND CONCLUSIONS

The results are summarized in Tables 1 and 2. There were no significant differences between the serum nickel concentrations in the three groups of subjects, or in Group 3 at stages B, C, or D of delivery. These findings do <u>not</u> confirm the occurrence of post-partum hypernickelemia.

Chemical Toxicology and Clinical Chemistry of Metals
ISBN 0.12.137520.X
Copyright © 1983 by IUPAC
All rights of reproduction in any form reserved

TABLE 1
Serum Nickel Concentrations in Parturient Women (Group 3)

Case	Age	Preg. week	Concentration of nickel in serum at the specified stage (μg/l) A	B	C	D	E
1[a]	30	40	2.1	2.1	2.0	2.1	3.1
2	24	38		2.3	2.3		
3[b]	35	37		1.7			
4	30	37		2.3	2.7	2.6	
5	34	41		1.0	2.2	2.1	
6	25	40		3.0		3.2	
7	29	38	2.0		2.8		
8	28	39			2.5		3.9
9	31	40		3.2	3.8		
10	29	41		3.7			
11	28	40		2.9			2.7[c]
12	26	42		2.1	3.4	4.1	
13	28	42		2.4		3.2	3.2
14	30	38		1.4		3.0	
15	30	40			4.4	2.5	
16	30	41			2.0		

[a]Caesarian section, [b]PG:prostaglandin treatment, and [c]Two days post-partum.

TABLE 2
Serum Nickel Concentrations in Groups 1, 2, and 3

Parameter	Group 1 (Non-Pregnant)	Group 2 (Third Trimester)	Group 3 Stages of Delivery A	B	C	D	E
$\bar{x}$	2.10	2.04	2.05	2.34	2.81	2.85	3.23
SD	0.90	0.77		0.77	0.81	0.67	
n	20	13	2	12	10	8	4

4. REFERENCES

G. Rubanyi, I. Birtalan, A. Gergely and A.G.B. Kovach. (1982) Am. J. Obstet. Gynecol. 15, 167-169.

S.S. Brown, S. Nomoto, M. Stoeppler and F.W. Sunderman Jr. (1981) Clin. Biochem. 14, 295-299.

REDUCTION OF THE BODY BURDEN OF ALUMINIUM IN RENAL PATIENTS BY DESFERRIOXAMINE CHELATION THERAPY

J. Philip Day, Peter Ackrill*, Fiona M. Garstang,
Kenneth C. Hodge, Philip J. Metcalfe,
Maureen O'Hara, Zully Benzo and Romer A. Romero-Martinez

Department of Chemistry, University of Manchester,
Manchester M13 9PL U.K.
**Artificial Kidney Unit, Withington Hospital,*
Manchester M20 8LR U.K.

1. INTRODUCTION

Patients in renal failure tend to accumulate aluminium in bone, blood and soft tissues, and as a result may develop a fatal neurological syndrome, dialysis encephalopathy, and/or a crippling bone condition, fracturing dialysis osteodystrophy (osteomalacia). These conditions are often associated with a patient's exposure to aluminium-contaminated dialysis solutions over a period of some years, and previously usually occurred in areas (e.g. the north of Britain) where water supplies used for haemodialysis contain abnormally high concentrations of aluminium (50 to 500 µg/l).

Recently, recommendations (Hodge et al., 1981) for an upper limit of 15µg/l for aluminium concentrations in dialysis solutions have been applied in the Manchester area, and the incidence of dialysis encephalopathy has greatly diminished. However, patients exposed to high levels of aluminium in the past may still present symptoms of encephalopathy or osteodystrophy, and a number of such patients have been, or are being, treated in our Renal Units to remove aluminium.

The iron-chelating agent, desferrioxamine (DFO), was introduced in 1979 to treat a case of severe dialysis encephalopathy (Ackrill et al., 1980), and later to treat dialysis osteodystrophy (Brown et al., 1982; Ackrill et al., 1982). Clinically, results have been dramatic. Several patients with encephalopathy, including the first one treated, have lost all symptoms of the disease (which previously had invariably proved fatal), and osteodystrophic patients have generally become fully ambulant and lost the bone pain associated with osteomalacia. This paper reports chemical studies carried out in relation to these clinical cases.

Chemical Toxicology and Clinical Chemistry of Metals
ISBN 0.12.137520.X
Copyright © 1983 by IUPAC
All rights of reproduction in any form reserved

2. CHEMICAL ACTION OF DESFERRIOXAMINE

Desferrioxamine is a well-known chelating agent which forms a stable complex with iron(III), and is used to reduce iron overload, for example in cases of acute poisoning, or chronic haemosiderosis. The compound forms a stable 1:1 complex with aluminium(III), and the stability constant and dialysance (dialysis molecular mass) of this complex have been determined (Table 1). Standard haemodialysis membranes are permeable to molecules of molecular mass up to at least 2000 daltons, so that the metal-DFO complexes would be expected to be cleared effectively by normal methods of dialysis.

TABLE 1

Stabilities and Molecular Masses of Desferrioxamine Complexes

Compound	Stability Constant	Molecular Mass
DFO	...	560
Fe-DFO	10^{31}	614*
Al-DFO	10^{22}	585*

* Molecular mass of single-charged complex cation.

DFO (6g "Desferal": Ciba) is usually administered by infusion into a blood line of the dialyser at one of the normal thrice-weekly dialysis sessions. Serum-Al rapidly rises from levels ca 100 to 200 μg/l in aluminium-loaded patients (cf. non-renal "normals", below 10 μg/l, and renal "normals", below ca 80μg/l), to levels up to 1000 and, exceptionally, 1600μg/l. The aluminium is presumably detoxified by complexation with DFO, as levels in this range would normally be fatal to the patient. Aluminium removal by haemodialysis has been measured directly in a number of cases. The rate of removal is proportional to serum-Al concentration, and up to 10mg Al may be removed in a single dialysis session. In vivo measurements of clearance indicate an effective molecular mass ca 600 daltons, corresponding to the formation of the 1:1 complex in vitro (Table 1).

We have concluded that, whatever may be the binding sites for aluminium within the body, the stability of the Al-DFO complex is sufficient to release stored aluminium and retain it in the serum, from where it may then be dialysed.

3. REDUCTION IN BONE ALUMINIUM

The bone aluminium levels in 5 patients treated with DFO are given in Table 2 (Day et al., 1983). Courses of treatment generally lasted from 6 to 10 months, and resulted in the estimated removal of 300 to 1500mg aluminium (from direct measurement of aluminium concentrations in dialysis solutions pre- and post-dialyser; Ackrill et al., 1982).

TABLE 2

Bone Aluminium Concentrations in Five Patients with Fracturing Dialysis Osteodystrophy, Treated with Desferrioxamine to Remove Aluminium.

Patient		Aluminium (mg/kg)		Bone histology[a]	
		Trabeculae	Marrow	OM	OF
GH	pre DFO	527	252	+2	0
	post DFO	213	91	0	+2
	post vit-D	149	57	0	+2
GO	pre DFO	245	488	+2	0
	post DFO (1)	192	225	+2	+1
	post DFO (2)	55	53	0	+2
WA	pre DFO	242	110	+1	+2
	post DFO	53	50	0	+2
RG	pre DFO	311	29	+2	+1
	post DFO	48	22	+1	0
DN	pre DFO	245	160	+1	0
	post DFO	206	79	+1	0

[a] OM = Osteomalacia; OF = Osteitis Fibrosa. Both conditions are graded histologically on a scale (0 - 2) of increasing severity.

The reductions in bone aluminium levels, if extended to the whole skeleton, would imply a total loss of about the same order as the estimated removal by dialysis. It is not yet clear whether significant quantities of aluminium are removed from other tissues concurrently, but it is known that the element also accumulates in the liver, kidneys and brain. Clinically, reduction in bone aluminium levels has been accompanied by a reduction or a total elimination of osteomalacia (OM), with the return of bone turnover, and in some cases the onset of osteitis fibrosa (OF).

4. CONCLUSIONS

Desferrioxamine (DFO) can be used effectively to treat cases of aluminium overload in renal patients, and can eliminate symptoms of both dialysis encephalopathy and osteomalacic fracturing dialysis osteodystrophy.

Chemically, DFO forms a stable aluminium(III) complex, which is much less toxic than free aluminium and which can be cleared by the normal methods of dialysis.

Body stores of aluminium can be greatly reduced by DFO therapy, and much of the reduction is from aluminium deposited in bone.

5. ACKNOWLEDGEMENTS

Grants from the Medical and Science Research Councils, and the Kidney Unit Endowment Fund, Withington Hospital, are gratefully acknowledged.

6. REFERENCES

P.Ackrill, A.J.Ralston, J.P.Day and K.C.Hodge (1980) Lancet,i, 692-693.

P.Ackrill, J.P.Day, F.M.Garstang, K.C.Hodge, P.J.Metcalfe, Z.Benzo, K.Hill, A.J.Ralston, J.Ball and J.Denton (1982) Proc. European Dialysis and Transplant Assocn, 19, 203-207.

D.J.Brown, J.K.Dawborn, K.N.Ham and J.M.Xipell (1982) Lancet,i, 343-346.

J.P.Day. P.J.Metcalfe, F.M.Garstang, K.C.Hodge, Z.Benzo and P.Ackrill (1983), These Proceedings.

K.C.Hodge, J.P.Day, M.O'Hara and P.Ackrill (1981) Lancet,ii,802-803.

PROTEIN-BINDING AND DIALYSATE CONCENTRATION OF ALUMINIUM - THE MAJOR DETERMINANTS OF ALUMINIUM KINETICS DURING HEMODIALYSIS

H. Graf, H.K. Stummvoll, R. Prager, A. Luger and J. Kovarik.

2nd Dept. of Int. Medicine, University of Vienna, A- 1090 Vienna, Austria

1. INTRODUCTION

It is generally agreed that accumulation of aluminum in patients on chronic hemodialysis is the cause of chronic toxicity syndromes as dialysis encephalopathy, microcytic anemia and a special form of osteomalacic bone disease (Sideman and Manor 1982). Whereas the water used for preparation of the dialysis bath solution has been recognized quite early as one major source for aluminum loading in these patients (Ward et al.1978) the quantitative aspects of the aluminum content of the dialysate on aluminum kinetics during hemodialysis remained unclear. We therefore evaluated the in vivo protein binding of aluminum kinetics during hemodialysis using different concentrations of aluminum in the dialysate. In a search for methods to enhance aluminum removal during hemodialysis anecdotal reports on an amelioration of dialysis encephalopathy by the administration of desferrioxamine (Ackrill et al. 1980) prompted us to study the effect of desferrioxamine on aluminum kinetics during hemodialysis.

2. METHODS

A detailed description of aluminum kinetic measurements during dialysis is given elsewhere (Graf et al. 1981). In brief: in vivo protein binding of aluminum was estimated by measuring an ultrafiltrate obtained at the beginning of a dialysis session using a hollow fiber dialysator with a molecular cut off of 10.000 Daltons; this concentration was compared with the simultaneously measured total plasma aluminum at the blood inlet site of the dialyser thus representing the free diffusible part of plasma aluminum, whereas the remaining fraction can be assumed as being protein bound. Aluminum removal during dialysis was estimated by measuring plasma aluminum at the beginning and at the end of a 6 hour dialysis. All aluminum measurements were done using flameless atomic absorption spectrophotometry (Fuchs et al. 1974). Aluminum kinetics were measured at dialysate alu-

Chemical Toxicology and Clinical Chemistry of Metals
ISBN 0.12.137520.X
Copyright © 1983 by IUPAC
All rights of reproduction in any form reserved

minum concentrations of 0.2 and 0.9 µmol/L. The effect of desferrioxamine was studied measuring aluminum kinetics at a dialysate aluminum of 0.9 µmol/l 48 hours after administration of 4 g desferrioxamine intravenously.

3. RESULTS AND DISCUSSION

Aluminum kinetics at 0.2 µmol/L dialysate aluminum are summarised in FIG. 1.

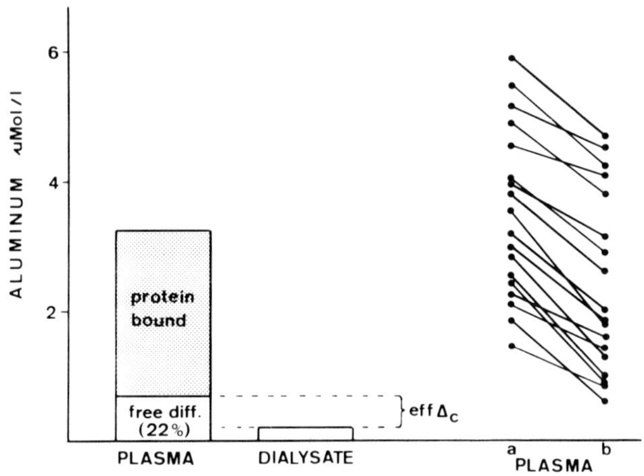

FIG. 1 Mean plasma and dialysate aluminum concentrations and individual plasma aluminum levels in 18 patients (eff Δ_c = effective concentration gradient , a= begin, b= end of dialysis).

Predialysis plasma aluminum levels in 18 patients were about 3 times the value found in normal controls and showed a significant correlation to the amount of daily ingested aluminum hydroxide used as phosphate binder (Graf et al. 1981). Free diffusible plasma aluminum was consistently about 20% and the concentration gradient between the free diffusible plasma aluminum and the dialysate aluminum was therefore always directed towards the dialysate. This resulted in a decrease of plasma aluminum levels during the 6 hour dialysis suggestive of a negative aluminum balance in all patients.

Aluminum kinetics at 0.9 µmol/L dialysate aluminum are summarised in FIG. 2.
Dialysing 18 patients against this relatively high aluminum concentration for a period of 4 weeks resulted in a significant increase of predialysis plasma aluminum levels compared to the very low content of 0.2 µmol/L. Again the free diffusible fraction was about 20% now giving a concentration gradient between free diffusible aluminum and dialysate aluminum directed to-

wards the dialysate only in patients whose predialysis aluminum level was above 5 μmol/L. During this 6 hour dialysis therefore only patients with a predialytic plasma aluminum of greater than 5 μmol/L showed a decrease of plasma aluminum levels whereas the others even showed an increase.

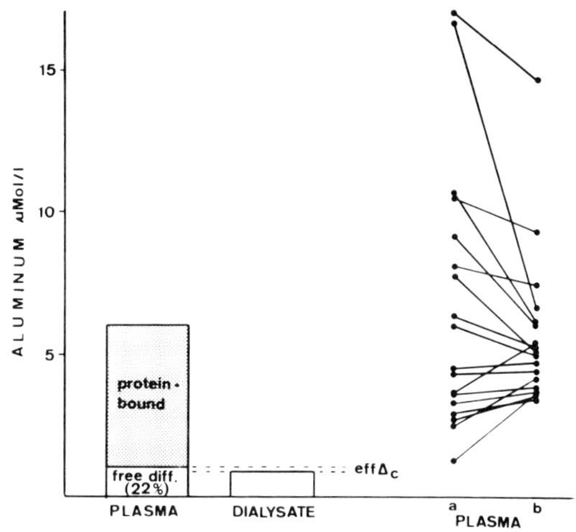

FIG. 2 Mean plasma and dialysate aluminum concentrations and individual plasma aluminum levels in 18 patients (eff Δ_c =effective concentration gradient, a= begin, b= end of dialysis).

Aluminum kinetics following administration of desferrioxamine are summarised in FIG. 3.

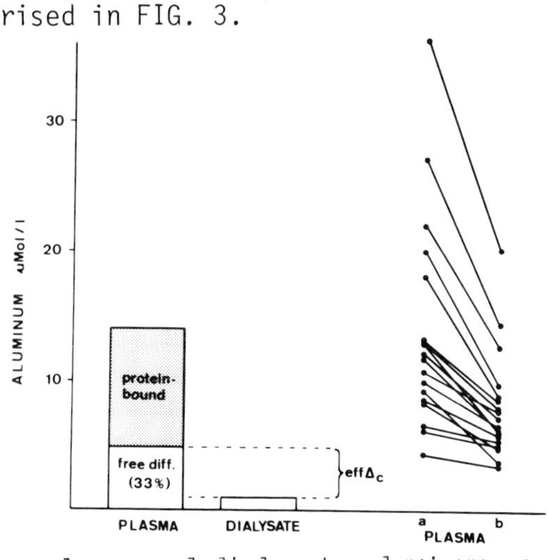

FIG. 3 Mean plasma and dialysate aluminum concentrations and individual plasma aluminum levels in 18 patients(eff Δ_c =effective concentration gradient, a= begin, b= end of dialysis).

48 hours after administration of 4 g desferrioxamine during the last hour of the preceeding dialysis there was a substantial increase of predialysis plasma aluminum levels in the 18 patients studied. As the aluminum hydroxide dosage had been unchanged it can be assumed that this increase reflects mobilisation of tissue aluminum by a chelating effect of desferrioxamine. Most strikingly, the free diffusible fraction of total plasma aluminum was altered to about 30%, resulting in a greatly enhanced concentration gradient between the free diffusible plasma aluminum and the dialysate aluminum directed towards the dialysate. Therefore a decrease of plasma aluminum levels during the 6 hour dialysis occured in all patients. Because of the absolute value of the free diffusible aluminum concentration, a substantial aluminum removal during this dialysis session can be assumed.

4. CONCLUSIONS

Under normal conditions, aluminum is protein bound to about 80%, so that 20% is free diffusible. Aluminum can be removed by hemodialysis depending on the concentration gradient between free diffusible plasma aluminum and the dialysate aluminum. An optimal aluminum concentration in the dialysate is below 0.2 umol/L as this concentration avoids aluminum loading even in patients with normal plasma aluminum levels. Desferrioxamine seems to mobilize tissue aluminum and greatly enhances the free diffusible fraction of total plasma aluminum thus facilitating increased aluminum removal during hemodialysis.

5. REFERENCES

P. Ackrill et al. (1980) Lancet II, 692-693.
C. Fuchs et al. (1974) Clin Chim. Acta, 52, 71-80.
H. Graf et al. (1981) Kidney International, 19, 587-592.
H. Graf, H.K. Stummvoll and V. Meisinger (1982) Lancet II, 46-47.
S. Sideman and D. Manor (1982) Nephron, 31, 1-10.
M.K. Ward et al. (1978) Lancet I, 841-845.

DYNAMICS OF CHROMIUM DETOXIFICATION FOLLOWING CUTANEOUS ABSORPTION OF CHROMIUM AND SUBSEQUENT RENAL FAILURE

Fiona M. Garstang and J. Philip Day[a]
Peter Ackrill and Peter Williams[b]

[a]*Department of Chemistry, University of Manchester, Manchester M13 9PL U.K.*
[b]*Artificial Kidney Unit, Withington Hospital, Manchester M10 8LR U.K.*

1. INTRODUCTION

Two contrasting cases of chromium poisoning have been studied. Severe burns from hot acidic solutions containing in case A, chromium(VI), and in case B, chromium(III), resulted in chromium absorption through the skin. Uptake of chromium and distribution to major body organs, accompanied by renal failure, occurred in both cases. However, the behaviour of chromium within the body showed marked differences between the two cases.

2. CHEMICAL METHODS AND RESULTS

Chromium uptake and distribution was monitored by analysis of blood (serum and red cells), urine, peritoneal dialysis solutions, skin, bone and other tissues. Tissues were ashed with boiling 10M nitric acid. Solutions were analysed by graphite furnace atomic absorption spectrometry.

(a) Patient A

Acute renal failure occurred 4 days after the patient received severe leg burns (ca 10% body surface) from a solution of chromium(VI) oxide (CrO_3) in 1M sulphuric acid at 60°C. Renal function (RF) started to return after 11 days of peritoneal dialysis (PD) and other treatment. The patient slowly recovered and was discharged from hospital on day 42 (70% RF). After one year, the patient remains well and has recovered normal renal function.

Serum chromium levels (Figure 1a) declined exponentially in two phases, with distinct half-lives of 3.5 days during the first 14 days, the period on PD, and 15 days subsequently. Red cell chromium (Figure 1b) reached a maximum concentration on day 8, and subsequently declined with half-life ca 28 days. The

Chemical Toxicology and Clinical Chemistry of Metals
ISBN 0.12.137520.X
Copyright © 1983 by IUPAC
All rights of reproduction in any form reserved

chromium concentration ratio (red cells:serum) increased to a limiting value of 10:1.

Chromium removal (Figure 1c) by PD was from 1 to 3mg/d (partly dependent on volume of PD fluid used), and by urinary excretion below 1mg/d (although by this stage serum chromium levels had dropped appreciably). Overall chromium removal was 14.8mg by PD, over 10 days, and 19.8mg through urine, over 50 days. After 125 days, the urine chromium concentration was still 77μg/l, compared to "normal" levels of below 1 μg/l.

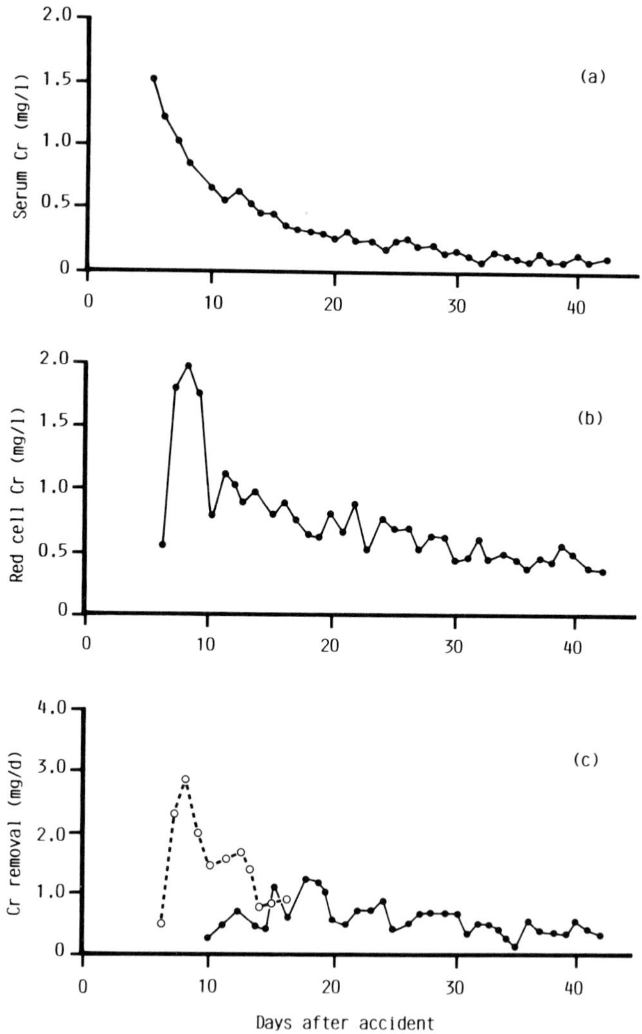

FIG. 1 Data for Patient A: (a) serum chromium concentrations; (b) red cell chromium concentrations; (c) chromium removal by PD (o) and urinary excretion (•).

Skin chromium adjacent to the burns was 58μg/g, and kidney chromium (biopsy on day 14) was 11μg/g. The initial body load of absorbed chromium was estimated at 80mg, based both on a comparision of the kidney chromium with that in patient B, for whom total body chromium was determined post mortem, and on back extrapolation of the serum chromium load.

(b) Patient B

Severe burns and shock followed immediately by acute renal failure resulted from the patient's immersion up to the neck (ca.90% body area) in chromium(III) sulphate in 1M sulphuric acid at 70°C. Death occurred after 48 hours. Blood chromium was measured hourly and tissue chromium, including bone and all major organs, after death. Uptake was by cutaneous absorption with no apparent ingestion (Kelly et al., 1982).

Serum chromium fell rapidly with a mean half-life ca 17 hours. Urine volumes were very small with negligable total chromium removal. Dialysis was not performed but exchange transfusion at 28 hours removed 12mg chromium. Total body chromium at death was ca.155mg, compared with total blood loads of 160mg and 18mg at 4 hours and death, respectively. The chromium concentration ratio (red cell:serum) at death was ca.1:10, in contrast with case A (10:1).

3. DISCUSSION

Both patients absorbed large quantites of chromium, ca 80mg and 160mg in the two cases A and B, respectively.

In patient B, almost total transfer of the initial blood load of chromium to tissues occured over 48 hours, with a serum half life ca 17 hours and little uptake by red blood cells. Assuming chromium(III) to have remained as such in the blood, we conclude that chromium in this oxidation state is rapidly removed from the serum, presumably by binding to protein in the tissues, and also does not readily cross the red cell membrane.

In patient A, a much slower loss of serum chromium occurred, and the effective removal by PD (half-life = 3.5 days) indicates facile transfer across the peritoneal membrane. We conclude that the major fraction of serum chromium was present in a different chemical form to that in case A, probably as chromium(VI) but just possibly as chromium(V). The delayed maximum in red cell chromium concentration indicates one-way transfer from serum, and the higher levels inside the cells could result from the passage of chromium(VI) into the cells, with subsequent reduction to chromium(III) and fixation by protein binding. The 28 day half-life for red cell chromium probably reflects increased turnover of red cells (normally ca 80 days) probably due to chromium-induced fragility.

The longer half life for serum chromium following the return of partial renal function (patient A) reflects the cessation of PD, after which chromium removal from the serum would be largely by a combination of 3 processes: (i) loss by excretion, (ii) continued transfer to new red cells, and (iii) reduction to chromium(III) followed by immediate removal to and binding in tissues.

The explanations suggested appear to relate plausibly to the chemical properties of chromium species in different oxidation states, and previous observations on the physiological behaviour of various chromium compounds (Baetjer et al., 1959; Lefrançois et al., 1983). Cationic chromium(III), either hydrated or complexed, has a high affinity for potential ligands, including protein molecules. In contrast, anionic chromium(VI) has little affinity for proteins or other ligands and readily crosses cell membranes . Alternatively, it is chemically possible that chromium(VI) would undergo reduction in serum to anionic chromium(V), which would also be expected to be physiologically mobile.

4. CONCLUSIONS

Both chromium(III) and chromium(VI) are readily absorbed through the skin, and chemical differences between the two forms persist in the body. Chromium(III) is rapidly removed to tissues, where it become firmly bound, but does not accumulate appreciably in the red cells. Chromium(VI) is more persistent in serum and readily enters the red cells where it becomes fixed, probably as chromium(III).

Chromium(VI) absorption may initially be treated by peritoneal dialysis but at later stages red cell chromium can only be removed by exchange of red cells. Chromium(III) absorption should, therefore, be treated immediately by exchange transfusion as dialysis is not likely to be effective.

5. ACKNOWLEDGEMENTS

The support of the Medical and the Science Research Councils, and the Kidney Unit Endowment Fund, Withington Hospital, is gratefully acknowledged.

6. REFERENCES

A.M. Baetjer, C. Damron and V. Budacz (1959) Arch. Ind. Health, 20, 136-150.

W.F. Kelly, P. Ackrill, J.P. Day, M. O'Hara and C.T. Tye (1982) British Journal of Industrial Medicine, 39, 397-400.

H. Lefrançois, J.F. Mercier, M.C. Herlant-Peers, H.F. Hildebrand and J.M. Haghenoer (1983) These Proceedings.

THE INFLUENCE OF FLUORIDE ON TISSUE ALUMINIUM AND BEHAVIOUR IN THE RAT

B.J. Stevens[1], G.L. Willis[2], T.J. Humphery[3], R.C. Atkins[3], G.C. Smith[2] and D.L. Copolov[2].

[1]*Department of Applied Biology, Royal Melbourne Institute of Technology,*
[2]*Monash University Department of Psychological Medicine, Prince Henry's Hospital,*
[3]*Department of Nephrology, Prince Henry's Hospital, Melbourne, Australia.*

1. INTRODUCTION

In patients with chronic renal failure, elevated levels of tissue aluminium have been reported in association with a progressive dementing neurological syndrome, and also a debilitating bone disease. The neurological disease, termed dialysis encephalopathy, is characterised by the gradual development of insidious psychological changes and deficits in motor reflex control (Alfrey, 1978).

Aluminium is an ubiquitous element, and renal failure patients treated by dialysis are exposed to the metal from a variety of sources - the water used for dialysis, the equipment, medications, foodstuffs, and inhaled dust.

Aluminium is known to have a high chemical affinity for fluorine, forming a number of stable complexes in aqueous solution (Matwiyoff and Wageman, 1970). Several studies (e.g. Lough *et al.*, 1975) have shown serum fluoride levels in uremic patients dialysed with fluoridated water to be about 10 times normal. Like aluminium, fluoride has been shown to be taken up by the blood during dialysis. The immediate post dialysis increase in serum fluoride concentration is followed by a subsequent fall, and this has been attributed to predominant uptake by bone (Taves *et al.*, 1965).

We postulated that the presence of increased fluoride in dialysate or other sources may enhance the uptake of aluminium by various tissues, and therefore contribute to both the neurological changes and the elevated bone aluminium levels seen in chronic renal disease.

2. EXPERIMENTAL

Eight groups of six male Sprague-Dawley rats were housed in controlled light and climate, and allowed food and water ad lib.

Chemical Toxicology and Clinical Chemistry of Metals
ISBN 0.12.137520.X
Copyright © 1983 by IUPAC
All rights of reproduction in any form reserved

Renal impairment was produced surgically in half the animals ('reduced renal mass', RRM) by 5/6 nephrectomy performed under ether anaesthesia and completed ten days before the treatment began. Control animals were designated 'intact'. There were four treatments within each division, consisting of daily subcutaneous injections for thirty days of (a) normal saline, (b) sodium fluoride in normal saline, (c) aluminium chloride in normal saline, and (d) aluminium chloride plus sodium fluoride in normal saline. These groups were designated 'saline', 'F', 'Al', and 'Al + F' respectively. Treatment groups (c) and (d) received 18.5 μmol Al/day and (b) and (d) received 55 μmol F/day. Thus in group (d) there was a fluorine: aluminium ratio of 3:1. The animals were weighed weekly.

On day 31 testing was carried out by measuring open field behaviour. The number of occasions on which animals reared in 5 minutes and also crossed squares during this period were observed.

On day 32 motor reflex activity was assessed by observing (a) the latency to retract an elevated forelimb, (b) latency to step up or down when hindquarters were placed on an elevated platform, and (c) latency to ambulate from a prescribed area.

These tests were selected on the basis of the motor impairments displayed by patients with dialysis encephalopathy (Kerr, 1980), and also on the basis of previous animal work (Bowdler *et al.*, 1979).

On day 33 all animals were decapitated, and trunk blood was collected in heparinised tubes. Haemotocrit and plasma levels of creatinine, aluminium, and β-endorphin were determined.

3. RESULTS

Autopsy was performed on all animals, and femur and soft tissues were sampled. The tissues were dissolved in 50 g/l aqueous tetramethyl ammonium hydroxide and the bone in nitric acid, and aluminium determined by graphite furnace atomic absorp-tion spectrophotometry.

At the end of the experiment, plasma creatinine concentrations were higher in RRM animals, but there were no significant differences between any of the chemically treated subgroups in the two major divisions. Haemotocrit values were significantly lower in both groups that were given Al + F.

Very severe macroscopic necrosis was observed in the region of the injection site in Al + F animals, whereas Al animals showed this to a much lesser degree. F and saline treated animals appeared unaffected.

The intact and RRM Al + F treated animals displayed severe deficits in open field behaviours (Fig I (a)). The number of squares crossed and the number of rearings in the open field were significantly reduced by Al + F treatment. There were instances when F animals and Al animals showed a significant reduction in open field performances, but this did not occur as reliably or to the same magnitude as that displayed by Al + F animals.

Limb retraction and ambulation were not affected in any group. Fluoride decreased the latency to step down ($P<0.05$) in intact rats but not those with RRM.

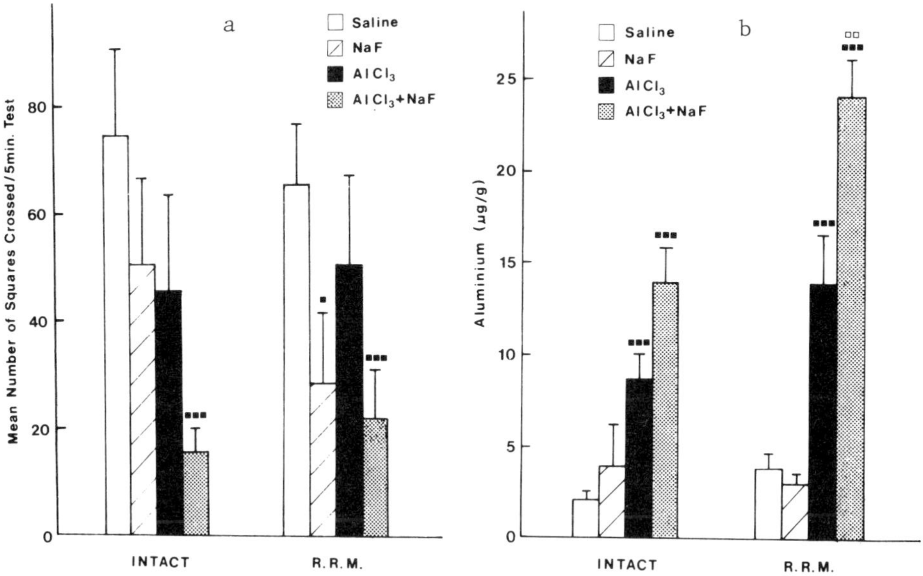

FIG. 1 Effects produced in rats by aluminium and fluoride separately and in combination (a) open field tests (b) aluminium concentrations in adrenal gland.

Plasma β-endorphin concentration was greater in RRM animals than in the intact group.

Tissue aluminium concentrations were significantly elevated in both the groups treated with aluminium. In the Al + F group, aluminium levels were higher than in the Al group in plasma, liver, spleen, adrenal (Fig 1(b)) and lung, but not in brain or muscle.

The body weights of both groups of animals treated with Al + F were significantly lower than that of their respective controls. Intact animals treated with F showed a significant increase in body weight over the thirty days of observation.

The major findings of the experiment are summarised as follows:

1. Treatment with Al + F produced significant impairment in open field behaviour and body weight regulation but not motor reflex function, and this is more severe than when aluminium alone is present.
2. When aluminium is injected, more is deposited in some tissue

compartments when fluoride is present than when fluoride is absent.

4. CONCLUDING REMARKS

From these findings, we make the following observations and speculations:

1. The total weight of injected aluminium (i.e. 15 mg) is comparable to the 3 g accumulated by uremic man.
2. We produced an impairment in locomotor behaviour which is similar to that seen in dialysis encephalopathy.
3. We altered the distribution of aluminium when fluoride was given simultaneously, and since fluoride is known to accumulate in chronic renal failure patients, we suggest that it may be responsible for influencing the pattern of toxicity at present linked solely to aluminium.
4. An anaemia has been attributed to aluminium toxicity in man and rat, and was seen in the rats given aluminium plus fluoride.

However, the assignment of this combination as the cause of the anaemia is complicated by the presence of the haemorrhagic necrosis seen at the injection site in this group. Nevertheless, it should be noted that a haemorrhagic tendency occurs in uremia which is not completely corrected by dialysis, and is thus in keeping with induction by substances absorbed during dialysis. Further work is proceeding to elucidate the relative significance and toxicity of aluminium fluoride complexes.

5. ACKNOWLEDGEMENTS

We thank the Rotary Club of Forest Hill, Victoria, Australia and the Percy Baxter Charitable Trust (Victoria, Australia) for providing funds for this study.

6. REFERENCES

A.C. Alfrey (1978) *Ann.Rev.Med.*, 29, 93-98.
N.C. Bowdler, D.S. Beasley, E.C. Fritze *et al.* (1979) *Pharmacol.Biochem.Behar.*, 10, 505-512.
D.N.S. Kerr (1980) *Adv.in Nephrol.*, 9, 109-132.
J. Lough, R. Noonan, R. Gagnon, *et al.* (1975) *Arch.Pathol.*, 99, 484-487.
N.A. Matwiyoff, W.E. Wageman (1970) *Inorg.Chem.* 9, 1031-1036.
D.R. Taves, R. Terry, F.A. Smith *et al.* (1965) *Arch.Int.Med.*, 115, 167-172.

EFFECTS OF CALCIUM AND MAGNESIUM ACETATES ON TISSUE DISTRIBUTION OF CADMIUM IN WISTAR RATS

Kazimierz S. Kasprzak and Lionel A. Poirier

Laboratory of Comparative Carcinogenesis, National Cancer Institute, FCRF, Frederick, MD 21701 U.S.A.

1. INTRODUCTION

The physiological divalent metals, manganese, zinc, magnesium, and calcium have been shown to antagonize the action of metallic carcinogens, cadmium, nickel and lead (Parizek, 1956; Sunderman et al., 1974, 1976, 1983; Poirier et al., 1980; Nordberg and Andersen, 1981). The mechanisms underlying such antagonism are not clear. In our recent study (Kasprzak et al., 1983), in which magnesium but not calcium, was found to prevent the occurrence of sarcomas at the site of s.c. injection of cadmium chloride, we have hypothesized that the preventive effects of magnesium might result from a decreased binding of cadmium by the target tissue. The present experiment attemoted (a) to test that hypothesis and (b) to furnish some new data on the influence of magnesium and calcium on the uptake, distribution and retention of cadmium injected subcutaneously into rats.

2. EXPERIMENTAL

Ninety male Wistar rats, 120-150 g, were divided into 6 equal groups and treated as follows: Three groups of rats were given a single s.c. injection of 0.02 mmol cadmium chloride-^{109}Cd ($CdCl_2$; 1 mCi/mmol)/kg, the other 3 groups received 0.04 mmol $CdCl_2$/kg. The rats, one group for each cadmium dose, were also given 3 s.c. injections of 4 ml saline/kg, 0.16 mmol calcium acetate (CaAc)/kg, or 4 mmol magnesium acetate (MgAc)/kg, on the day before, the day of and the day after the $CdCl_2$ dosing. Five rats from each group were sacrificed by decapitation 4,15 and 45 days after the $CdCl_2$ injection. The tissue cadmium was determined by counting preweighed samples in a well-type gamma counter.

3. RESULTS

The results are shown in Fig. 1. The amounts of cadmium taken up at the injected site on day 4 equalled approx. 7-15 % of the injected $CdCl_2$ dose, and its concentration there was practically

Chemical Toxicology and Clinical Chemistry of Metals
ISBN 0.12.137520.X
Copyright © 1983 by IUPAC
All rights of reproduction in any form reserved

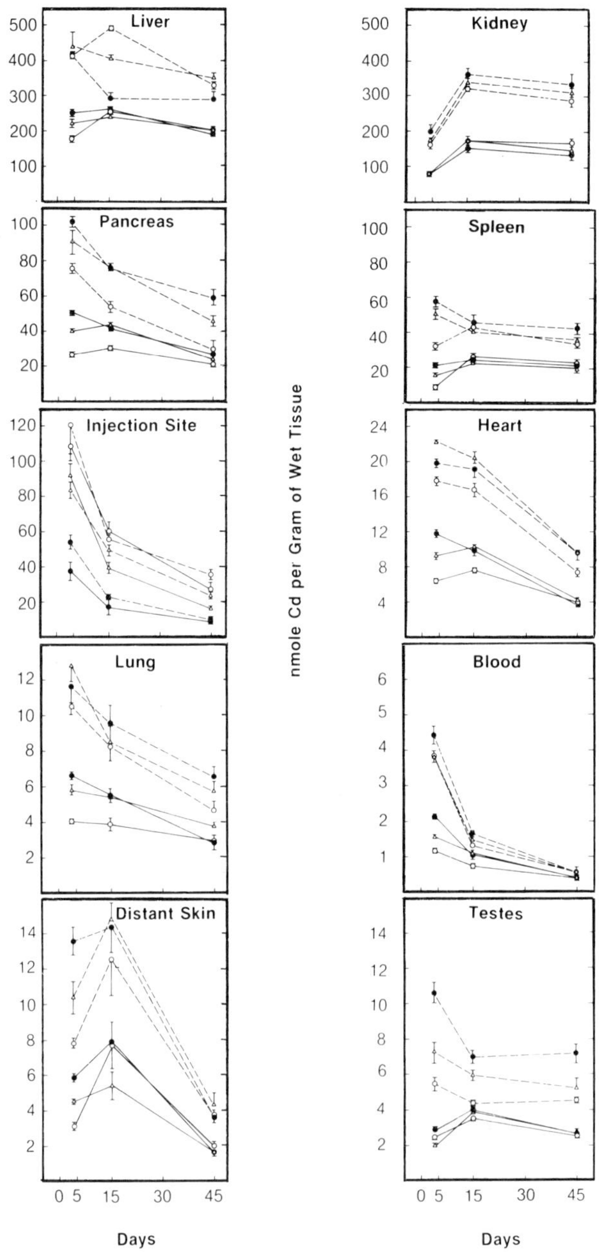

FIG. 1 Cadmium concentrations in tissues at different times following s.c. injection of 0.02 mmoles of $CdCl_2$/kg (solid lines) or 0.04 mmoles $CdCl_2$/kg (dashed lines) alone (△), with 3x0.16 mmoles CaAc/kg (○) or with 3x4.0 mmoles MgAc/kg (●). Vertical bars indicate the standard error limits.

equal for both doses. On day 4 the concentrations of cadmium ranked the tissues as follows: liver > kidney > the injection site skin > pancreas > spleen > heart > lung > distant skin > testes > blood. On day 45 the ranking was different: liver > kidney > pancreas > spleen > the injection site skin > heart > lung > testes > distant skin > blood.

Administration of CaAc or MgAc along with $CdCl_2$ markedly influenced the uptake of cadmium by different tissues. At the injection site CaAc caused significant increase (over 20%) while MgAc produced a significant decrease (over 30%) in cadmium uptake. In the distant skin, pancreas and spleen at either $CdCl_2$ dose and in liver, heart, lung and blood at the low $CdCl_2$ dose, the effects were reversed: CaAc decreased and MgAc increased the uptake of cadmium. The differences became less evident at later times. In the testis, liver, heart, lung and blood the effects of CaAc and MgAc were less uniform; they depended on both the tissue and the dose of $CdCl_2$ administered. No effects of CaAc or MgAc were noticed in the kidney.

4. DISCUSSION

The uptake of cadmium by tissues, its distribution and concentration levels found in this study are consistent with previous findings (Lucis et al., 1972; Nordberg, 1972; Kotsonis and Klaassen, 1977). The initial tissue concentration of cadmium was not always proportional to the injected dose. When the double dose was administered, correspondingly increased concentrations of cadmium were observed in blood, liver, kidney, pancreas, heart, lung and skin distant from the injection site, but not at the injection site itself or in spleen and testes. At the injection site administration of either dose of $CdCl_2$ resulted in practically identical cadmium uptake. In spleen and testes the initial 4-day uptake of the high $CdCl_2$ dose was more than doubled; both organs are known to be especially sensitive to cadmium toxicity (Bondia et al.,1980; Suzuki et al.,1981). The concentration of cadmium at the injection site decreased from day 4 to day 45 in both high and low dose groups. In the low dose rats, but not in the high dose groups, this decrease was generally accompanied by a rise in the cadmium concentration in distant tissues from day 4 to day 15. This observation suggests that cadmium is slowly released from the injection site and is taken up by other tissues whose binding capacity has not been reached.

CaAc and MgAc administered along with $CdCl_2$ significantly affected cadmium uptake. However, almost all of the effects seem to result indirectly from the interactions between the metals and the tissues at the injection site, rather than from interactions in each separate organ or tissue. Thus, CaAc, which enhanced the retention of cadmium at the site of injection, decreased cadmium accumulation in other tissues. On the other hand, the administration of MgAc inhibited the accumulation of cadmium at the injection site which led to increased retention in distant tissues. Quite possibly, the different cadmium uptake patterns at the injection site resulted from different effects exerted by CaAc and MgAc on blood vessels. Calcium by its contracting action upon the blood vessel smooth muscles (Aikawa, 1981) would hinder blood flow, slowing the mobilization of cadmium. Magnesium which is antagonistic to calcium in this respect (Aikawa, 1981) would produce opposite effects.

The results of this study generally support our recent suggestion that the prevention of local cadmium carcinogenesis by MgAc may result from the inhibition of binding of the carcinogen in the target tissue (Kasprzak et al., 1983). However, the suppression of cadmium binding by MgAc is incomplete and the effect of the dose of cadmium on injection site sarcoma formation has not been established. Thus the observed decrease in cadmium uptake is a possible, but not necessarily the sole mechanism by which magnesium prevents the carcinogenic activity of cadmium.

5. REFERENCES

J.K. Aikawa (1981) in Magnesium: Its Biologic Significance, CRC Press, Boca Raton, Fla., pp. 72-78.

S. Bondia, A. Jaudi, B. Ribas, A. Santos-Ruiz and M.I. Sanchez (1980) in Trace Element Analytical Chemistry in Medicine and Biology, W. de Gruyter and Co., Berlin, New York, pp. 37-46.

K.S. Kasprzak, K.L. Hoover, L.A. Poirier and M.L. Wenk (1983) Proc. Am. Assoc. Cancer Res., 24, 84.

F.N. Kotsonis and C.D. Klaassen (1977) Toxicol. Appl. Pharmacol., 41, 667-680.

O.J. Lucis, R.Lucis and K. Aterman (1972) Oncology, 26, 53-67.

G.F. Nordberg (1972) Environ. Physiol. Biochem., 2, 7-36.

G.F. Nordberg and O. Andersen (1981) Environ. Health Perspect., 40, 65-81.

J. Parizek (1956) Nature, 177, 1036-1037.

L.A. Poirier, J.C. Theiss, L.J. Arnold, M.B. Shimkin (1980) Proc. Am. Assoc. Cancer Res., 21, 87.

F.W. Sunderman, Jr., T.J. Lau and L.J. Cralley (1974) Cancer Res., 34, 92-95.

F.W. Sunderman, Jr., K.S. Kasprzak, T.J. Lau, PP. Minghetti, R.M. Maenza, N. Becker, P.J. Goldblatt and C. Onkelinx (1976) Cancer Res., 36, 1780-1800.

F.W. Sunderman, Jr. and K.S. McCully (1983) Carcinogenesis, 4, 461-465.

K.T. Suzuki, J.K. Yamada and F. Shimizu (1981) Biochem. Pharmacol., 30, 1217-1222.

EFFECTS OF CALCIUM, MAGNESIUM AND SODIUM ACETATES ON TISSUE DISTRIBUTION OF NICKEL(II) IN STRAIN A MICE

Kazimierz S. Kasprzak and Lionel A. Poirier

Laboratory of Comparative Carcinogenesis, National Cancer Institute, FCRF, Frederick, MD 21701 U.S.A.

1. INTRODUCTION

It has been recently shown by Poirier et al. (1980) that the essential divalent metals, calcium and magnesium inhibit the tumorigenicity of nickel and lead in the lung of strain A mice. The authors hypothesized that the mechanisms responsible for that inhibition might consist of a prevention of binding of the carcinogenic metals by the target tissue. The present paper attempts to verify that hypothesis by investigating the effects of calcium and magnesium on the uptake of nickel by the lung and other organs of strain A mice under the same experimental conditions in which both essential metals inhibited lung tumor formation. Some effects of sodium are also shown.

2. EXPERIMENTAL

Eighty male, strain A mice, 16-24 g, were divided into 8 equal groups. Five mice per group received 1 or 24 (3x/week) i.p. injections, of 42.5 µmol/kg of nickel acetate-^{63}Ni (NiAc; 1 mCi/mmol) alone or combined with 42.5, 127.5, or 425 µmol/kg of calcium acetate (CaAc), magnesium acetate (MgAc), or with 850 µmol/kg of sodium acetate (NaAc) - all in 0.155 M saline. The mice were sacrificed by decapitation 48 hr after the last injection. Nickel in tissue samples was determined by liquid scintillation counting (Kasprzak and Sunderman, 1979).

3. RESULTS

The results are shown in Fig. 1. The concentration of nickel in blood remained below 0.1 nmol/ml and was not significantly affected by the administered essential metals. In most organs the multiple dosing of NiAc alone resulted in greater retention of nickel than did the administration of a single dose. The greatest increase in nickel concentration was noticed in the pancreas (over 10-fold), followed by spleen (2-fold), brain (2-fold), lung (1.5-fold) and liver (1.5-fold). Retention of

Chemical Toxicology and Clinical Chemistry of Metals
ISBN 0.12.137520.X
Copyright © 1983 by IUPAC
All rights of reproduction in any form reserved

nickel in kidney and heart after 24 injections was about 50% lower than after one injection. The nickel contents after one injection of NiAc alone ranked the tissues as follows: kidney > lung > pancreas > spleen > liver > brain > heart > blood. After 24 injections the ranking changed to: pancreas > lung > kidney > spleen > liver > brain > heart > blood. Simultaneous administration of CaAc assisted the uptake of nickel in all the tissues investigated, except heart and blood. MgAc either enhanced or

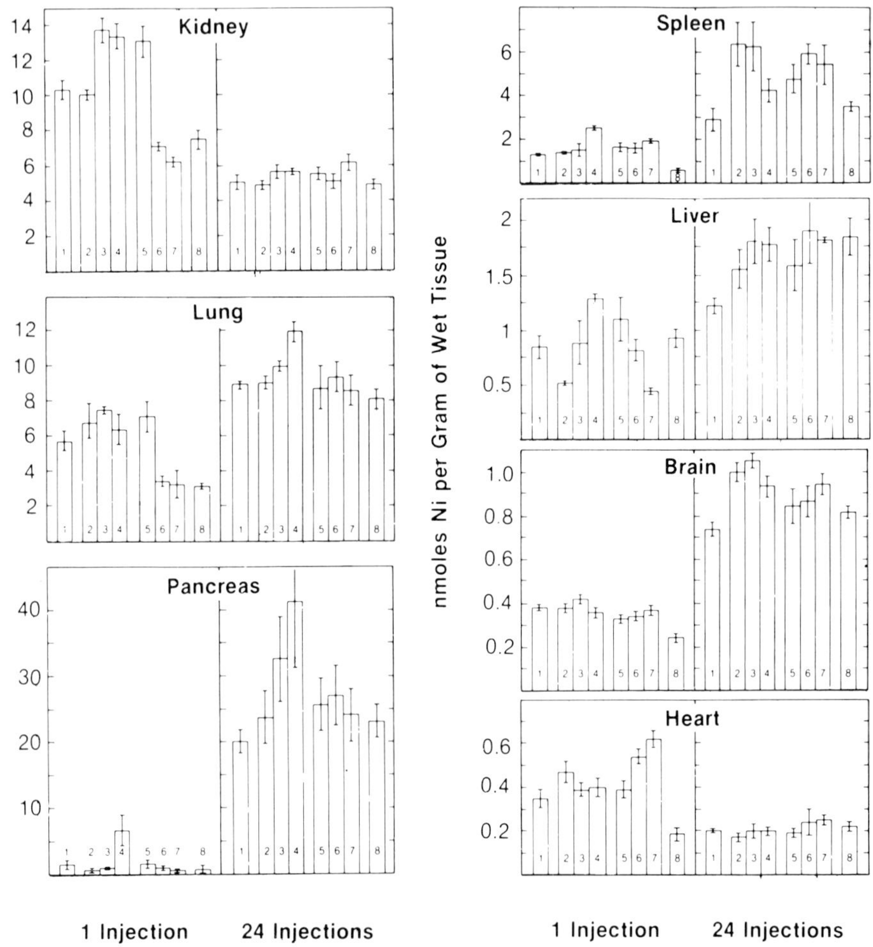

FIG. 1 Concentrations of nickel (nmol/g wet tissue ± S.E.) in organs of mice after a single or 24 i.p. injections of NiAc alone (42.5 μmol/kg - bar 1) and together with different proportions of CaAc (42.5 μmol/kg - bar 2; 127.5 μmol/kg - bar 3; 425 μmol/kg - bar 4), MgAc (42.5 μmol/kg - bar 5; 127.5 μmol/kg - bar 6; 425 μmol/kg - bar 7), or NaAc (850 μmol/kg - bar 8)

inhibited nickel uptake, depending on the tissue, dose and number of injections. NaAc inhibited nickel uptake after one injection in all tissues, except pancreas, liver and blood, but had no influence on the nickel levels after 24 injections, except for an increase in the liver. No consistent effects of the doses of CaAc or MgAc on nickel uptake were noticed.

4. DISCUSSION

The present study confirms that kidney and lung are the main targets for parenterally administered nickel in rodents (Smith and Hackley, 1968; Oskarsson and Tjalve, 1979; Herlant-Peers et al., 1983). The uptake of this metal by liver, spleen, pancreas and heart is ranked very differently by different authors and seems to depend strongly on the nickel compound administered, dose, time, strain of animals and route of administration (Clary, 1975; Mathur et al., 1978; Oskarsson and Tjalve, 1979). The most striking findings of the present study are the high accumulation of nickel in pancreas and the decreasing nickel accumulation in kidney and heart (with no significant change in organ weights) after multiple i.p. injections of NiAc. The first finding would link nickel with zinc and insulin metabolism and relate the commonly observed Ni-induced elevation of serum glucose to that interaction (Clary, 1975; Sunderman et al., 1976). The second finding would suggest the development of some detoxifying mechanisms in kidney and heart during the prolonged exposure to nickel.

CaAc, MgAc and NaAc exerted some significant effects on the amount of uptake of nickel. Generally, CaAc tended to enhance and NaAc tended to inhibit nickel uptake. MgAc either enhanced or inhibited that uptake. The effects of NaAc were more pronounced after a single injection than after 24 ones. Biochemical interactions among the essential divalent metals and their non-essential, toxic counterparts have already been described by Whanger (1973), Fine et al.(1976), Washko and Cousins (1977), Meredith et al. (1977), Sugawara (1977) and Walker and Shelburne (1982). The theoretical and experimental bases of such interactions have been reviewed by Jacobson and Turner (1980) and Nordberg and Andersen (1981). The ionic radii of Ni^{2+} and of Mg^{2+} are very similar, 0.69 Å and 0.66 Å, respectively, while that of Ca^{2+} is 0.99 Å. The arrangement of biocomplexes of Ni^{2+} is also much closer to that of Mg^{2+} than to the arrangement of the Ca^{2+} biocomplexes (Whanger, 1973). Hence, stronger biochemical interactions may be anticipated between Ni^{2+} and Mg^{2+} than between Ni^{2+} and Ca^{2+}. Unfortunately, published experimental evidence of such interactions is sparse and indirect. Nickel - magnesium interactions have not been reported so far. Clary (1975) observed a sharp decrease in the excretion of calcium in urine following a 4 month exposure of rats to nickel chloride in drinking water. Gitlitz et al. (1975) noticed a decrease in urine calcium in rats after a single i.p. injection of nickel chloride; plasma calcium was not affected. An increased accumulation of iron and chromium in organs of nickel-treated rats was observed by Whanger (1973) and Clary (1975). It seems possible that nickel and each of the other metals mutually enhance their tissue retention. Such effects are also described for calcium and cadmium at the site of their s.c. injection (Kasprzak et al., 1983). The mechanisms of these effects remain, however, obscure.

The results of the present investigation clearly demonstrate that the inhibition by CaAc and MgAc of lung tumor induction by multiple i.p. injections of NiAc (Poirier et al., 1980) was not caused by a prevention or a decrease of nickel uptake in lung. The anti-tumorigenic action of CaAc and MgAc would thus depend on some more complex, molecular interactions of nickel and the essential divalent metals. Further studies on such interactions are in progress.

5. REFERENCES

J.J. Clary (1975) Toxicol. Appl. Pharmacol., 31, 55-65.

B.P. Fine, A. Barth, A. Sheffet and M.A. Lavenhar (1976) Environ. Res., 12, 224-227.

P.A. Gitlitz, F.W. Sunderman, Jr. and P.J. Goldblatt (1975) Toxicol. Appl. Pharmacol., 34, 430-440.

M.C. Herlant-Peers, H.F. Hildebrand and J.P. Kerckaert (1983) Carcinogenesis, 4, 387-392.

K.B. Jacobson and J.E.Turner (1980) Toxicology, 16, 1-37.

K.S. Kasprzak, K.L. Hoover, L.A. Poirier and M.L. Wenk (1983) Proc. Am. Assoc. Cancer Res., 24, 84.

K.S. Kasprzak and F.W. Sunderman, Jr. (1979) Pure Appl. Chem., 51, 1375-1389.

A.K. Mathur, T.S.S. Dikshith, M.M. Lal and S.K. Tandon (1978) Toxicology, 10, 105-113.

P.A. Meredith, M.R. Moore and A. Goldberg (1977) Biochem. J., 166, 531-537.

G.F. Nordberg and O. Andersen (1981) Environ.Health Perspect., 40, 65-81.

A. Oskarsson and H. Tjalve (1979) Ann. Clin. Lab. Sci., 9, 47-59.

L.A. Poirier, J.C. Theiss, L.J. Arnold and M.B. Shimkin (1980) Proc. Am. Assoc. Cancer Res., 21, 87.

J.C. Smith and B. Hackley (1968) J. Nutr., 95, 541-546.

N. Sugawara (1977) Arch. Environ. Contam. Toxicol.,5, 167-175.

F.W. Sunderman, Jr., K.S. Kasprzak, E. Horak, P. Gitlitz and C. Onkelinx (1976) Toxicol. Appl. Pharmacol., 38, 177-188.

S.R. Walker and J.D. Shelburne (1982) Fed. Proc., 41, 924.

P.W. Washko and R.J. Cousins (1977) J. Nutrition, 107,920-928.

P.D. Whanger (1973) Toxicol. Appl. Pharmacol., 25, 323-331.

CHELATION OF LEAD WITH SODIUM DIMERCAPTOPROPANE SULPHONATE AND ESTIMATION OF RENAL LEAD BURDEN

Teresa Twarog and M. George Cherian

Department of Pathology, University of Western Ontario
London, Ontario, CANADA

1. INTRODUCTION

Chelation has been used for both treatment of lead poisoning, and for the estimation of body burden of lead in exposed people. Presently EDTA and BAL are considered to be the drugs of choice, although their use is limited due to their toxicity (Chisolm, 1968). In addition, the body burden of lead in exposed people has been estimated by determination of the urinary mobilization yield of lead following chelation with EDTA (Araki and Ushio, 1982).

A water soluble derivative of BAL, the sodium salt of dimercaptopropane sulfonic acid (DMPS) has been found to be effective in the treatment of various types of metal poisoning, such as mercury and arsenic (Gabard, 1976); (Tadlock and Aposhian, 1980). It has been observed that administered DMPS, unlike BAL, is not easily metabolized and is excreted unchanged (Gabard and Walser, 1979). DMPS has low toxicity with only few side effects.

A preliminary study showed that DMPS could increase the urinary excretion of lead and preferentially remove lead from the kidneys of lead exposed rats (Twarog and Cherian, 1983). Since renal disease is one of the major consequences of chronic lead exposure, a drug that selectively removes lead from the kidneys would be of value not only for treatment, but also for measurement of renal lead burden in excessive lead exposure. Thus it was of interest to examine the correlation between renal lead burden and urinary lead excretion after a single oral administration of DMPS.

2. MATERIALS AND METHODS

Four groups of rats were injected (i.p.) daily for 7 days, each with a different dose of lead (.5 to 4 mg Pb/kg as lead acetate) to obtain different body burdens of lead. A fifth

Chemical Toxicology and Clinical Chemistry of Metals
ISBN 0.12.137520.X
Copyright © 1983 by IUPAC
All rights of reproduction in any form reserved

Concurrent with the decrease in renal lead was an increase in urinary lead excretion. A linear relationship was observed between urinary excretion of lead and renal lead concentrations following DMPS treatment (Y=.38X+.18, r=.94, p<0.01; Figure 1).

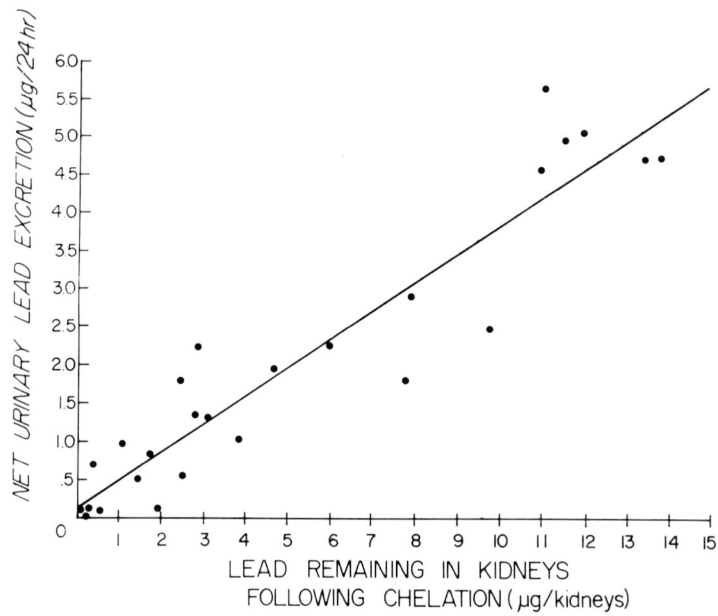

FIG. 1 Correlation between net urinary excretion of lead and renal lead burden following treatment with DMPS. Net excretion is defined as the difference between the amount of metal excreted during the first 24 hours after DMPS treatment and the amount excreted in the preceding 24 hours.

3. DISCUSSION

DMPS is a non-toxic water soluble chemical analogue of BAL and is absorbed intact from the G.I. tract (Gabard, 1978). About 30-40% of ingested DMPS is absorbed, which can effectively chelate lead from the kidney. The major excretory route of DMPS is urine, where it is excreted intact without any degradation or metabolic transformation (Gabard and Walser, 1979). Metabolic studies using ^{14}C-DMPS have shown that only a small amount of administered DMPS is deposited in tissues; the highest amount being in the kidneys (Gabard, 1978). Due to the hydrophilic property of DMPS, it cannot normally enter the cell and thus may mainly chelate metals from the interstitial space.

The organ specificity in chelation of lead by DMPS is related to this hydrophilic property of DMPS. The water soluble DMPS is unexposed group served as a control. Five days after the last injection of lead, the animals were kept in individual metabolic cages for urine collection. One day 6, half the animals in each group were given 150 μmol DMPS/kg orally by gastric tubing. Urine samples were collected from each rat 24 hours before and after DMPS treatment and were analyzed for lead. The rats were sacrificed on day 7 and various organs were removed (liver, kidneys, spleen and bone) for lead determination by flameless atomic absorption spectrophotometry.

3. RESULTS

A single oral administration of DMPS resulted in a significant decrease of renal lead in rats previously injected with more than .5 mg Pb/kg. There was no change in lead concentration in blood or other tissues (Table 1)

TABLE 1

Tissue and Blood Lead Concentrations Following Oral Chelation with DMPS

Dose of Pb (mg/kg)	Organ	Pb alone (μg/g)	Pb + DMPS (μg/g)
0.5	Blood (ug/dl)	6.88 ± 1.74	9.60 ± 3.28
	Kidney	1.09 ± .224	.854 ± .255
	Liver	.240 ± .140	.259 ± .090
	Bone	4.28 ± 1.18	6.78 ± 2.24
	Spleen	.556 ± .403	.416 ± .124
1.0	Blood	15.96 ± 3.71	15.22 ± 5.02
	Kidney	2.10 ± .250	*1.23 ± .176
	Liver	.493 ± .097	.393 ± .124
	Bone	10.79 ± 5.23	11.40 ± 4.45
	Spleen	1.14 ± .825	.848 ± .263
2.0	Blood	18.96 ± 2.96	16.27 ± 4.08
	Kidney	7.07 ± 2.56	*3.49 ± .881
	Liver	2.46 ± .303	2.18 ± 1.50
	Bone	69.98 ±16.90	70.75 ±40.31
	Spleen	5.50 ± 2.78	5.75 ± 1.47
4.0	Blood	22.64 ± 3.62	21.64 ± 5.91
	Kidney	10.67 ± 2.48	*5.56 ± .942
	Liver	3.61 ± 1.03	3.86 ± .975
	Bone	65.91 ±12.17	75.91 ±23.94
	Spleen	9.00 ± 5.97	10.74 ± 4.46

* Significant difference from Pb alone group ($p<0.05$)

actively filtered by the glomerulus with little reabsorption from the tubular cells. Although the mechanism of chelation of metals by DMPS is not yet understood, the renal handling of DMPS and lead suggests that DMPS may chelate only easily filterable fractions of lead in the kidney, such as the interstitially deposited lead. In addition DMPS may also act by preventing the reabsorption of lead from the tubular cells.

In the present study, we have shown that specific chelation of lead from the kidney is possible after oral administration of 150 μmol DMPS/kg to lead exposed rats. It was also observed that DMPS could complex a constant percent of renal lead independent of the total renal lead content. Thus there is a direct relationship between the DMPS induced increase in urinary lead excretion and the renal content of lead. The slope of the linear regression equation suggests that the maximum chelatable renal lead by DMPS is approximately 38%.

4. CONCLUSION

In conclusion it seems that there are at least two major pools of lead in the kidney; an easily chelatable fraction (such as the DMPS chelatable fraction) and a more intracellularly located fraction which is not easily available to certain chelating agents. If the chelating agent can enter the cell by pinocytosis or by virtue of its lipophilicity, it may be able to chelate lead from both these fractions. DMPS may be special in this case because it has access only to certain areas in the kidney from which it may chelate metals.

The results of this preliminary study show the potential use of DMPS as a diagnostic drug for the estimation of renal lead content in excessive exposure to lead.

5. ACKNOWLEDGEMENT

This work was supported by NIH grant ES-01535.

6. REFERENCES

S. Araki and K. Ushio (1982) Brit. J. Ind. Med. 32, 157-160.
J. Chisolm (1968) J. Pediat. 73, 1-38.
B. Gabard (1976) Arch. Toxicol. 35, 15-24.
B. Gabard (1978) Arch. Toxicol. 39, 289-298.
B. Gabard and R. Walser (1979) J. Toxicol. Environ. Health 5, 759-764.
C. Tadlock and H. Aposhian (1980) Biochem. Biophys. Res. Comm. 94, 501-507.
T. Twarog and M. G. Cherian (1983) Bull. Environ. Contam. Toxicol. 30, 165-169.

INFLUENCE OF DIET ON METHYLMERCURY EXCRETION AND DEMETHYLATION IN MICE - ROLE OF THE INTESTINAL MICROFLORA

Ian R. Rowland*, Roy D. Robinson and Richard A. Doherty

**British Industrial Biological Research Association, Woodmansterne Road, Carshalton, Surrey 2M5 4DS U.K. and Environmental Health Sciences Center, University of Rochester School of Medicine, NY 14642 U.S.A.*

1. INTRODUCTION

Methylmercury (MeHg), unlike mercuric Hg, is rapidly absorbed from the mammalian gut (Walsh 1982) and hence the retention of mercury after an oral dose of MeHg is related to the rate of excretion, the main route of which is the faeces. Major differences in Hg excretion and whole-body retention, after MeHg exposure, are exhibited by mice fed different diets. In particular, a milk diet appears to increase, and a synthetic liquid diet to decrease, the retention of MeHg by mice, in comparison to animals fed a stock diet (Landry et al. 1979). Two possible explanations for the diet-related effects are examined in the present communication:

(i) Differences in concentration of thiols in the diets and in liver and gut contents of mice fed the diets. Mercurials bind strongly to sulphydryl groups and feeding low molecular weight thiols has been shown to influence the tissue levels of mercury compounds (Magos et al. 1978; Thomas & Smith 1982).

(ii) Diet-related differences in the rate of demethylation of MeHg. In mice MeHg undergoes enterohepatic circulation, being absorbed rapidly from the gut and subsequently secreted in the bile (Norseth 1973). Conversion of MeHg to the poorly absorbed mercuric form would interrupt the enterohepatic circulation and so increase Hg excretion in faeces. The intestinal microflora is the most likely site of this demethylation (Rowland et al. 1978; Rowland et al. 1980).

2. METHODS AND RESULTS

(a) Methods

Adult female BALB/c mice were fed a pelleted rodent diet

Chemical Toxicology and Clinical Chemistry of Metals
ISBN 0.12.137520.X
Copyright © 1983 by IUPAC
All rights of reproduction in any form reserved

(RMH3000, Agway Inc.) evaporated whole milk, or a synthetic liquid diet (GIBCO 116 E.C.). In 4 animals from each dietary group the intestinal microflora was suppressed by giving neomycin, streptomycin and bacitracin (3 mg per mouse per day). After seven days on the diets all the mice were given a single p.o. dose of radiolabelled $Me^{203}HgCl$ (0.6 mg Hg/kg body weight). Faeces and urine were collected separately and counted for radioactivity. Six days after dosing with MeHgCl, the animals were killed and the ^{203}Hg content of the tissues was measured. The proportions of organic and mercuric Hg in the tissues and excreta were determined by the method of Cappon & Smith (1977). Total and non-protein bound thiols were measured (Sedlak & Lindsay, 1968) in the three diets, and in liver and gut contents of mice given the diets for 10 days.

(b) Diet-related differences in Hg retention and thiol concentration

Mice given the three diets exhibited differences in whole-body retention of Hg after exposure to MeHgCl (Table 1), the milk-fed mice having the highest and those fed the synthetic diet the lowest Hg retention after 6 days. The major route of Hg excretion was the faeces, with the highest and lowest rates of excretion being seen in the mice fed synthetic and milk diets respectively. The differences in whole-body retention could not be related to the thiol concentration in the diets which were similar (0.4-0.5 μmole/g diet). Concentration of thiols in the gut contents were highest in mice fed the milk diet (Table 2) although it seems unlikely that this would affect the Hg retention since in all dietary groups the intestinal thiol concentrations were in excess of the Hg concentration to which the mice were exposed. Few diet-related differences were seen in the total, or non-protein bound, thiol concentrations in the liver, in particular mice fed milk or synthetic diet had almost identical hepatic thiol concentrations (Table 2).

Whole-body Hg retention appeared to be related to the proportion of the total Hg present as mercuric Hg. The milk-fed mice, having the highest Hg retention, had the lowest proportion of mercuric Hg in the body, whereas the mice fed the synthetic diet, having a low total Hg body-burden, had the highest proportion of mercuric Hg (Table 1). These data suggest that demethylation of MeHg is an important factor determining whole body retention of Hg.

(c) Effects of antibiotics on Hg retention

Whole-body retention of Hg after MeHgCl exposure was greatly increased by treatment of the mice with antibiotics which suppressed the intestinal microflora (Table 1). In addition, the differential effects of diet on Hg retention were virtually

TABLE 1

Total and mercuric Hg body burden of conventional and antibiotic-treated mice exposed to $Me^{203}HgCl$ (0.6 mg Hg/Kg).

	Hg body-burden (fraction of initial dose)					
	Conventional mice			Antibiotic-treated mice		
Diet	Total Hg	Hg++	(% BB)	Total Hg	Hg++	(% BB)
RMH3000	0.637	0.066	(10.4%)	0.942	0.065	(6.9%)
Milk	0.861	0.057	(6.6%)	1.005	0.064	(6.4%)
GIBCO 116EC	0.428	0.151	(35.5%)	0.995	0.070	(7.0%)

Total Hg body burden for each mouse was calculated from the sum of the total Hg in each organ.

Body burden (BB) of mercuric (Hg++) mercury was obtained calculating the sum of total Hg in each organ or tissue X % mercuric Hg (determined by the method of Cappon & Smith, 1978).

% BB is calculated from $\frac{\text{Hg++}}{\text{Total Hg}}$ x 100%.

Values shown are the means for 4 animals.

TABLE 2

Total and non-protein bound (free) thiols in livers and gut contents of mice fed different diets for 10 days.

	Thiol concn. (μmol/g gut contents or liver)					
	Small intestine		Caecum		Liver	
Diet	Total/SH	Free/SH	Total/SH	Free/SH	Total/SH	Free/SH
RMH3000	5.0,6.3	5.0,7.2	2.9,3.0	0.7,0.7	18.6±1.3	5.1±0.8
Milk	13.2,12.3	7.7,10.4	4.3,4.4.	1.0,1.3	17.6±2.1	3.4±1.0
GIBCO 116EC	6.8,7.1	6.8,8.3	2.5,2.7	0.9,1.0	18.7±1.8	3.5±0.5

*Each value is for the pooled caecal or intestinal contents of 4 mice.

+Values shown are the means ± 95% confidence intervals for 6 mice.

eliminated by antibiotic treatment.

The proportion of the Hg body burden present as mercuric Hg was decreased by antibiotics to similar levels in all dietary groups (Table 2) suggesting that less demethylation of MeHg occurred in antibiotic-treated animals, and implicating the intestinal microflora in demethylation.

3. CONCLUSIONS

The diet-related differences in whole body retention of Hg after MeHg exposure did not appear to be related to the concentrations of thiols in the diet, liver or gut contents of the mice. Retention of Hg was, however, inversly related to the proportion of the total Hg body-burden present as mercuric Hg, suggesting that demethylation of MeHg was an important determinant of Hg excretion in mice. The effect of antibiotics in increasing the whole-body retention of Hg and decreasing the proportion of the Hg body-burden present as mercuric Hg, implicates the intestinal microflora in MeHg demethylation. The results indicate that diet-related changes in the gut flora and its ability to demethylate MeHg (Rowland et al. 1978) may have profound consequences for mercury retention and hence toxicity after MeHg exposure in mammals.

4. ACKNOWLEDGEMENTS

We thank Donna Oberst and Frederich Genett for technical assistance. The work was funded by NIEHS (ES01247, ES01248).

5. REFERENCES

C.J. Cappon and J.C. Smith (1978). *J. Anal. Toxicol.*, 2, 114-120.

T.D. Landry, R.A. Doherty and A.H. Gates (1979). *Bull. Environ. Contam. Toxicol.* 22, 151-158.

L. Magos, T.W. Clarkson and J. Allen (1978). *Biochem. Pharmacol.*, 27, 2203-2208.

T. Norseth (1973). *Acta Pharmocol. Toxicol.*, 33, 280-288.

I.R. Rowland, M.J. Davies and J.G. Evans (1980). *Arch. Environ. Health*, 35, 155-160.

I.R. Rowland, M.J. Davies and P. Grasso (1978). *Xenobiotica*, 8, 37-43.

J. Sedlak and R.H. Lindsay (1968). *Anal. Biochem.*, 25, 192-205.

D.J. Thomas and J.C. Smith (1982). *Toxicol. Appl. Pharmacol.*, 62, 104-111.

C.T. Walsh (1982). *Environ. Res.*, 27, 412-420.

EFFECTS ON THE POSTNATAL DEVELOPMENT OF THE MOUSE OF PRECONCEPTION, POSTCONCEPTION AND/OR SUCKLING EXPOSURE TO MANGANESE VIA MATERNAL INHALATION EXPOSURE TO MnO_2 DUST

Bradley A. Lown[1], John B. Morganti[1], Rosemary D'Agostiono[2], Carl H. Stineman[3], and Edward J. Massaro[4]

[1]*Department of Psychology, SUNY-College at Buffalo, NY*
[2]*West Seneca Developmental Center, West Seneca, NY*
[3]*Department of Biochemistry, SUNY at Buffalo, NY*
[4]*Center for Air Environment Studies, The Pennsylvania State University, 226 Fenske Laboratory, University Park, PA 16802 U.S.A.*

1. INTRODUCTION

Many substances well tolerated by the adult organism have deleterious effects on the embryo, fetus, and/or neonate (Wilson, 1973). In addition, toxic effects may be manifested as subtle disturbances of behavior before obvious physical symptoms of toxicity become apparent (Weiss and Spyker, 1974). And, during the late fetal and early neonatal periods (times of rapid neurological development, growth, and metabolism), the developing organism is highly vulnerable to toxic neurological insult (Davison and Dobbing, 1968). Little information is available on the effects of excess exposure to manganese (Mn) during development (Deskin et al., 1980, 1981; Gray and Laskey, 1980), especially exposure via maternal inhalation. Therefore, employing the mouse as a model mammalian system, we have investigated effects on the development and behavior of offspring of mothers exposed prior to conception and/or during gestation to MnO_2 dust via inhalation. To differentiate prenatal from postnatal (suckling) effects, the offspring were subjected to fostering/cross-fostering procedures.

2. MATERIALS AND METHODS

Random bred female Swiss mice (ICR strain), approximately one month old at the beginning of the study, were exposed (in Rochester-type chambers) either to MnO_2 dust (mass median aerodynamic diameter = 1.5 μm) or filtered air (controls), 7 hr/day, 5 days/week for 16-32 weeks. For the first 12 weeks, exposure was to 49.1 ± 2.3 mg Mn/m^3 and, subsequently, to 85.3 ± 15.6 mg Mn/m^3.

Chemical Toxicology and Clinical Chemistry of Metals
ISBN 0.12.137520.X
Copyright © 1983 by IUPAC
All rights of reproduction in any form reserved

Following 16 weeks of exposure to experimental or control conditions, groups of 5 females were bred to a single male of the same strain (which was not part of the study except for its role as a breeder). On the day after detection of the copulatory plug (day 1 of gestation), the pregnant MnO_2-exposed and control females were randomly assigned either to experimental or control postconception exposure conditions and were exposed either to MnO_2 dust or filtered air for 7 hours/day, 5 days/week until day 17 of gestation. This procedure generated 4 preparturient groups: Mn(MnO_2)-Mn, C(filtered air)-Mn, Mn-C, and C-C; (designating the type of exposure the female received prior to and postconception). Parturition occurred on day 19 postconception.

When not in the inhalation chambers, the experimental and control animals were maintained in clear plastic home cages in environmental chambers maintained at $21 \pm 1°C$ on a 12 hr light cycle (on at 0730 hr). Preparturient females were housed 5/cage while lactating females were housed individually with their litters. Food (Charles River RMH 3000 diet) and water were provided (ad libitum) only when the subjects were in their home cages. Weaned offspring were housed individually in suspended steel cages in the environmental chamber with free access to water and food (Charles River RMH 1000 maintenance diet).

Immediately after parturition (day 0 postpartum), and before nursing was initiated, litters were removed from the mothers, pooled by sex, and individual pups were randomly assigned to newly parturient mothers of the same treatment group (fostering) or to newly parturient mothers of a different treatment group (cross-fostering). The newly formed litters were composed of 6 pups, 3 of each sex. All pups involved were born on the same day. These procedures produced 4 primary groups and 4 secondary groups of fostered or cross-fostered offspring based on the exposure history of the biological mother (before slash) and foster mother (after slash) which have been designated: Mn-Mn/Mn-Mn, C-C/Mn-Mn, Mn-Mn/C-C, and C-C/C-C, and Mn-Mn/Mn-C, Mn-Mn/C-Mn, C-C/C-Mn, and C-Mn/C-C, respectively.

As an index of rate of growth, the offspring were weighed on postpartum days 3, 7, and 12, and at weaning (day 21).

On day 7 postpartum, gross locomotor activity of the pups was evaluated on an activity field (Cone, 1974; D'Agostino et al., 1982). On day 12 postpartum, gross locomotor activity was assessed employing a circular runway (D'Agostino et al., 1982). Maternal retrieval latency was measured employing the standard litter scatter technique (D'Agostino et al., 1982).

On day 45 post-partum, one male and one female from each litter of the 4 primary groups of fostered/cross-fostered offspring were randomly selected for assessment of open-field behavioral parameters, passive avoidance learning, "hole-in-board" exploratory behavior, and rotarod performance (Lown et al., 1980; D'Agostino et al., 1982).

Six weeks after the completion of behavioral testing, all subjects were sacrificed and cerebral and cerebellar + brain stem mitochondrial Mn levels were measured by atomic absorption spectroscopy.

3. RESULTS AND CONCLUSIONS

The mean number of pups born to females exposed to Mn prior to conception was significantly larger than the number born to control females ($F=4.02$, d.f.=1/69, $p<0.05$). Postconception exposure to Mn had no significant effect on litter size ($F=3.60$, d.f.=1/69, $p<0.10$) although it approached significance.

ANOVA of the maternal retrieval latency data revealed that offspring of mothers exposed to Mn during gestation were retrieved significantly faster than offspring of mothers exposed to filtered air ($F=9.03$, d.f.=1/32, $p<0.01$). There was no effect of foster mother exposure history on offspring retrieval latency ($F=0.28$, d.f.=1/32, NS).

A significant multivariate effect (across the dependent measures) of prenatal exposure condition was found ($F=2.62$, d.f.=5/123, $p<0.003$). Univariate analysis revealed that prenatal exposure to Mn resulted in reduced offspring weight and activity on day 7 postpartum compared to prenatal exposure to filtered air ($F=6.77$, d.f.=1/127, $p<0.001$ and $F=0.32$, d.f.=1/127, $p<0.003$, respectively).

A significant multivariate interaction (across the dependent measures) of preconception (Mn or C) and postconception (Mn or C) exposure history of the foster mother also was found ($F=3.54$, d.f.=5/123, $p<0.005$). A significant ($F=4.70$, d.f.=1/127, $p<0.03$) univariate test of day 7 weight was obtained which was due to the reduced weight of offspring reared by foster mothers exposed to Mn prior to conception and filtered air after conception in comparison to offspring reared by foster mothers exposed to filtered air both prior to and after conception ($t=2.04$, d.f.=64, $p<0.05$. Another factor contributing significantly to the interaction was the higher day 12 activity ($F=8.42$, d.f.=1/127, $p<0.004$) of offspring reared by foster mothers exposed to Mn prior to conception and filtered air postconception compared to offspring reared by foster mothers exposed to filtered air prior to conception and to Mn after conception (Mn-C vs C-Mn; $t=3.00$, d.f.=64, $p<0.01$). In addition, offspring reared by foster mothers exposed to Mn prior to conception and filtered air postconception exhibited significantly higher day 12 activity than offspring reared by foster mothers exposed to filtered air both prior to and after conception (Mn-C vs C-C; $t=2.33$, d.f.=36, $p<0.05$).

A significant multivariate interaction of the preconception exposure history of the biological mother (Mn or C) and the postconception exposure history of the foster mother (Mn or C) was obtained (F=2.73, d.f.=5/123, p<0.02). Post hoc analysis revealed that, regardless of maternal postconception exposure history, offspring of biological mothers exposed to Mn prior to conception and reared by foster mothers exposed to Mn after conception, regardless of their preconception exposure history, were significantly less active than offspring of mothers exposed to filtered air prior to conception and reared by foster mothers exposed to Mn after conception (F=11.182, d.f.=1/127, p<0.001).

MANOVA of the growth and behavioral data for the sexually mature (day 45) offspring of the four primary groups of fostered/cross-fostered offspring revealed a significant multivariate effect of maternal exposure condition (F=2.05, d.f.=10/51, p<0.05). Underlying this effect was the significantly reduced day 45 weight of offspring of mothers exposed to MnO_2 dust prior to and postconception compared to offspring of mothers exposed to filtered air both prior to and postconception ($\bar{x}$= 31.14 gm vs. $\bar{x}$=34.12 gm, respectively: F=12.05, d.f.=1/60, p<0.001).

A significant multivariate interaction of the exposure history of the biological mother (Mn or C) and the foster mother (Mn or C) was found across behavioral measures (F=2.32, d.f.=10.51, p<0.02). Univariate tests revealed that the behavioral measures underlying this interaction were: rearing scores (F=7.86, d.f.=1/60, p<0.007), exploration scores (F=4.76, d.f.=1/60, p<0.03), and trial-2 scores on the rotarod (F=8.20, d.f.=1/60, p<0.006). Post hoc tests showed that offspring of Mn-exposed mothers fostered to Mn-exposed mothers exhibited significantly reduced rearing scores compared to offspring of filtered air-exposed mothers fostered to Mn-exposed mothers (t=4.60, d.f.=30, p<0.05). Analysis of the exploratory behavior scores revealed that mean differences in explorations between offspring groups exposed both prenatally and postnatally to Mn and offspring groups exposed to Mn either prenatally (t=-1.14, d.f.=35, p<0.08) or postnatally (t=-1.24, d.f.=33, p<0.07), approached, but did not reach statistical significance. For rotarod scores, group means were not significantly different. Passive avoidance and open-field measures did not contribute to the multivariate interaction.

MANOVA of brain Mn levels in the 4 primary groups of fostered/cross-fostered offspring showed a significant multivariate effect of exposure history of the foster mother (F=5.84, d.f.=6/52, p<0.0002). Univariate tests revealed that, independent of their in utero exposure history, sexually mature (day 45 post partum) offspring reared by mothers exposed to Mn had significantly lower concentrations of Mn in mitochondria obtained from the cerebellum plus brain stem

($F=14.05$, d.f.=1/57, $p<0.02$). In addition, there was a significant multivariate interaction of the exposure history of the biological mother and the foster mother on cerebral mitochondrial Mn levels ($F=3.82$, d.f.=6/58, $p<0.003$). Offspring of mothers exposed only to filtered air that were reared by foster mothers exposed only to MnO_2 dust had lower cerebral mitochondrial Mn levels than offspring of mothers exposed only to filtered air that were reared by foster mothers exposed only to filtered air ($t=-2.13$, d.f.=30, $p<0.05$). Other pair-wise comparisons were not significantly different.

4. REFERENCES

A.L. Cone (1974) Bull. Psychon. Soc., 4, 233-234.

R.B. D'Agostino, B.A. Lown, J.B. Morganti and E.J. Massaro (1982) J. Toxicol. Environ. Health, 10, 449-458.

A.N. Davison and J. Dobbing (1968) in Applied Neurochemistry (Eds, A.N. Davison and J. Dobbing), F.A. Davis Co., Philadelphia.

R. Deskin, S.J. Bursion and F.W. Edens (1980) Neurotoxicology, 2, 65-73.

R. Deskin, S.J. Bursion and F.W. Edens (1981) Gen. Pharmac., 12, 279-280.

L.E. Gray and J.W. Laskey (1980) J. Tox. Env. Health, 6, 861-867.

B.A. Lown, J.B. Morganti, C.H. Stineman, R.B. D'Agostino and E.J. Massaro (1980) Environ. Health Perspect., 34, 203-212.

J.B. Morganti, B.A. Lown, C.H. Stineman, R.B. D'Agostino and E.J. Massaro (in press) Neurotoxicology.

B. Weiss and J.M. Spyker (1974) Pediatrics, 53, 851-856.

J.G. Wilson (1973) in Environment and Birth Defects, Academic Press, New York.

VARIABLE ABSORPTION AND EXCRETION OF BISMUTH AND ITS POTENTIAL FOR TOXICITY

D.W. Thomas, S. Sobecki, T.F. Hartley, P. Coyle and M.H. Alp

Institute of Medical and Veterinary Science and
The Queen Elizabeth Hospital, Adelaide 5000, Australia

1. INTRODUCTION

Bismuth containing pharmaceuticals are numerous and have been used by the medical profession for many years for a variety of purposes. The greatest use has been for gastro-intestinal tract disturbances such as gastritis, peptic ulcer, constipation and even diarrhoea.

2. BISMUTH NEUROTOXICITY

In 1974 a previously undescribed reversible form of encephalopathy was observed in some patients with colostomies who had been ingesting bismuth subgallate to reduce colostomy odour, Burns *et al.* (1974). Almost simultaneously an identical syndrome was observed in France in a number of patients taking bismuth subnitrate for constipation and other gastrointestinal tract disorders, Buge *et al.* (1974).

This striking neurotoxic effect has been reported in only one other circumstance, where two patients developed an organic brain syndrome thought to be due to bismuth absorbed from a skin cream, Kruger *et al.* (1976).

Complete reversal of these symptoms was obtained when the bismuth medication was stopped. It was, therefore, considered that bismuth was associated with this new toxic effect. Laboratory investigation of symptomatic patients who were taking bismuth subgallate produced no significant abnormality except for an elevated concentration of whole blood bismuth.

The detection of bismuth in blood of both symptomatic and asymptomatic patients indicates that both groups were absorbing bismuth from the gastrointestinal tract, with the possibility that the magnitude of absorption and/or retention is higher in those who develop symptoms.

Chemical Toxicology and Clinical Chemistry of Metals
ISBN 0.12.137520.X
Copyright © 1983 by IUPAC
All rights of reproduction in any form reserved

3. IN VITRO ASSESSMENT OF BISMUTH SOLUBILITY

To assess the amount of bismuth available for absorption from the gastrointestinal tract, experiments were conducted with bismuth subgallate, bismuth subnitrate and bismuth subcitrate, simulating first gastric and then duodenal conditions.

Amounts of the various bismuth salts, in proportion to their normal daily doses, were added to hydrochloric acid pH 1.9 and incubated for several hours at 37°C. Duodenal fluid was then added which resulted in a change in pH to between 6.7-8.9. The mixtures were incubated for a further 2 hours at 37°C.

At various stages the mixtures were centrifuged, and "soluble" bismuth measured in the supernatant fluids using the technique of flameless atomic absorption spectrophotometry Allain (1975).

TABLE 1.

'IN VITRO' ESTIMATES OF 'SOLUBLE' BISMUTH AVAILABLE FOR ABSORPTION

COMPOUND	AVERAGE DAILY DOSE	TOTAL Bi (mmol)	Bi RENDERED 'SOLUBLE' BY HCl (μmol)	Bi RENDERED 'SOLUBLE' BY HCl & DUODENAL ASPIRATE (μmol)
BISMUTH SUBNITRATE	10g	32.8	29.9	51.7 – 64.1
BISMUTH SUBCITRATE	15ml	4.6	0	2.7 – 3.6
BISMUTH SUBGALLATE	1g	2.4	0.4	0.7 – 2.3

The largest amount of "soluble" bismuth available for absorption would seem to occur with bismuth subnitrate ingestion, and smaller amounts with bismuth subcitrate and bismuth subgallate ingestion. The actual forms in which bismuth is available for absorption following ingestion of these compounds is not known.

4. IN VIVO ESTIMATION OF BISMUTH ABSORPTION

Since "soluble" bismuth may be present in the gastrointestinal tract and presumably available for absorption, attention was focussed on estimating the extent of absorption that occurred with each of these three compounds. This was assessed by measuring urinary bismuth excretions, making the assumptions that when subjects and patients were ingesting constant amounts of bismuth containing compounds and were in a steady state, without any change in body content or significant excretion into the gastrointestinal tract, then urinary excretion would approximate gastrointestinal tract absorption.

Both symptomatic and asymptomatic patients taking bismuth subgallate, and asymptomatic patients taking bismuth subcitrate and bismuth subnitrate excreted approximately similar amounts of bismuth. Symptomatic patients taking bismuth subnitrate excreted considerably more.

TABLE 2.

BISMUTH IN BLOOD AND URINE

COMPOUND	Bi RENDERED SOLUBLE BY HCl & DUODENAL ASPIRATE (μmol/L)	URINARY EXRECTIONS OF BISMUTH (μmol/day)		BLOOD Bi (μmol/L)
BISMUTH SUBNITRATE	51.7 – 64.1	ASYMPTOMATIC[a]	0.97	0.17
		SYMPTOMATIC[a]	7.74	4.27
BISMUTH SUBCITRATE	2.7 – 3.6	ASYMPTOMATIC	1.20	0.05
BISMUTH SUBGALLATE	0.7 – 2.3	ASYMPTOMATIC	1.61	0.18
		SYMPTOMATIC	1.52	1.00

a) FROM LITERATURE SOURCES

In patients exhibiting neurotoxicity, those ingesting bismuth subnitrate appear to increase absorption of "soluble" bismuth to 12-15% of that available, whereas those ingesting bismuth subgallate appear to absorb the same amount of "soluble" bismuth as asymptomatic patients.

5. ASSESSMENT OF BISMUTH DISTRIBUTION

Symptomatic patients ingesting bismuth subgallate had significantly higher blood bismuth concentrations than asymptomatic patients, even though their urinary excretions of bismuth were similar. Patients taking bismuth subnitrate who developed symptoms had even higher concentrations of blood bismuth. The mean blood bismuth concentration in patients taking bismuth subcitrate was much less, even though their urinary excretions were quite substantial.

These observations could be explained by the formation of large molecular weight complexes of bismuth in those patients ingesting bismuth subnitrate or bismuth subgallate. Under certain circumstances the formation of these complexes may increase, accumulate within the body and lead to the development of neurotoxicity. Such conditions in patients ingesting bismuth subnitrate could be increased absorption from the gastrointestinal tract and in patients ingesting bismuth subgallate a change in the nature and/or amount of large molecular weight complexes.

6. GEL FILTRATION EXPERIMENTS

A study was undertaken to investigate the possibility of bismuth salts binding to plasma proteins, and then to identify the various proteins and other plasma species which may potentially bind bismuth. Bismuth subgallate of known concentration was added to human serum and incubated at 37°C. The distribution of bismuth was then studied under equilibrium conditions using a gel filtration technique. A 40x900 mm column maintained at 4°C was packed with Sephacryl S-200 super fine. The column eluent contained sodium (140 mM) in triethanolamine (20 mM) adjusted to pH 7.4. Fractions were collected and analysed for bismuth and protein content.

Bismuth was found associated with high molecular weight proteins collected in the first elution peak. This peak was shown to contain α-2-macroglobulin, IgM, β-lipoprotein and haptoglobin. Bismuth was not detected in any other fractions. This indicates that bismuth subgallate may bind "in vivo" to proteins with molecular weights greater than or equal to 200,000.

7. CONCLUSION

From the information presented, it can be concluded that (1) the amount of "soluble" bismuth within the gastrointestinal tract available for absorption, (2) the actual amount of bismuth absorbed, (3) the probable formation of large molecular weight complexes within the vascular compartment and subsequent transport into the central nervous system and (4) the amount of bismuth excreted in the urine are all dependent on the nature of the bismuth containing compound ingested.

8. REFERENCES

P. Allain (1975) *Clin.Chim.Acta.*, 64, 281-286.

A. Buge, G. Rancurel, M. Poisson and H. Dechy (1974). *Nouv. Presse. Med.*, 3, 2315-2320.

R.J. Burns, D.W. Thomas and V.J. Barron (1974) *Brit.Med.J.* 1, 220-223.

G. Kruger, D.J. Thomas, F. Weinhardt and S. Hager (1976). *Lancet*, ii, 485-487.

THE EFFECTS OF SUBSTITUTED DITHIOCARBAMATES ON THE NEPHROTOXICITY AND INTESTINAL TOXICITY OF CISPLATIN

Ernest M. Walker Jr.[1], William B. Greene[2], Glen R. Gale[3], James H. Nicholson[2], Josef Vesely[2], and Gordon R. Hennigar[2]

[1]Department of Laboratory Medicine, [2]Department of Pathology, [3]Veterans Administration Medical Center (20403)
Medical University of South Carolina
Charleston, South Carolina 29425 U.S.A.

1. INTRODUCTION

Cisplatin, a widely used antineoplastic agent, produces several associated side effects, chiefly a dose-limiting nephrotoxicity and deleterious GI effects including severe nausea and vomiting and intestinal toxicity. Reduction of cisplatin nephrotoxicity has been attempted by a number of methods which have been recently reviewed (Walker and Gale, 1981). One promising approach appears to be the use of diethyldithiocarbamate (DDTC). Intraperitoneal injections of DDTC at 500 mg per kg has been shown to protect mice when given 30 minutes before or up to five hours after LD_{100} doses of cisplatin without concurrently diminishing the antitumor efficacy of cisplatin against L1210 leukemia in the same mice (Gale *et al.*, 1982). Administration of DDTC markedly reduced the nephrotoxicity of toxic doses of cisplatin in F344 rats without diminishing its therapeutic efficacy against a subcutaneously implanted mammary tumor in the same rats (Borch *et al.*, 1980).

This study was undertaken to evaluate the ability of selected substituted dithiocarbamates to antagonize the toxic effects of cisplatin, a widely used antitumor chemotherapeutic agent, to the kidneys and small intestine. The agents evaluated in the study include DDTC and dimethyldithiocarbamate (DmDTC), a close structural analogue of DDTC.

2. METHODS AND RESULTS

(a) *Effects of DDTC and DmDTC on Cisplatin Nephrotoxicity*

Cisplatin at doses approaching the LD_{100} level produces extensive proximal tubular destruction of kidneys in mice. Light microscopic evaluation of proximal tubules in mouse renal sections fixed in Tellyesniczky's fixative and stained by the

Chemical Toxicology and Clinical Chemistry of Metals
ISBN 0.12.137520.X
Copyright © 1983 by IUPAC
All rights of reproduction in any form reserved

alcian blue-periodic acid-Schiff method demonstrates a toxic type of acute tubular necrosis, specifically extensive degeneration or necrosis of proximal tubular epithelial cells followed by desquamation, and filling of many lumina with granular cellular debris with preservation of tubular basement membranes (Gale *et al.*, 1982). Approximately 60 percent of proximal tubules, on an average, appear to be severely damaged in renal sections obtained from mice given cisplatin at doses approaching the LD_{100} level when examined microscopically.

TABLE I

Evaluation of Renal Proximal Tubules by Computerized Image Analysis

Group		Ratio[1] (Cumulative Percent)		
Cisplatin (mg per kg)	DDTC (mg per kg)	<2:1	<3:1	<4:1
0	0	--	--	7
18	0	42	65	72
18	500 (-30 minutes)[2]	29	46	63
18	500 (+30 minutes)	24	34	43
18	500 (+60 minutes)	12	34	62
18	500 (+120 minutes)	19	40	58
18	500 (+240 minutes)	41	55	70

[1](Total area of the proximal tubule/area of the tubular lumen) ratio. A total of 2,000 control proximal tubules and 1,000 proximal tubules from each test group were evaluated.

[2]Represents time relationship of DDTC administration with respect to that of cisplatin. Both agents were given by i.p. injection.

This observation has been confirmed by analysis of the same or similar renal sections by use of a Zeiss Videoplan Image Analysis Computer in which total proximal tubular/luminal area ratios were computed for individual proximal tubules. Necrosis and desquamation of tubular epithelium into lumina in cisplatin-treated mice causes the luminal area ratio to decrease. Analysis of over 2,000 proximal tubules in renal sections made from nontreated, control mice reveals that the total tubular/luminal area ratio is almost never less than 3:1 in normal proximal tubules, and that the ratios are within a 3:1 to 4:1 range in seven percent of the normal proximal tubules. (Table 1). Similar analysis of a total of 1,000 proximal tubules in renal sections taken from mice four days after receiving 18 mg per kg of cisplatin showed that about 65 percent of the ratios are less than 3:1 indicating moderate to complete destruction of nearly two-thirds of the proximal tubules and confirming the results obtained by light microscopic methods. A total of 42 percent of the proximal tubules had total tubular/luminal area ratios of less than 2:1 indicating severe or complete destruction of those proximal tubules. The administration of DDTC 30 minutes before or up to two hours after cisplatin appears to dramatically reduce the number of severe or completely destroyed proximal tubules.

Electron microscopic examination of proximal tubular epithelial cells not desquamated into lumina revealed a generalized disruption of cellular architecture with mitochondrial changes of high amplitude swelling with disruption of cristae, condensation, and clumping of mitochondria. Autophagocytic vacuoles were observed containing mitochondria and other cellular debris. Hydropic and fatty changes were apparent. Basal cytoplasmic infoldings were sometimes disrupted and swollen remnants of endoplasmic reticulum were seen. Large primary lysosomes were present and clearing of nuclear chromatin was noted in some cells. Basement membranes appeared to remain intact. No prominent glomerular changes were noted. The relative percentage of extensively damaged proximal tubules as well as the toxic cellular changes are greatly reduced by the administration of DDTC or DmDTC at doses of 2.2 mmol per kg 30 minutes before or 30 minutes after cisplatin. DDTC afforded appreciable protective effects when given as long as two hours after cisplatin as determined from the results of light microscopic (LM), electron microscopic (EM), and computerized image analysis techniques and reduced the number of severely damaged proximal tubules by nearly 50 percent when given 30 minutes or 60 minutes after cisplatin.

(b) *Effects of DDTC and DmDTC on Cisplatin Gastrointestinal Toxicity*

Cisplatin appears to produce sustained pyloric constriction in mice causing retention of undigested food. The effect appears to be markedly reduced by the administration of 500 mg per kg of DDTC as long as six hours after 18 mg per kg cisplatin (Gale *et al.*, 1982).

Cisplatin appears to be toxic to both columnar and goblet cells in the small intestine. Light microscopic evaluation of hematoxylin and eosin (H & E) stained sections of intestine obtained from mice given 18 mg per kg cisplatin reveals swollen lamina propria of some villi and the transition of the tall columnar epithelium of many villi toward a more cuboidal type of epithelium. Specific cellular degenerative changes include condensation or vacuolization of cytoplasmic elements and conversion of normally elongated cell nuclei into rounded or bizarre shaped nuclei, in many cases containing prominent and irregular nucleoli. Most cells exhibit decreased nucleotytoplasmic ratios. Electron microscopic evaluation reveals a number of dead columnar and goblet cells with the latter cells containing marked reduction in numbers of mucous granules. Altered columnar cells contain phagocytic vacuoles, disrupted mitochondria or mitochondrial inclusions, occasional swollen Golgi bodies, and occasional myelin bodies (whorls). Fewer dead or abnormal cells are seen in sections of small intestine from mice sacrificed four days after treatment with cisplatin and DDTC or DmDTC. Administration of DDTC or DmDTC 30 minutes before or 30 minutes after cisplatin dramatically reduces the intestinal toxicity of cisplatin as determined by LM and EM evaluation. Some protection is produced by DDTC when given up to four hours after cisplatin.

3. CONCLUSIONS

The agents DDTC and DmDTC appear to be protective in mice against the nephrotoxic and gastrointestinal toxic effects produced by cisplatin. Both agents provide appreciable protection when given 30 minutes before or 30 minutes after cisplatin and prevent the lethal effects produced by LD_{100} doses of cisplatin without diminishing the chemotherapeutic efficacy of the agent. Appreciable protective effects against cisplatin toxicity are conferred by DDTC when given as long as five or six hours after cisplatin.

4. ACKNOWLEDGEMENTS

This study was supported in part by the Veterans Administration.

5. REFERENCES

R.F. Borch *et al.* (1980) *Proc. Nat. Acad. Sci.* 77, 5441-5444.
G.R. Gale *et al.* (1982) *Ann. Clin. Lab. Sci.* 12, 345-355.
E.M. Walker, Jr. and G.R. Gale (1981) *Ann. Clin. Lab. Sci.* 11 397-410.

SODIUM DIETHYLDITHIOCARBAMATE ADMINISTRATION IN NICKEL-INDUCED RHABDOMYOSARCOMA

F. William Sunderman Sr., Henry P. Schneider, and George Lumb*

*Institute for Clinical Science and *Department of Pathology, Hahnemann University School of Medicine, Philadelphia, PA 19102 U.S.A.*

1. INTRODUCTION

Studies were undertaken to ascertain the possible anti-tumorigenic effectiveness of sodium diethyldithiocarbamate (dithiocarb) on the development of tumors in rats following the muscular implantations of nickel subsulfide (Ni_3S_2). The tumors that developed have been designated rhabdomyosarcoma on the basis of (1) site of origin, (2) morphologic appearances, including spindle and strap-like cells with nuclei, and (3) tendency to metastasize. Histologically, the tumors show elongated spindle cells and strap-like cells with fibrillar longitudinal striations.

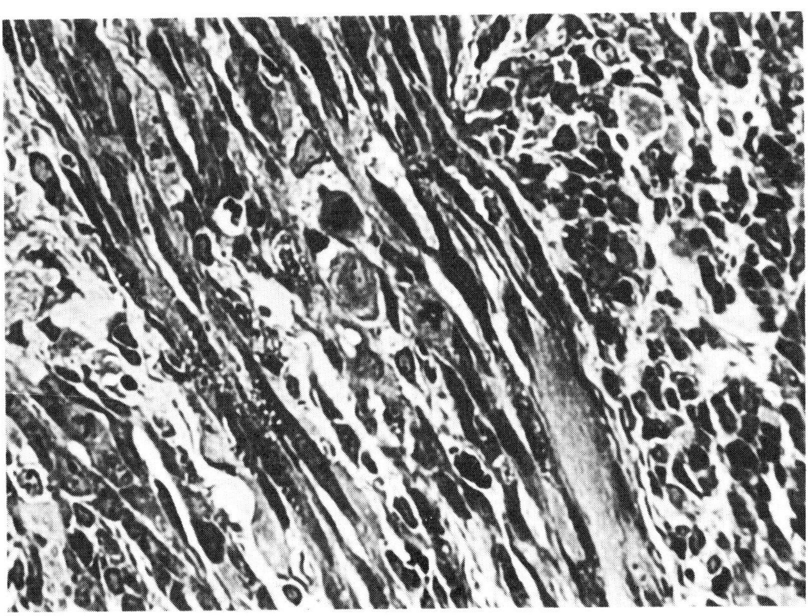

FIG. 1 Pleomorphic malignant cells with elongated strap-like cells with fibrillar markings in some cytoplasm. H & E x 400.

Chemical Toxicology and Clinical Chemistry of Metals
ISBN 0.12.137520.X
Copyright © 1983 by IUPAC
All rights of reproduction in any form reserved

Transverse striations were difficult to find in tumor cells, but occasional examples were seen. Once the tumors began to grow, they increased rapidly in size and underwent necrosis. Extreme anaplasia was common, and in these areas no specific features could be identified. Metastases were always anaplastic.

2. METHODS AND RESULTS

In two separate studies over a period of two years, a total of 100 four-month old Fischer rats received muscular implantations of Ni_3S_2. Of this total, 50 rats (25 males and 25 females) were treated with dithiocarb for a period of four to six weeks. Seventy-eight percent of the untreated rats developed sarcomas as compared to 32 percent of the treated rats.

Analyses of the data revealed a striking difference in the sex responsiveness to dithiocarb. Of the 25 female rats with Ni_3S_2 implantations and treated with dithiocarb, only 12 percent developed sarcomas; of the male rats similarly treated, 52 percent developed sarcomas. The difference in sex response is statistically significant ($p = 0.005$). The mechanisms by which dithiocarb causes an increased inhibition of tumorigenic activity in female rats as compared to males are not readily apparent and are obviously speculative. However, it is well known that the metabolism of metals, such as copper, in animals is influenced by pregnancy, ovariectomy, castration, administration of estrogens and progesterone, and various hormonal factors. Furthermore, sex-linked differences in the concentrations of metallothionein have been reported in rat kidneys, - the concentrations being greater in females than in males.

3. CONCLUSION

Nickel-containing alloys have been implanted into patients in a variety of therapeutic devices, including stainless steel wire suture material, orthopedic prosthetic implants, heart valve prostheses, nickel-containing intrauterine contraceptive devices, nickel-chrome mesh for nasal prostheses, implantable nickel-cadmium batteries for cardiac pacemakers, dental inlays, fillings, etc. It is recognized that patients with implanted prostheses may, on occasion, develop malignant tumors, osteonecrosis, osteomyelitis, hemangioendotheliomas, and allergenic conditions, and it seems probable that their development is related to the presence of nickel. Under such circumstances, chelation therapy with the dithiocarbamates might be considered.

LONG-TERM TOXICITY OF THALLIUM IN THE RAT

L. Manzo, R. Scelsi, A. Moglia, P. Poggi, E. Alfonsi,
R. Pietra, F. Mousty and E. Sabbioni

*Institutes of Pharmacology, Neurology and Pathology,
University of Pavia Medical School, 27100 Pavia, Italy;
Nuclear and Radiochemistry Division, Commission of the
European Communities JRC, 21020 Ispra, Varese, Italy*

1. INTRODUCTION

Thallium compounds are highly toxic to most biological systems (Sabbioni and Manzo,1980). Severe accidental poisonings by thallium rodenticides occur rather frequently in many parts of the world.Chronic occupational intoxications showing serious neurological disorders and alopecia have also been sporadically described (Smith and Carson,1977).

In the last few years, attention has been paid to potential human health hazards related to environmental thallium (Zitko, 1975). Thallium compounds are released by smelters and coal mines. In addition, coal fired power plants emit thallium in fly ash and the concentration of thallium was found to increase from 4-45 μg Tl/g in large particles to 29-76 μg Tl/g in air-born fly ash (Davison et al.,1974).

Recently, health disorders related to excessive thallium absorption have been described in subjects living around a thallium emitting cement plant. These subjects showed accumulation of thallium in hair and thallium levels in urine 50-100 times as high as the upper reference limits (Brockhaus et al., 1981).

While the effects of large doses of thallium have been extensively studied in laboratory animals (Herman and Bensch,1967) no experimental data are available on the toxicity potential of thallium after continuous low-level exposure.

In this report, the results of a long-term toxicity study showing cutaneous and peripheral nervous system disorders in rats treated with low levels of thallium are summarized.

2. EXPERIMENTAL STUDIES

Eighty female Sprague-Dawley rats (180-200 g) were given thallium via the drinking water (10 mg Tl/l, 24 μM) as thallium sulfacte, and the effects were investigated over a 36-week period.

Chemical Toxicology and Clinical Chemistry of Metals
ISBN 0.12.137520.X
Copyright © 1983 by IUPAC
All rights of reproduction in any form reserved

(a) *General findings*

Changes consisting of poor hair luster, periorbital redness and irritability were observed as early as 20-25 days after starting thallium treatment. Hair loss appeared after 32 days and within a few days thereafter involved about 20 per cent of the treated animals. Over the 36-week period of exposure, several rats presented almost complete hair loss or focal alopecia mainly in the neck region and periorbital area. In other animals, hair loss did not occur at all in spite of intake of a total amount of thallium as high as 80 mg Tl/rat. The overall mortality rate was 15 and 21 per cent after 40 and 240-280 days of treatment, respectively.

TABLE 1

Toxicological Findings in Thallium Treated Rats

	Days of treatment	
	40	240
Thallium Dose		
total, mg Tl/rat	10.31 ± 0.77[†]	72.74 ± 9.12
daily intake, μg/rat	258.10 ± 21.6	303.10 ± 34.8
Mortality	12/80	17/80
Cutaneous Findings		
mild	7/80[§]	33/80
severe	1/80	19/80
Electrophysiological Abnormalities	0/8[§]	10/16

[§]Animals showing abnormal findings. [†]Mean ± S.D.

(b) *Electrophysiological findings*

Serial measurements of electrodiagnostic parameters of caudal nerve (Moglia et al.,1981) were performed at various time intervals in groups of control and thallium treated rats.

No change related to thallium exposure could be observed after 40 days of treatment. However, abnormal parameters were seen after 240 days in up to 60 per cent of the treated rats (Table 1). The motor action potential (MAP) and sensory action potential (SAP) amplitudes were decreased by 44 and 30 per cent respectively. MAP latency was increased by 25 per cent as compared to control values. No fibrillation activity was observed in any of the treated rats examined by needle-electrode in the tibialis anterior muscle.

(c) *Neuromorphological data*

Sciatic nerve samples from six thallium treated rats and three controls were examined by light and electron microscopy at the 8th month of the study.

Morphological changes confined to some large and medium-sized myelinated fibres were seen in one half of the thallium treated rats. The changes were characterized by Wallerian degeneration of scattered fibres as well as vacuolization and lamination of the myelin sheath of about 10 per cent of nerve fibres (Fig.1).

On electron microscopic examination, the affected fibres showed vacuolar changes and fragmentation of the myelin sheath mainly confined to the internodal region. The pictures of Wallerian degeneration were characterized by complete destruction of the axon with mitochondrial disintegration, neurofilamentous clustering and abundant lysosomal activity, as well as myelin degeneration with formation of rows of ovoids or round lamellar osmiophilic inclusions.

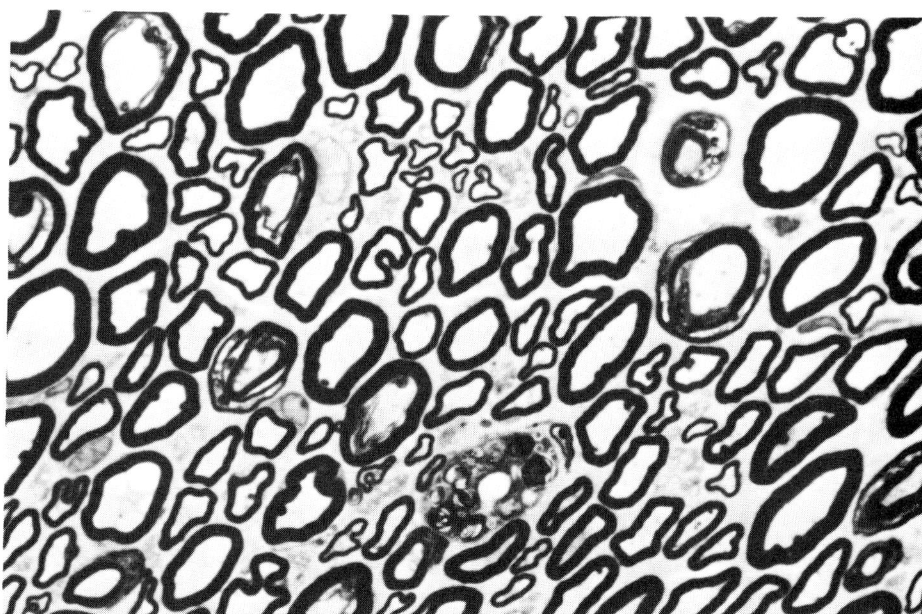

Fig. 1. Wallerian degeneration with myelin debris and myelin vacuolization. Groups of small nerve fibres surrounded by a thin myelin sheath are present (X1200).

TABLE 2

Thallium Levels in Representative Tissues

Tissue	Tl-rats†	control	ratio
	μg/g		
Blood	0.67±0.08	0.045±0.006	15
Brain	6.11±1.03	0.088±0.011	69
Bone	7.71±0.88	0.119±0.023	65
Skin	2.57±0.22	0.040±0.013	63
Heart	6.89±0.60	0.070±0.010	98
Kidney	17.15±1.23	0.038±0.011	451

†Total intake: 78 mg Tl/rat over a 280-day period

(d) *Thallium levels in tissues*

Thallium levels were measured by atomic absorption spectrometry (Mousty et al.,in preparation) in tissues of control and thallium treated rats after 280 days of exposure. As shown in TABLE 2, ppb levels of thallium were detected in tissues of control (non-exposed) rats. In the chronically treated rats the tissue concentrations of thallium were strikingly increased and were of the same order of magnitude as those found in a previous study in rats given a single sub-lethal dose of thallium sulfate (Sabbioni et al.,1982). Most of the tissues examined exhibited a 60-100 fold increase in thallium concentrations after the long-term treatment. A greater accumulation occurred in the kidney. The increase in blood thallium levels was relatively low suggesting that the blood is a poor indicator of thallium tissue content following repeated dosing.

3. CONCLUSIONS

The prolonged absorption of a low dose of thallium in rats resulted in tissue accumulation of the element at levels high enough to induce the appearance of overt signs of thallotoxicosis namely hair loss and peripheral nervous system damage.

A large variability in the response of individual animals to the toxic effects of thallium was apparent. Certain rats succumbed to a cumulative dose of 10 mg thallium while several animals survived the 36-week period of exposure in spite of intake of amounts as high as 80 mg Tl/rat, i.e. 250-300 mg Tl/kg.

In humans, disorders of the peripheral nervous system associated with acute or chronic thallium intoxication generally appear 2 to 4 weeks before alopecia (Bank et al.,1972). In our study, neurological changes only appeared in a limited number of the treated rats at the late stage of the chronic exposure period. This suggests that the rat is less susceptible than man to thallium neurotoxicity.

The mode of action of thallium compounds is still unclear. It is generally accepted that thallium substitutes for potassium in many physiological processes. This would account for the cumulative behaviour of thallium in tissue cells as well as of certain manifestations of thallium toxicity such as cardiac and gastrointestinal disorders. However,this mechanism is unlikely to be involved in the development of cutaneous and neurological disorders following chronic thallium exposure (Sabbioni and Manzo,1980).

There is considerable evidence suggesting that the neurotoxicity of thallium intoxication may be related to impairment of energy supply to the nerve (Cavanagh, in press). The energy deprivation mechanism has also been claimed as a causative factor of thallium-induced inhibition of hair follicle mitosis (Cavanagh and Gregson, 1978).

Cutaneous and neurological disorders like those found in thallium-exposed subjects are known to occur in non-toxic conditions, i.e. beri-beri, which have been related to impaired energy production in nervous tissue.

5. REFERENCES

W.J. Bank, D.E.Pleasure, K.Suzuki, M.Nigro and R.Katz (1972) *Arch.Neurol.*, 26, 456-464.

A.Brockhaus, R.Dolgner, U.Ewers, U.Krämer, H.Soddemann and H.Wiegand (1981) *Int.Arch.Occup.Environ.Health*, 48,375-389.

J.B.Cavanagh, in *Neurotoxicology* (Eds, K.Blum and L.Manzo), Marcel Dekker, New York, in press

J.B.Cavanagh and M.Gregson (1977) *J.Pathol.*, 125,179-191.

R.L.Davison, D.F.S.Natusch, J.R.Wallace and C.A.Evans,Jr., (1974) *Environ.Sci.Technol.*, 8,1007-1113.

M.M.Herman anf K.G.Bensch (1967) *Toxicol.appl.Pharmacol.*, 10,199-222.

A.Moglia, A.Tartara, A.Arrigo, P.Poggi, M.Scelsi and R.Scelsi (1981) *Il Farmaco*, 36,419-424.

E.Sabbioni and L.Manzo (1980) in *Advances in Neurotoxicology* (Ed, L.Manzo), Pergamon Press, Oxford-New York, pp.249-270

E.Sabbioni, C.Gregotti, J.Edel, E.Marafante, A.Di Nucci and L.Manzo (1982) *Arch.Toxicol.*, suppl.5, 220-223.

I.C.Smith and B.L.Carson (1977) *Trace Metals in the Environment*, Ann Arbor Press, Ann Arbor.

V.Zitko (1975) *Sci.Total Environ.*, 4,185-192.

AUTHOR INDEX

SUBJECT INDEX